1,000,000 Books

are available to read at

www.ForgottenBooks.com

Read online
Download PDF
Purchase in print

ISBN 978-1-333-02382-9
PIBN 10453727

This book is a reproduction of an important historical work. Forgotten Books uses state-of-the-art technology to digitally reconstruct the work, preserving the original format whilst repairing imperfections present in the aged copy. In rare cases, an imperfection in the original, such as a blemish or missing page, may be replicated in our edition. We do, however, repair the vast majority of imperfections successfully; any imperfections that remain are intentionally left to preserve the state of such historical works.

Forgotten Books is a registered trademark of FB &c Ltd.
Copyright © 2018 FB &c Ltd.
FB &c Ltd, Dalton House, 60 Windsor Avenue, London, SW19 2RR.
Company number 08720141. Registered in England and Wales.

For support please visit www.forgottenbooks.com

1 MONTH OF
FREE
READING

at
www.ForgottenBooks.com

By purchasing this book you are eligible for one month membership to ForgottenBooks.com, giving you unlimited access to our entire collection of over 1,000,000 titles via our web site and mobile apps.

To claim your free month visit:
www.forgottenbooks.com/free453727

_{* Offer is valid for 45 days from date of purchase. Terms and conditions apply.}

English
Français
Deutsche
Italiano
Español
Português

www.forgottenbooks.com

Mythology Photography **Fiction** Fishing Christianity **Art** Cooking Essays Buddhism Freemasonry Medicine **Biology** Music **Ancient Egypt** Evolution Carpentry Physics Dance Geology **Mathematics** Fitness Shakespeare **Folklore** Yoga Marketing **Confidence** Immortality Biographies Poetry **Psychology** Witchcraft Electronics Chemistry History **Law** Accounting **Philosophy** Anthropology Alchemy Drama Quantum Mechanics Atheism Sexual Health **Ancient History** **Entrepreneurship** Languages Sport Paleontology Needlework Islam **Metaphysics** Investment Archaeology Parenting Statistics Criminology **Motivational**

EXPERIMENTAL HISTORY OF THE MATERIA MEDICA,

OR OF THE

NATURAL AND ARTIFICIAL SUBSTANCES

MADE USE OF IN

MEDICINE:

CONTAINING A COMPENDIOUS VIEW OF THEIR

NATURAL HISTORY,

An ACCOUNT of their PHARMACEUTIC PROPERTIES,

And an Eſtimate of their

MEDICINAL POWERS,

So far as they can be aſcertained by EXPERIENCE, or by RATIONAL INDUCTION from their SENSIBLE QUALITIES.

By WILLIAM LEWIS, M.B. F.R.S.

IN TWO VOLUMES.

VOL. II.

THE FOURTH EDITION,

With numerous ADDITIONS and CORRECTIONS

By JOHN AIKIN, M.D.

Rationalem quidem puto Medicinam eſſe debere: inſtrui vero ab evidentibus. CELSUS.

LONDON:

Printed for J. Johnſon, in *St. Paul's Church-Yard*; R. Baldwin, in *Pater-noſter-Row*; J. Sewell, in *Cornhill*; and S. Hayes, in *Oxford-Street*.

MDCCXCI.

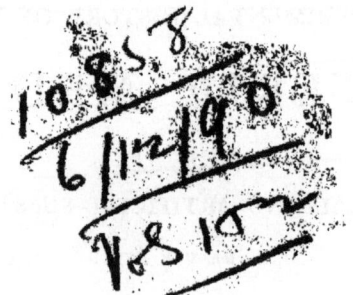

THE MATERIA MEDICA.

JACOBÆA.

JACOBÆA Pharm. Parif. Jacobæa vulgaris laciniata C. B. Senecio major five flos fancti jacobi Matthiol. Senecio Jacobæa Linn. RAGWORT: a plant with a firm round ftalk generally purplifh; oblong dark green leaves, deeply jagged almoft to the rib, and the fections jagged again and fomewhat crumpled: on the tops of the ftalks grow umbel-like clufters of yellow flowers, of the radiated difcous kind, followed by fmall oblong feeds winged with down. It is perennial, common in uncultivated fields and by road fides, and flowers in July.

THE leaves of ragwort have a roughifh bitterifh fubacrid tafte, extremely naufeous, far different from that of the herbaceous groundfel to which they have by fome been accounted fimilar. Simon Paulli relates, that they were found

found of great service in an epidemic camp dysentery, many soldiers having been cured of disease by drinking a decoction of them; expresses some concern, that a medicine of so much efficacy should be at the same time so very disgustful to the palate. This however is an inconvenience that may be easily palliated: the active matter of ragwort, whatever its virtue may be, is dissolved both by water and by rectified spirit, and on inspissating the filtered tinctures, remains concentrated in the extracts; which may be taken, without offence to the palate, in the form of a bolus or pill. The spirituous extract is in less quantity than the watery, and proportionably stronger in taste; though rather less nauseous than the herb in substance.

JALAPIUM.

JALAPIUM Pharm. Lond. Jalapa Pharm. Edinb. Machoacanna nigra. JALAP: the dried root of the *mirabilis peruviana* or marvel-of-peru, *mirabilia peruviana* Gerard. *solanum mexicanum flore parvo* C. B. *Convolvulus Jalapa Linn.*[a] a plant with thick, fleshy radish-like roots; jointed stalks and branches; acuminated somewhat oval leaves set in pairs; and elegant, numerous, monopetalous, funnel-shaped flowers, purple, yellow, white, or diversely variegated, standing in double cups, of which the innermost incloses the flower and the outer surrounds its basis: each flower is followed by a wrinkled, roundish,

---

[a]. The later botanists are not perfectly agreed concerning the genus of the plant producing jalap. Linnæus first made it a *mirabilis*; and Bergius now gives it as the *mirabilis dichotoma Linn.* from the resemblance between the root of that plant and the jalap of the shops.

pentagonal umbilicated fruit, about the size of a pepper corn, including a white kernel. It is perennial, a native of the West Indies, and cultivated in our gardens on account of the beauty and duration of its odoriferous flowers, which open only during the night, and of which it produces continual successions from June or July till checked by frosts; at which time the roots, which do not endure our winter, are taken up, and preserved in sand till spring. Whether the roots produced here are equivalent in virtue to those which are brought from abroad, has not, that I know of, been tried.

The officinal jalap roots come from the province of Xalapa in New Spain; in thin transverse slices, solid, hard, weighty, of a blackish colour on the outside or cortical part, internally of a dark greyish with several black circular striæ: the hardest, darkest coloured, and those which have the most of these resinous veins, are the best. Slices of bryony root, which are said to be sometimes mixed with them, may be distinguished by their whiter colour and less compact texture.

This root has scarcely any smell, and very little taste upon the tongue: swallowed, it affects the throat with a slight kind of pungency and heat. Taken in doses of a scruple or half a dram, it proves an effectual and in general a safe purgative; very rarely occasioning any severe gripes or nausea, which too frequently accompany the other strong cathartics. Some have prohibited the use of this cathartic to children; probably on no very good foundation. Young children, from the laxity of their solids, and the soft lubricating quality of their food, generally bear these kinds of medicines better than adults,

adults, and adults of a spungy, lax, or weak habit, better than the rigid or robust. Few, if any, of the strong resinous purgatives are in either case more innocent than jalap.

Jalap root, digested in as much rectified spirit as will cover it to the height of about four fingers, gives out greatest part of the resinous matter in which its activity resides, and tinges the menstruum of a yellowish brown colour. On inspissating the filtered tincture to about one half, and adding to the remainder a proper quantity of water, the liquor becomes milky, and on standing deposites the pure resin. This preparation, given by itself, irritates and gripes violently, without proving considerably purgative: thoroughly triturated with testaceous or other powders, or with soap; or ground with almonds or powdered gum-arabic, and made into an emulsion with water; or dissolved in rectified spirit, and mixed with a proper quantity of syrup, that the solution may bear being diluted with watery liquors without precipitation; it purges, in doses of eight or ten grains, as effectually, and for the most part as mildly, as the jalap in substance.

The jalap remaining after sufficient digestion with spirit, has no cathartic virtue: boiled in water, it gives out a mucilaginous substance, which operates only by urine. Water applied at first takes up a portion of the resin along with the gum, and hence the watery decoction and extract prove weakly cathartic as well as diuretic: the root still retaining great part of its resin, so as to purge considerably. The resinous and gummy parts may be united into one extract, by first drawing a tincture from the powdered root with rectified spirit, then boiling the residuum in fresh quantities of water, evaporating

the

JALAPIUM.

the decoctions till they begin to grow thick, mixing in by degrees the tincture infpiffated to a like thicknefs, and continuing a gentle heat till the whole is reduced to a due confiftence. This extract may be taken by itfelf in dofes of twelve grains or more: the gummy matter of the jalap being fufficient to divide the refin and prevent its too violent irritation. [Extractum jalapii *Ph. Lond. & Ed.*]

The proportion of active matter differs greatly in different parcels of the jalap; fixteen ounces of fome forts yielding hardly two of refin, while the fame quantity of others affords three or four. Hence the extracts of jalap appear preferable to the root in fubftance, not only on account of the dofe being rendered fmaller by the rejection of the woody parts, but likewife as being more uniform and certain in ftrength. Tinctures of jalap made in proof fpirit are nearly fimilar in quality to the gummy refinous extract, this menftruum taking up both the refinous and gummy parts of the root: thefe preparations, made from different kinds of jalap, will vary in ftrength fomewhat more than the folid extract or refin, but not fo much as fome have fufpected, or as the roots in fubftance; for in the proportions ufually employed, the proof fpirit does not take up the whole of the virtue of any kind of jalap, and perhaps it does not extract much more from one kind than from another, provided the jalap be of moderate goodnefs. If three† or four‡ ounces of jalap be digefted in a pint of proof fpirit, the refiduum will ftill give out a portion of refinous matter to rectified fpirit, and this refin will be in greater quantity in proportion as the root itfelf was the more refinous. [Tinct. jalap. † *Ph. Ed.* ‡ *Ph. Lond.*]

ICHTHYOCOLLA.

ICHTHYOCOLLA Pharm. Lond. Isinglas or Fish-glue: a solid glutinous substance, prepared from a fish of the sturgeon kind, caught in the rivers of Russia and Hungary. The skin, fins, &c. are boiled in water, the decoction inspissated to a due consistence, and then poured out so as to form thin cakes; which are either exsiccated in that form, or cut while soft into slices and rolled up into spiral, horse shoe, and other shapes. The best is in thin, clear, and almost transparent pieces.

* A different account of the formation of isinglass is given by Mr. Jackson in *Philos. Trans.* vol. lxiii. part I. He asserts that the solution of animal substances of every kind gives *glue*, not *isinglass*; that this last is nothing more than certain membranous parts of fishes, as the air-bladder, intestines, peritonæum, &c. in their entire state, only freed from their natural mucus and adhering matters, and rolled and twisted into the forms in which we get it.

Ichthyocolla is one of the purest and finest of the animal glues, of no particular smell or taste. Beaten into shreds, it dissolves pretty readily in boiling water or milk, and forms a gelatinous substance, which yields a mild nutriment, and proves useful medicinally in some disorders arising from a sharpness and colliquation of the humours. A solution of it in water, curiously spread, whilst hot, upon silk, affords an elegant sticking plaster for slight injuries of the skin, not easily separable from the part by water, and scarcely inferiour to the more compounded one sold under the name of the ladies black plaster,

ILLECEBRA.

plaster, in which different balsams and resins are joined to the ichthyocolla.

ILLECEBRA.

ILLECEBRA, *Vermicularis*, *Piper murale*, *Sedum minus*. *Sempervivum minus vermiculatum acre C. B. Sedum acre Linn.* WALLPEPPER or STONECROP: a small plant, having its stalks covered with little fleshy conical leaves set thick together in the manner of scales: on the tops appear pentapetalous yellow flowers, each of which is followed by several pods full of small seeds. It is annual, grows on old walls and dry stony grounds, and flowers in July.

THIS plant has a very acrid taste, and no remarkable smell: applied externally, it vesicates the part: taken internally, in no great quantity, it proves strongly emetic. Its active matter appears, from the accounts given by authors, to be in great part forced out along with the watery juice by expression; to dissolve both in water and fermented liquors by infusion; and not to be dissipated, or not soon, by boiling. It is said to have been used with success in sundry chronical disorders (a), but its durable acrimony, and the great vehemence of its operation, have prevented its being received in practice.

IMPERATORIA.

IMPERATORIA *Pharm. Edinb. Imperatoria major C. B. Imperatoria astrutium Lob. & Linn. Astrantia Dod. Smyrnion hortense Trag.* Stru-

(a) Below, *Eph. nat. curiof. dec.* i. ann. vi. & vii. *obf.* 22. Boerhaave, *hist. plant.* p. 369.

thium hodie vocatum Cord. Masterwort: an umbelliferous plant, with large winged leaves divided into three indented fegments; producing thick oblong, ftriated feeds furrounded with a narrow leafy margin: the roots are oblong, thick, knobby, jointed, with feveral lateral fibres, brown on the outfide and whitifh within. It is perennial, a native of the Alps and Pyreneans, from whence we are fupplied with roots fuppofed to be fuperiour to thofe which are raifed in our gardens.

The root of imperatoria is a very warm and moderately grateful aromatic, nearly of the fame nature with that of angelica. Infufed in water, or digefted in rectified fpirit, it impregnates both menftrua ftrongly with its fragrant fmell; the former weakly, the latter ftrongly, with its warmth, pungency, and bitterifhnefs; the former with a muddy brownifh, the latter with a bright yellow colour. On infpiffating the fpirituous tincture, very little of its flavour exhales with the fpirit: the remaining deep yellow extract fmells moderately of the root, and impreffes on the organs of tafte a confiderable bitternefs and glowing pungency. Water carries off in evaporation nearly all the fpecific flavour of the mafterwort, leaving, in the dark brown extract, a naufeous bitternefs with a flight degree of warmth or acrimony.

IPECACOANHA.

IPECACUANHA Pharm. Lond. & Edinb. Hipecacuanna; Radix brazilienfis. Pfychotria emetica

emetica Linn.*(a) Ipecacoanha: a slender root, brought from the Spanish West Indies, in short pieces, variously bent and contorted, full of wrinkles and deep circular fissures, which reach quite down to a small whitish woody fibre that runs in the middle of each piece: the cortical part is compact, brittle, and looks smooth and resinous on breaking. Two sorts of this root are met with in the shops, one brought from Peru, the other from Brazil; usually denominated from their external colour, the first *whitish, grey,* or *ash-coloured,* the other *brown* ipecacoanha.*(b) The first is generally preferred, being found to operate with the greatest certainty and mildness.

A root has been brought over under the name of white ipecacoanha, which has little or nothing of the virtues of the two foregoing: this is readily distinguished by its yellowish white colour, woody texture, and having no fissures or wrinkles. More dangerous abuses have sometimes been committed, by the substitution or mixture of the roots of an American *apocynum,* which have been found to operate with great violence both upwards and downwards, and in some instances, as is said, to prove fatal: these may be known by their being larger than the true ipecacoanha, the fissures more distant, the intermediate spaces smoother, and more particularly by the colour of the medullary fibre; which in the poisonous roots is a deep reddish yellow, in the true ipecacoanha a whitish or pale greyish.

*(a) This is the name given it in the supplement to Linnæus: it was formerly reckoned by him a species of *Lonicera.*

*(b) Both kinds have been found in the neighbourhood of Rio di Janeiro. *Lond. Med. Journ.* ix. 69.

MATERIA MEDICA.

Ipecacoanha has scarcely any smell, unless during its pulverization or infusion in liquors, in which circumstances it emits a faint nauseous one: in chewing, the wrinkled cortical part proves bitterish and subacrid, and covers the tongue as it were with a kind of mucilage; the medullary woody fibre is nearly insipid, and gives out to menstrua very little active matter. Geoffroy observes, that in pulverizing considerable quantities, the finer powder that flies off, unless great care be taken to avoid it, is apt to affect the operator with a difficulty of breathing, a spitting of blood, a bleeding at the nose, or a swelling and inflammation of the eyes and face, and sometimes of the throat; and that these symptoms go off in a few days, either spontaneously, or by the assistance of venæsection. *In the Philosophical Transactions, vol. lxvi. part I. is a remarkable case of violent asthmatic fits in a lady caused by the effluvia of powdered ipecacoanha.

This root is the mildest and safest emetic that has yet been discovered; and may be ventured on almost in the lowest circumstances where the stomach requires to be unloaded. The common dose is from ten grains to a scruple and upwards: in the medical observations and inquiries published by a society of physicians in London, a great number of cases are mentioned, in which two grains operated sufficiently: in constitutions which bore vomiting ill, and which were greatly ruffled by the usual doses, two or three grains operated with great ease. Where it fails of operating upwards, it commonly purges, and sometimes considerably: in this intention it may be employed, in several cases, to advantage, in conjunction with other purgatives,

to

to determine its action downwards: I have found fifteen grains of jalap, with two or three of ipecacoanha, purge more than twice the quantity of jalap by itself.

The ipecacoanha was first introduced, about the middle of last century, as a specific in dysenteries; and repeated experience has confirmed its efficacy in this distemper, not only when used as an emetic, but likewise when given in such small doses as scarcely to affect the grosser emunctories. In common dysenteric fluxes, it frequently performs a cure in a very short space of time; not by its exerting an astringent power, as some have supposed, for it does not appear to have any real astringency; nor by its mucilaginous substance covering the intestines and incrassating thin humours, as others with more plausibility, have inferred both from its mucilaginous taste, and from the ropiness and fliminess which it manifestly communicates to the contents of the stomach; but apparently by promoting perspiration, the freedom of which is in these cases of the utmost importance, and an increase of which, even in a state of health, is generally observed to diminish the evacuation by stool. In common dysenteries, the skin is for the most part dry and tense, and perspiration obstructed: and indeed this obstruction, and the conversion of the perspirable matter upon the intestines, is very frequently the immediate cause of the disease. Most of the common diaphoretics pass off, in these cases, without effect: but ipecacoanha, if the patient, after a puke or two be covered up warm in bed, brings on a free diaphoresis or a plentiful sweat, by which I have often known the distemper terminated at once.

In putrid or malignant dysenteries, or where the patient breathes a tainted air, it has not been found equally succesful: it requires here to be continued for several days, or repeated as an evacuant, with the further assistance of rhubarb, cordial antiseptics, and mild opiates or astringents. Where plentiful evacuation is necessary, or the offending matter lodged deep, and the operation can be borne without inconvenience, the ipecacoanha, as Dr. Pringle observes, is most advantageously given in small quantities at a time, and repeated at proper intervals, till a vomiting or purging comes on.

*In the spasmodic asthma, Dr. Akenside remarks, that where nothing contraindicates repeated vomiting, he knows no medicine so effectual as ipecacoanha. In violent paroxysms, a scruple procures great and immediate relief. For habitual indisposition, from three to five grains every morning, or from five to ten every other morning, may be given for a month or six weeks. It is equally useful where it does not vomit, as where it does. The relief seems owing to its general antispasmodic or relaxing property, of which its emetic operation is probably a particular consequence *(a)*.

*In the *Stockholm acts* 1770, are several cases of uterine hæmorrhages cured by one third or half a grain, rubbed with sugar, given every four hours or oftener. In one case, the hæmorrhage returned on discontinuing the medicine, and ceased on repeating it. These small doses had good effects in catarrhal coughs, even in those which attend consumptions; and if not beneficial, are at least not hurtful, in bloody coughs, in which vomiting has several times

(a) Med. Transact. i. 93.

IPECACOANHA.

been obferved to come on, without any increafe of the hæmorrhage. They may be ufeful in peripneumony and pleurify, in which cough is often the moft troublefome fymptom, and in which Seneka root (which in increafed dofes proves alfo emetic) has been fo much recommended.

The emetic virtue of ipecacoanha refides in its refinous parts. By digefting the root in frefh quantities of rectified fpirit, and infpiffating the filtered tinctures, a refinous extract is obtained, to the quantity of about three ounces from fixteen, which, by itfelf, vomits ftrongly, and with great irritation: the refiduum yields to water nearly four ounces of a foft tenacious mucilage, which has fcarcely any fenfible operation. If only a part of the refin be extracted, by flight digeftion in a little highly rectified fpirit, the remaining root proves more gentle, and rather purgative than emetic: in this ftate it is recommended by fome in dyfenteries accompanied with a confiderable fever, where the root with its natural quantity of refin might irritate too much; but as fmall dofes of the root itfelf operate with all the safe and gentlenefs that can be wifhed for, this precarious method of weakening it does not appear advifable.

By boiling it in water, a part of the refin is taken up with the mucilage; the extract amounting to about fix ounces from fixteen, and proving mildly emetic. The beft menftruum for extracting the entire virtue of the root appears to be a mixture of one part of pure fpirit with two or rather three parts of water: after fufficient digeftion in this menftruum, neither water nor fpirit took up any thing confiderable from the remainder. In the fhops wine is employed: an ounce of the root

is

<small>Vinum ipe-cacuanhæ
† *Ph. Lond.*
‡ *Ph. Ed.*</small>

is macerated or digested in a pint† or fifteen ounces of mountain‡. These tinctures, in doses of from half an ounce or less to an ounce and upwards, prove mildly emetic.

I R I S.

IRIS: a perennial plant with long narrow sword-like leaves, standing edgewise to the stalk; and large naked flowers, divided deeply into six segments, of which, alternately, one is erect and another arched downwards, with three smaller productions in the middle, inclosing the stamina and pistil: the roots are tuberous, irregular, and full of joints.

1. IRIS *vulgaris germanica sive silvestris C. B. Iris germanica Linn.* Flower-de-luce, common iris or orrice: with blue flowers, whose arched segments are bearded with a yellowish matter, standing several on one stalk higher than the leaves. It is a native of the mountainous parts of Germany, common in our gardens, and flowers in June.

THE roots of this plant have, when fresh, a disagreeable smell, and an acrid nauseous taste. They are a strong irritating cathartic; in which intention, their expressed juice has been given in hydropic cases, from one or two drams to three or four ounces, diluted largely with watery or vinous liquors, to prevent its inflaming the throat. The remarkable differences in the dose, as directed by different practical writers, appear to have proceeded from hence; that some employed the juice in its recent turbid state, loaded with the acrimonious cathartic matter of the root; others, such as had been depurated

depurated by fettling, and which had depofited, along with the feculencies, a great fhare of the active parts. By gently infpiffating the juice, it is rendered lefs violent in cathartic power, and lefs liable to irritate and inflame; but becomes at the fame time too precarious in ftrength to be depended on: by infpiffation to perfect drynefs, its purgative virtue is almoft, if not altogether, deftroyed. The root itfelf lofes alfo, in drying, its offenfive fmell, and its naufeous acrimony, and along with thefe its cathartic quality: in this ftate, it difcovers a flight and not difagreeable pungency and bitter- ifhnefs, accompanied with a kind of aromatic flavour, nearly of the fame kind with that of the following fpecies, but weaker and lefs grateful.

The bluifh expreffed juice of the flowers changes on being infpiffated, efpecially if a little lime-water is added, to a fine green; and in this form is directed, in foreign pharma- copœias; for tinging fome of the unctuous compofitions called odoriferous or apoplectic balfams.

2. IRIS *Pharm. Lond. Iris florentina Pharm. Edinb. & Linn. Iris alba florentina C. B.* Florence orrice; fuppofed to be only a variety of the foregoing occafioned by difference of climate; diftinguifhable from it in our gardens, by the flowers being white, and the leaves inclining more to bluifh. The fhops are fup- plied from Italy with dried roots fuperiour to thofe of our own growth; in oblong flattifh pieces freed from the fibres and brownifh bark, externally of a whitifh colour with brownifh fpecks, internally inclining to yellowifh, eafily
reducible

reducible into a farinaceous, yellowish white powder.

This root, in its recent state, does not seem to differ much from the preceding; being, like it, naufeous, acrimonious, and purgative, though not quite in fo great a degree; and lofing thefe qualities on being dried. The dry root, as met with in the fhops, has an unctuous, bitterifh, pungent tafte, not very ftrong, but very durable in the mouth: and a light agreeable fmell, approaching to that of violets. It is ufed in perfumes; in fternutatory powders; for communicating a grateful flavour, fomewhat like that of rafpberries, to wines and to fpirits; and medicinally in diforders of the breaft, for attenuating vifcid phlegm, and promoting expectoration. Its fmell and tafte are extracted both by water and rectified fpirit, moft perfectly by the latter. In diftillation, it gives over with water the whole of its peculiar flavour, its bitternefs and a flight acrimony remaining in the infpiffated extract: the diftilled water fmells very agreeably, but no effential oil is obtained though fome pounds of the root be fubjected to the operation at once. Rectified fpirit brings over a part of its violet fmell, but little or nothing of its warmth or tafte: the infpiffated extract is a pungent, bitterifh, balfamic mafs, glowing in the mouth like pepper; its quantity is about one fifteenth of the weight of the root.

3. Iris *paluftris* Pharm. Edinb. *Iris paluftris lutea* Ger. *Gladiolus luteus. Acorus vulgaris* Pharm. Auguftan. *Acorus adulterinus* C. B. *Pfeudoacorum* Matth. *Pfeudoris* Dod. *Butomon* Cluf. *Iris Pfeud-Acorus* Linn. Yellow water-flag,

IRIS.

flag, baftard acorus, fedge: with reddifh roots, yellow unbearded flowers ftanding feveral on one ftalk, and the middle ribs of the leaves prominent. It is common by the fides of rivers and marfhes, and flowers in June.

THE roots of this fpecies are, when frefh, rather more acrid, and more ftrongly cathartic, than either of the preceding. The expreffed juice, given to the quantity of eighty drops every hour or two, and occafionally increafed, has, in fome inftances, produced plentiful evacuations, after jalap, gamboge, and mercurials had failed *(a)*: but however fuccefsful it may have fometimes been as a draftic purgative, it is accompanied, like the other irifes, with a capital inconvenience; its ftrength being fo precarious, or fo variable in different ftates, that it is by no means fit for general ufe. The juice, both of this and of the other kinds of iris, has been employed alfo externally for clearing the fkin of ferpiginous eruptions; and fometimes fnuffed up the nofe as a ftrong errhine: even for thefe purpofes it is to be ufed with caution, being fubject, by its great acrimony, to inflame or veficate the parts.

The dry roots are much weaker and lefs agreeable than thofe of either of the preceding fpecies of iris. They have fcarcely any fmell; and when chewed in fubftance, difcover very little tafte. An extract made from them by rectified fpirit is likewife weaker and more naufeous, though its quantity is lefs, amounting only to one twenty-fourth of the weight of the root: it has nothing of the flavour or aromatic warmth of thofe of the other two, but an

---

(a) Edinburgh medical effays, vol. v. art. 8.

ungrateful

ungrateful auftere bitterifhnefs and a kind of faline pungency. It is the root in this dry ftate that the writers on medicines mean, when they fpeak of the yellow water-flag root as being aftringent and ftomachic: it does not, however, appear to have any great claim to thefe virtues, and among us is no otherwife made ufe of than as an ingredient in the officinal arum powder, in which it is faid to be employed only in deference to the original of Birckmann firft publifhed by Quercetanus.

* JUGLANS.

JUGLANS Pharm. Lond. Nux juglans C. B. Juglans regia Linn. WALNUT: a large tree commonly cultivated in this and moft other countries of Europe for its fruit, which is a flefhy drupe, becoming hufky when ripe, and inclofing a nut with an edible kernel. The unripe fruit (which is the part fpecified in the London catalogue) has a fharp acerb tafte, and when handled, tinges the fkin with ruft-coloured durable fpots. Infufed in water it imparts a bitter harfh tafte to the fluid, which becomes blackifh on the addition of vitriol of iron. An extract prepared from it is fubfaline, lightly acerb and ftyptic, and fufficiently grateful to the fmell. This extract is accounted an excellent anthelmintic, given twice or thrice a day in the dofe of a tea fpoonful to children. It proves purgative, and expels the worms with the ftools. A fyrup made with a ftrong decoction of green walnuts and brown fugar, is much ufed in fome parts of England as a domeftic aperient medicine. The outer covering and fhell of the fruit have been joined with guaiacum and farfaparilla as ingredients for fudorific

decoctions

decoctions in rheumatic and venereal cases. Green walnuts enter an antivenereal decoction, the formula of which is given in a Treatise on the Venereal Disease by Dr. Swediaur, edition second and third. This decoction is by some supposed to be the genuine LISBON *diet drink* which has acquired considerable reputation among the nostrums for this malady.

The culinary use of unripe walnuts as a pickle is well known.

JUJUBÆ.

JUJUBÆ *Pharm. Paris. Jujubæ majores oblongæ* C. B. *Zizyphus Dod. Rhamnus Zizyphus Linn.* JUJUBES: a half-dried fruit of the plum kind, about the size and shape of an olive: consisting of a pretty thick reddish yellow skin, a whitish fungous pulp, and a wrinkled stone pointed at both ends: the produce of a prickly tree, with three-ribbed leaves, and herbaceous or yellowish flowers, sometimes found wild, and commonly cultivated in the southern parts of Europe.

This fruit, when in perfection, has an agreeable sweet taste; and in those countries where it is common, makes an article of food in its recent state, and of medicine when half dried; decoctions of it being used, like other glutinous sweets, as incrassants, and demulcents in defluxions on the breast. Among us, it has long stood neglected, and is now become a stranger to the shops; the tree not producing fruit in this climate; and that, which we received from abroad, being commonly mouldy or carious.

ANOTHER fruit of the same kind, of a dark blackish hue, furnished with an ash-coloured cup at the bottom, from which it eafily parts, is fometimes brought from the eaftern countries, under the names of *febeften, myxa,* or *myxaria.* It is produced by the *Cordia Myxa* of Linnæus. It is more glutinous than the jujube; to which it has been commonly joined in pectoral decoctions; and along with which it is now difcarded by the colleges both of London and Edinburgh.

JUNCUS ODORATUS.

JUNCUS ODORATUS five aromaticus C. B. *Schænanthus, fquinanthum, fœnum camelorum, & palea de mecha quibufdam. Andropogon Schænanthus Linn.* SWEET RUSH or CAMEL'S HAY: a dried herb, of the grafs kind, brought from Turkey and Arabia, in bundles about a foot long; confifting of fmooth ftalks, in fhape and colour fomewhat refembling barley ftraws, full of a fungous pith like thofe of rufhes; and leaves like thofe of wheat furrounding the ftalk with feveral coats: towards the tops of the ftalks are fometimes found fhort woolly fpikes of imperfect red flowers, fet in double rows.

THE fweet rufh, when in perfection, has an agreeable fmell, and a warm, bitterifh, not unpleafant tafte. Diftilled with water, it yields a fmall quantity of a yellowifh, fragrant, and very pungent effential oil: the remaining decoction, thus divefted of the aromatic matter of the plant, proves unpleafantly roughifh, bitterifh, and fomewhat acrid. A tincture made in rectified fpirit, in colour greenifh yellow, yields, on being infpiffated, a tolerably grateful, bitterifh, aromatic extract. This plant, formerly employed as a warm ftomachic and deobftruent,

appears

JUNIPERUS.

appears from the above experiments to be of no inconfiderable activity; but in this country, more common aromatic vegetables have now fuperfeded its ufe. It has been kept in the fhops only as an ingredient in the mithridate and theriaca; and the two colleges, having at laft expunged thofe compofitions, have dropt the *juncus odoratus*.

JUNIPERUS.

JUNIPERUS: *juniperus vulgaris fruticofa* C. B. *Juniperus communis Linn.* JUNIPER: an evergreen tree or bufh, clothed with flender narrow ftiff fharp leaves, like prickles, which ftand generally three together: the flowers are a kind of fmall fcaly catkins growing on one plant; the fruit, round berries, growing on a different one, containing, each, three oblong irregular feeds. It is common on heaths in different parts of Europe; and is found, at all feafons of the year, both with unripe green or red berries, and with ripe bluifh black ones.

The BERRIES, *baccæ juniperi Pharm. Lond. & Edinb.* are brought chiefly from Holland and Italy: they fhould be chofen frefh, not much fhrivelled, and free from mouldinefs, which they are very fubject to contract in keeping. They have a moderately ftrong not difagreeable fmell, and a warm pungent fweetifh tafte, which if they are long chewed or previoufly well bruifed, is followed by a confiderable bitternefs. The fweetnefs appears to refide in the juice or foft pulpy part of the berry: the bitternefs, in the feeds; and the aromatic flavour, in oily veficles, fpread throughout the fubftance both of the pulp and of the feeds, and diftinguifhable even by the eye. The frefh berries yield, on expreffion, a rich, a rich, sweet, honey-like aromatic juice: if previously powdered, so as to thoroughly break the seeds, which is not done without difficulty, the juice proves tart and bitter. The same differences are observable also in tinctures and infusions made from the dry berries, according as the berry is taken entire or thoroughly bruised.

They give out nearly all their virtue both to water and rectified spirit, and tinge the former of a brownish yellow, the latter of a bright orange colour. Distilled with water, they yield a yellowish essential oil, very subtile and pungent, in smell greatly resembling the berries, in quantity (if they have been sufficiently bruised) about one ounce from forty: the decoction, inspissated to the consistence of a rob or extract, has a pleasant, balsamic, sweet taste, with a greater or less degree of bitterishness. A part of the flavour of the berries arises also in distillation with rectified spirit: the inspissated tincture consists of two distinct substances; one oily and sweet: the other tenacious, resinous, and aromatic.

Ol. e baccis juniperi Pb. Lond. & Ed.

These berries are useful carminatives, detergents, and diuretics. The distilled oil is a very stimulating diuretic, approaching in quality to that of turpentine, like which, it impregnates the urine with a violet smell: the spirituous extract gives the same kind of smell; as does likewise the berry in substance, in a lower degree; but the watery extract or rob, as being divested of the oil, has no such effect. This last may be used with advantage in cases where the more stimulating preparations would be improper; as in catarrhs, debilities of the stomach and intestines, and difficulties of the urinary excretions, in persons of an advanced age. Among the aromatics that have been tried in composition

with

with juniper berries, sweet fennel seeds and caraway seeds seem the best adapted to improve their flavour: a cordial water is prepared in the shops by drawing off a gallon of proof spirit from a pound of the berries and an ounce and a half of each of the seeds. The water is strongly impregnated with the volatile virtue of the berry; to which the more fixt ones may in many cases be usefully superadded, by mixing with it a proper quantity of the rob. *Spiritus juniperi comp. Ph. Lond.*

The WOOD, *lignum juniperinum, cedrinum lignum Pharm. Parif.* has been recommended as a sudorific, and by some accounted similar to guaiacum or sassafras, to either of which it is greatly inferiour. It has a weak not unpleasant smell, and very little taste: decoctions and extracts, made from it with water, are disagreeably bitterish, subastringent, and balsamic: the spirituous tinctures are weaker than the watery, and yield, on being inspissated, an almost insipid resin. The quantity of watery extract, according to Cartheuser's experiments, is about one twelfth the weight of the wood; of spirituous extract, one eighth.

IN the warmer climates, particularly on the coasts of Africa, there exudes, from a larger species of juniper, a resinous juice, which concretes into semipellucid pale yellowish tears or glebes, resembling mastich, but larger; the *sandaracha* of the Arabians, and *gummi juniperinum* of the shops called by some, from the use to which it has been principally applied, *vernix*. This resin has a light agreeable smell, and no considerable taste: it dissolves in rectified spirit, and in oils both expressed and distilled, but gives out little or nothing to watery liquors, and thus discovers that it is nearly a pure resin. It is is supposed to be similar in quality, as in appearance, to mastich; and has been sometimes given internally, against hemorrhagies, old fluxes, and ulcerations; but principally employed externally in corroborant, nervine, traumatic applications. Among us, it is scarcely ever made use of for any medicinal purposes; other resinous substances, more common in the shops, being apparently superiour to it.

KALI.

KALI majus cochleato semine C. B. *Salsola quibusdam. Salsola Soda Linn.* SNAIL SEEDED GLASSWORT or SALTWORT: a plant with spreading, reddish, pretty thick branches; oblong, narrow, pointed, fleshy leaves like those of the houseleeks; and imperfect flowers in the bosoms of the leaves, followed each by one seed spirally curled and inclosed in the cup. It is annual, and grows wild on the sea coasts in the southern parts of Europe, particularly of the Mediterranean.

This herb is very juicy, in taste bitterish and remarkably saline. The expressed juice, and infusions or decoctions of the leaves, are said to be powerfully aperient and diuretic, and in this intention have by some been greatly recommended in hydropic cases: half a dram of the juice is reckoned a sufficient dose. But the kali is principally regarded, on account of its yielding copiously, when burnt, the fixt alkaline salt called *soda* or *soude:* an impure soda is prepared from it about Montpelier, where the plant is said to be cultivated for this use in the salt marshes; and a purer kind at Alicant in Spain the *barilla Pharm. Lond.* from a somewhat

KERMES.

a somewhat different species of kali *(a)*. The salt called *kelp*, prepared among ourselves from different marine plants, contains an alkali of the same kind, but more impure.

The soda is much milder in taste than the common vegetable alkalies, and is in several other respects also very considerably different from them, being of the same nature with the mineral alkali or basis of sea salt (see *Natron*). It promises to be an useful article of the materia medica, and has for some time past been received in practice in this country, as it has long been among the French, both by itself, and combined with tartar into the neutral salt called *sal rupellense*. *The Edinburgh college have received a purified salt of this kind, under the title of *sal alcalinus fixus fossilis purificatus*; and the London, under that of *natron præparatum*.

KERMES.

GRANUM TINCTORIUM & coccus baphica quibusdam. KERMES: round reddish-brown grains, about the size of peas: found in Spain, Italy, and the southern parts of France, adhering to the branches of the scarlet oak. These grains appear, when fresh, full of minute reddish ova or animalcules, of which they are the nidus, and which in long keeping change to a brownish red powdery substance. They are cured by sprinkling with vinegar before exsiccation: this prevents the exclusion of the ova, and kills such of the animals as are already hatched; which would otherwise become

---

(a) Kali hispanicum supinum annuum sedi foliis brevibus, *Mem. de l'acad. des scienc. de Paris, pour l'ann.* 1717. & *Pharm. Parif. p.* lxiv. *Salsola sativa Linn.*

winged

winged insects, and leave the grain an empty husk.

Fresh kermes yield upon expression a red juice, of a light pleasant smell, and a bitterish, subastringent, somewhat pungent taste: this juice, or a syrup made from it, are brought from the south of France, and sometimes made use of as mild restringents and corroborants. An elegant cordial confection, for these intentions, is prepared in the shops, by dissolving, in the heat of a water bath, six ounces of fine sugar in six ounces by measure of damask rose water, then adding three pounds of the juice of kermes warmed and strained, and after the whole has grown cold, mixing in half a scruple of oil of cinnamon: this confection is taken from a scruple to a dram or more; either by itself, or in juleps, with which it mingles uniformly without injuring their transparency. The dried grains, if they have not been too long kept, give out, both to water, and to rectified spirit, the same deep red colour, and nearly the same kind of smell and taste, with those of the expressed juice. The watery tinctures lose nearly all their smell in evaporation: the spirituous retain nearly the whole of their smell as well as of their taste. The inspissated extracts are considerably bitter, astringent, and of a kind of mild balsamic pungency: the spirituous is stronger and in somewhat smaller quantity than the watery, but the difference in 'strength is more considerable than that of the quantity, spirit seeming to extract the active matter more completely than water.

Confectio alkermes.

KINO.

KINO.

GUMMI rubrum aftringens gambienfe D. Fothergill in *med. obf. Lond.* vol. i. 1757. *Kino, Pharm. Lond. & Edinb.* Red aftringent gum from Gambia; fuppofed to exude from incifions made in the trunks of certain trees, called *pau de fangue*, growing in the inland parts of Africa.

It is very friable, fo as to be crumbled in pieces by the hands; of an opake dark reddifh or almoft black colour in the mafs, and when reduced to powder, of a deep brick red: fmall particles of it, viewed with a magnifying glafs, appear of a femitranfparent red like bits of garnet. In chewing, it firft crumbles, then fticks together a little, and in a fhort time feems wholly to diffolve, impreffing a very confiderable aftringency accompanied with a flight fweetifhnefs. It has no fmell.

To oils it gives little or no tincture. On a red-hot iron, it glows for a long time like a bit of burning charcoal, without fhewing any difpofition to melt: it yields, during a little while, a flight dull flame hovering about the furface, and leaves at length a large proportion of greyifh afhes.

Both rectified fpirit and water diffolve, each, about two thirds of it, the fpirit fomewhat more than the water. Both folutions, when made with the fame quantities of the two menftrua, as twenty or thirty times the weight of the powdered gum, appear of the fame deep bright red colour, the fpirituous rather deepeft: with folution of chalybeate vitriol, they both produce inky mixtures, from which the black matter fpeedily concretes, and fettles to the bottom,

leaving

MATERIA MEDICA.

leaving the liquors colourlefs (a). The watery folution fuffers no apparent change from the addition of alkalies fixt or volatile; but acids render it turbid, and occafion a copious precipitation.

The part, which water leaves undiffolved, feems as dark-coloured as the gum at firft: it gives the fame deep red tincture to fpirit, and this tincture ftrikes the fame black with folution of vitriol. The part which fpirit leaves undiffolved is much paler than the original gum, gives no tincture to water, and produces no change with the vitriolic folution.

It appears therefore that both the colouring and aftringent matter are more completely taken up by fpirit than by water; though water extracts readily enough a great fhare of both.

ONLY a little quantity of this drug has hitherto been brought over. Dr. Fothergill, the firft perfon, as far as I can find, who gave notice of it to the public, and who favoured me with the fpecimens on which the above experiments were made, informs us, that he had the firft intimation of it from a phyfician, who had met with good effects from it in obftinate chronical diarrhœæ; and that a parcel was afterwards fhewn to him, which had been received from a Guinea fhip, and taken for a

(a) The black matter in thefe kinds of mixtures appears to confift of the iron of the vitriol, difengaged from its acid folvent, and combined with the vegetable aftringent fubftance; the acid ferving only as a neceffary intermedium for procuring this union. The above black precipitates, after repeated ablutions with water, retained their blacknefs; and the clear liquors from which they had fettled, being examined with alkaline falt on the principles to be mentioned hereafter under the article *fales alkalini*, feemed to contain as much acid as the quantity of vitriol employed in them.

fine

LABDANUM.

fine kind of dragons blood, which it pretty much resembles in appearance, though in quality essentially different. He observes, that from the trials which have been made, and from its sensible qualities, it promises to be an article worth inquiring after, and to become in time a valuable addition to the materia medica. In disorders from laxity and acrimony, it may, doubtless, be of great advantage; nor do I recollect any other drug, that is so much of a gummy nature, and at the same time so astringent. Terra japonica comes the nearest to it, but is manifestly less astringent. The terra japonica differs likewise, in its watery solutions suffering no considerable separation of their parts from the addition of acids; and in the black matter, which they produce with vitriol, being little disposed to concrete and precipitate. Whether the cause, on which these kinds of diversities depend, be sufficient to influence also their medicinal powers, our knowledge, both in the chemical composition of bodies, and in the operation of medicines, is as yet too imperfect to permit us to judge. *The London and Edinburgh colleges have now received this gum as an officinal, and the latter have directed a tincture, in which two ounces of it are dissolved in a pound and a half of proof spirit.

*Tinctura e Kino *Ph. Ed.*

LABDANUM.

LADANUM Pharm. Lond. LABDANUM: a resinous juice, exuding upon the leaves of a small shrub, *cistus ladanifera cretica flore purpureo Tourn. Cistus creticus Linn.* which grows plentifully in Candy and some of the other islands of the Archipelago, and bears the winters of our own climate. The juice is said to be collected,

lected, by lightly brushing the shrub, in the summer heats, with a kind of rake having several straps or thongs of leather fixed to it instead of teeth *(a)*: the unctuous juice adheres to the thongs, and is afterwards scraped off with knives. The shrub is said to be very plentiful also in Spain *(b)*, but it does not appear that any labdanum is brought from thence.

Two sorts of labdanum are met with in the shops. The best, which is very rare, is in dark-coloured black masses, of the consistence of a soft plaster, growing still softer on being handled: the other is in long rolls coiled up, much harder than the preceding, and not so dark. The first has commonly a small, and the last a very large admixture of fine sand, which, in the labdanum examined by the French academy, amounted to three fourths of the mass. It is scarcely indeed to be collected pure, independently of designed abuses; the dust, blown on the plant by winds from the loose sands among which it grows, being retained by the tenacious juice.

LABDANUM has been sometimes exhibited as a resinous corroborant and restringent, but principally employed in external applications and perfumes: the soft kind makes an useful ingredient in the cephalic and stomachic plasters of the shops. This sort has an agreeable smell, and a lightly pungent bitterish taste: the hard is much weaker, and the common means of purifying these kinds of substances, though they

---

(a) Belon, *(Bellonius) observations des choses memorables trouvées en Grece*, &c. *l.* i. *c.* vii.

(b) Clusius, *Rariorum stirpium per Hispanias observatarum historia*, *l.* i. *c.* v.

separate

L A C.

separate the sandy matter mixed with it, render it weaker still. Rectified spirit of wine dissolves nearly the whole of the pure labdanum into a gold-coloured liquor: on inspissating the filtered solution, the finer part of the labdanum rises with the spirit, and the remaining resin proves both weaker and less agreeable than the juice at first. On infusing the labdanum in water, it impregnates the liquor considerably with its smell and taste: in distillation with water, there comes over a fragrant essential oil; and there remains in the still a brittle almost insipid resin, with a pale coloured liquor, which, inspissated, yields a weakly bitterish extract. The specific flavour of this juice seems to be sooner dissipated by heat than that of almost any of the other officinal resins or gummy resins.

L A C.

LAC; *lac asininum, caprinum, muliebre, ovillum, vaccinum.* MILK: asses, goats, human, sheeps, and cows milk: a fluid prepared and secreted in the bodies of animals, but not completely elaborated into an animal nature. On a chemical analysis, it yields the same general principles with substances of the vegetable-kingdom.

MILK is a mild nutritious balsamic fluid; when taken freely, an excellent obtunder of acrid and deleterious substances, and of over-doses of the stronger cathartics and emetics; one of the best restoratives in emaciated habits; a palliative, whilst its use is continued for the only aliment, in gouty cases not inveterate, and in some rheumatic pains; the medicine principally depended on in hectics and consumptions;
- prejudicial

prejudicial in acute difeafes, bilious fluxes and dyfenteries, fwellings of the præcordia, and obftructions of the abdominal vifcera.

It fometimes happens, that when the body ftands moft in need of this medicinal nutriment, the inteftines are too flippery to retain it. In fuch cafes it may be advantageoufly boiled with gentle aftringents, as granate peel, balauftines, red rofes; about an equal quantity of water being added, by a little at a time as the milk boils up, fo as that all the water may be wafted in the boiling *(a)*.

It may be prefumed that milk thickens in a found ftomach, before its digeftion, nearly in the fame manner as it is thickened by the runnet or infufion of the ftomach of a calf; and that, where the gaftric juices are too inert to produce this change, or fo acid as to produce it in too great a degree and to feparate a firm curd from the ferous part; the milk will be difficult of digeftion. Debilities of the ftomach are endeavoured to be corrected by the medication above-mentioned, or by the interpofition of proper ftomachics; acidities by the abforbent earths. The abforbent earths, however, are in this intention commonly infufficient, unlefs affifted by ftomachics; for as they abforb only the acid already generated, and have no power of remedying the weaknefs or indifpofition which tends to produce more, they afford only a temporary and palliative relief: and indeed it may be queftioned, whether they are capable of fo far deftroying the force, even of the acid they are mixed with, as to prevent its curdling milk in the ftomach.

(a) Mead, *monita & præcepta medica*, p. 49.

Milk

Milk is curdled by all acids; by moſt, perhaps by all, of the combinations of acids with earthy and metallic bodies ; by alkaline ſalts both fixt and volatile; by ſome vegetables that have no acidity or alkaline quality, as muſtard ſeed; and by ſtrong vinous ſpirits. The concentrated acids produce a ſtrong curd immediately on mixture: moſt of the other ſubſtances ſcarcely have their full effect without a boiling heat. The coagulum made by acids falls to the bottom of the ſerum: that made with alkalies ſwims on the ſurface, forming, eſpecially if the alkali is of the volatile kind, a thick coriaceous ſkin. The ſerum, with alkalies, proves of a greeniſh hue: that made with the other ſubſtances is nearly of the ſame appearance with the whey which ſeparates ſpontaneouſly.

The perfect neutral ſalts, or thoſe compounded of an acid and an alkali, produce no coagulation, either with or without heat: ſome of them, particularly nitre and ſal ammoniac, make the milk leſs coagulable, and, if added to the boiling mixture when already curdled by vegetable acids, render nearly the whole fluid again (a). Sugar retards the ſpontaneous coagulation, and impedes likewiſe the ſeparation of the cream from milk, and of the butyraceous part from cream. Lime-water and animal gall rediſſolve the coagula.

MILK, diſtilled with a gentle warmth, gives over a colourleſs and taſteleſs liquor, which ſeems to be mere water, but is found to differ from the ſimple element in growing ſour upon keeping. The reſiduum is a grumous, unctu-

---

(a) Willis, *Pharmaceutice rationalis*, parſ i. ſect. iv. cap. i. § 8.

ous, yellowish or brownish mass; which, on being boiled in water, partially diffolves. This folution contains the fweet fubftance of the milk, freed from the groffer unctuous cafeous matter; and proves an elegant whey, more agreeable in tafte, and which keeps better, than thofe prepared in the common manner. Thefe forts of liquors are very ufeful, cooling, diluent, aperients and detergents; in hypochondriacal complaints, impurities of the humours, acute difeafes, &c. They promote the natural excretions in general, and remarkably increafe the action of the purgative fweets, cafia and manna. The faline matter of thefe liquors may be obtained in a pure folid cryftalline ftate, by clarifying the whey with whites of eggs, and, after due evaporation, fetting it to fhoot, in the fame manner as other faline folutions.

Saccharum lactis Ph. Parif.

Thus milk is refolved into a watery fluid; a grofs fubftance indiffoluble in water, which appears to contain the directly nutrimental part; and a fweet aperient falt. The milks of different animals differ remarkably in the proportions of thefe ingredients, and in the quality of the falt.

Breaft milk and affes milk are very nearly alike: twelve ounces leave on evaporation, according to Hoffman's experiments, eight drams of folid matter, of which boiling water diffolves fix drams: the folution, infpiffated or cryftallized, yields a falt of a rich honey-like or faccharine fweetnefs. The fame quantity of cows milk leaves thirteen drams of folid matter, from which water extracts only about a dram and a half: the falt obtained from this folution is much lefs fweet, when purified is almoft infipid, diffolves very difficultly, and feems to have little claim to the pectoral and antiphthifical virtues

tues vulgarly afcribed to it. All the other milks that have been examined are of an intermediate nature between the two firft and the laft: goats milk approaches more to that of the afs than fheeps milk does, though both of them come nearer to that of the cow than of the afs.

There are confiderable differences in the milk of one and the fame animal according to its different aliment. Diofcorides relates, that the milk of goats, which fed on the fcammony plant and fpurges, proved cathartic; and inftances are given, in the *Acta Hafnienfia*, of bitter milk from the animal having eaten wormwood. It is a common obfervation, that cathartics, fpirituous liquors, &c. taken by a nurfe, affect the child; that the milk of animals, feeding on green herbs, is more dilute than when they are fed on dry ones; and that many of the common plants, which are eaten by cattle, give a particular tafte to their milk. Hoffman is of opinion, that, on this principle, milk may be ufefully impregnated with the virtues of different medicinal fubftances.

LACCA.

LAC, STICK-LAC, improperly called GUM-LAC: a concrete brittle fubftance, of a dark red colour; brought from the Eaft Indies incruftated on pieces of fticks; internally divided into feveral cells; faid to be the refinous juice of certain trees, collected by winged red infects of the ant kind, impregnated with the tinging matter of the infects, and by them depofited either on the branches of the trees, or on fticks faftened in the earth for that purpofe. In the cells

cells are often obferved fmall red bodies, which appear to be the young infects *(a)*.

* A curious account by Mr. Kerr of the infect producing this gum, is contained in the *Philof. Tranf.* vol. lxxi. part ii. From this it appears, that thefe infects are inhabitants of four trees; the *Ficus religiofa Linn.* the *Ficus Indica Linn.* the *Plafo Hort. Malabar.* and the *Rhamnus Jujuba Linn.* The lac is however rarely found upon this laft, and of an inferiour quality. The two fpecies of *Ficus* yield a milky juice when wounded, which inftantly coagulates into a vifcid fubftance. The *Plafo* tree by incifion gives out a red gum very fimilar to the lac. Hence the infect feems to have little trouble in animalizing the juices of thefe trees fo as to make its cell, which is the ftick-lac. It is found in very great quantities on the uncultivated mountains on both fides the Ganges; and is of great ufe to the natives in various works of art, as varnifh, painting, dying, &c.

The tinging red animal matter of the fticklac diffolves both in water and in rectified fpirit, and appears to be of the fame general nature with that of cochineal; like which it is made dull by alkalies, and brighter by acids, and turned to a fcarlet by folution of tin. If the lac be broken in fmall pieces, or grains, and infufed in warm water, till it ceafes to give any tincture to the liquor; the remainder appears of a transparent yellowifh or brownifh colour, and, on raifing the heat fo as to make the water boil, melts and rifes to the furface. The grains, or the plates formed from them by liquefaction,

Seed lac *of the fhops.*

Shell lac *of the fhops.*

(a) See the *Memoires de l'acad. roy. des fciences de Paris, pour l'ann.* 1714.

thus

thus robbed of great part of the animal tincture, seem to be of an intermediate nature between that of wax and resins, or to partake of the nature of both: they crumble on chewing, and do not soften or stick together again: laid on a red-hot iron, they instantly catch fire, and quickly burn off, with a strong and not disagreeable smell: distilled, they yield, like wax, an acid spirit and a butyraceous oil: alkaline lixivia, and volatile alkaline spirits, dissolve them into a purplish liquor: they dissolve also, by the assistance of heat, in rectified spirit of wine, and communicate to it a yellowish or brownish red colour, an agreeable smell, and a bitterish, subastringent, not unpleasant taste. The lac in substance, whether entire, or freed from so much of its colouring matter as boiling water is capable of extracting, has no manifest taste or smell.

A spirituous tincture of stick-lac has been sometimes given as a mild restringent and corroborant in female weaknesses, and in rheumatic and scorbutic disorders. But the principal medicinal use of this concrete is as a topical corroborant and antiseptic, in laxities and scorbutic bleedings and exulcerations of the gums: some employ for this purpose a tincture of the lac in alum water; others, a tincture made in vinous spirits impregnated with the pungent antiscorbutics.

LACTUCA.

LETTUCE: a plant with slender but firm stalks, which yield, as do the leaves, a milky juice on being wounded: the flower consists of a number of flat flosculi set in a small scaly cup, followed by short flat seeds, which are pointed at both ends and winged with down.

1. Lactuca *sativa* C. B. & *Linn*. Garden lettuce: with oblong, broad, rounded, uncut leaves; and numerous flowers ftanding on long pedicles in the form of an umbel. It is annual, and raifed at different times of the year in culinary gardens.

The young leaves of the feveral fpecies or varieties of garden lettuce are emollient, cooling, in fome fmall degree laxative and aperient, eafy of digeftion, but of little nourifhment; falubrious in hot, bilious, indifpofitions; lefs proper in cold phlegmatic temperaments. In fome cafes, they tend to procure fleep; not as being poffeffed of any ftrictly hypnotic power; but by virtue of their refrigerating and demulcent quality. When the plant is grown up, it proves confiderably bitter, though lefs fo than moft of the others of the lactefcent kind, to which it is fimilar in its general virtues.

The feeds, which in the common lettuce are of a grey or afh colour, in the cabbage lettuce black, unite with water, by trituration, into an emulfion or milky liquor, which has nothing of the aperient bitternefs of the milky juice of the leaves, and is nearly fimilar to the emulfions made with almonds. The lettuce emulfions have been fuppofed to be more refrigerant than thofe of the almond, and hence have fometimes been preferred in heat of urine and other diforders from acrimony or irritation.

2. Lactuca silvestris *Medicorum. Lactuca filveftris & Scariola Pharm. Parif. Lactuca filveftris cofta fpinofa* C. B. *Lactuca Scariola Linn.* Wild lettuce: with the leaves cut almoft to the rib into indented triangular fegments; and the ftalks and the ribs prickly. It is
. biennial,

biennial, grows wild in hedges, and flowers in June.

This species is considerably bitterer than the garden lettuces, and more aperient and laxative. It is nearly similar, in virtue as in taste, to endive unblanched.

3. LACTUCA GRAVEOLENS: *lactuca silvestris odore viroso* C. B. *Lactuca virosa Linn.* Strong scented lettuce, by some erroneously supposed to be the wild lettuce of medical writers: with the lower leaves entire, the upper jagged, the stalks and leaves prickly. It is biennial, found in hedges and by the sides of ditches, and flowers in June.

This species differs greatly in quality from the two preceding, though reckoned by botanists to be only a variety of the second. It smells strongly of opium, and appears to partake, in no small degree, of the virtues *(a)* of that narcotic drug. The opiate power of the lettuce, like that of the poppy-heads, resides in its milky juice, but whether the milk of the lettuce is of equal safety, or its virtue precisely of the same kind, with that of the poppy, is not known.

* Dr. Collin of Vienna has written a tract recommending the use of this plant in the cure of dropsies. The preparation he employs, is an extract from the expressed juice, first sufficiently clarified, and evaporated by a very gentle heat. He begins with small doses; but in dropsies of long standing, originating from visceral obstructions, he rises to the quantity of from one to three drams in twenty-four hours. He has constantly found it a mild re-

---

(a) Ray, *Historia plantarum*, i. 222. Boerhaave, *Hist. hort. Lugd. Bat. p.* 127.

medy, agreeing perfectly with the stomach. It usually kept the body open, but without exciting a purging. It seldom failed of proving powerfully diuretic, and at the same time mildly diaphoretic. The patient's thirst is said to have been totally extinguished by its use; but at the same time we are told that they were allowed to drink freely of diluting liquors during the course. Dr. Collin asserts, that out of twenty-four dropsical cases, all but one were cured by the use of this medicine; a degree of success that certainly entitles it to the further notice of the faculty.

LAMIUM.

LAMIUM ALBUM Linn. Lamium album non fœtens folio oblongo C. B. Galeopsis & archangelica, & urtica mortua sive alba quibusdam. WHITE ARCHANGEL or DEAD NETTLE: a plant with square stalks; oblong indented acuminated leaves, like those of the stinging nettle, set in pairs at the joints; and clusters, in the bosoms of the leaves, of white labiated flowers, whose upper lip is entire, arched, and hairy, the lower lip cloven. It is perennial, common in hedges and about the borders of fields, and found in flower from April to near the end of summer.

INFUSIONS of this plant, drank as tea, are said to be beneficial in uterine hemorrhagies and the fluor albus: the flowers are supposed to be more efficacious than the leaves, and hence those only are directed by the college of London. The sensible qualities, either of the one or the other, afford little foundation to expect from them any considerable virtues. The flowers have only a slight mucilaginous sweetishness, without any remarkable smell or flavour:

the leaves have a weak not unpleafant fmell, and a fmall degree of roughnefs,. which may entitle them to a place among the milder corroborants.

LAMPSANA.

LAMPSANA, *Lapfana, Napium, Papillaris herba. Soncho affinis lampfana domeftica* C. B. *Lapfana communis Linn.* DOCK CRESSES, NIPPLEWORT : a roughifh plant ; bearing fmall yellow flofculous flowers, fet in form of an umbel on the top of the ftalk, followed by little crooked naked feeds : the lower leaves are deeply cut, towards the pedicle, into generally two or four oppofite fections ; the upper are oblong, narrow, undivided, and have no pedicles. It is annual, grows wild by road fides, and flowers greateft part of the fummer.

THIS is one of the bitter lactefcent plants, nearly fimilar in virtue to dandelion, endive, cichory, and the others of that clafs. It has been employed chiefly for external purpofes, againft wounds and ulcerations, particularly of the nipples, whence its names *nipplewort* and *papillaris*.

LAPATHUM.

DOCK : a perennial plant bearing numerous imperfect flowers fet in double cups : the outermoft cup confifts of three fmall green leaves ; the inner of three larger reddifh ones, which become a covering to a gloffy triangular feed.

1. OXYLAPATHUM. *Lapathum acutum folio plano* C. B. *Rumex acutus Linn.* Sharp-pointed wild dock: with long acuminated leaves, not curled

curled about the edges, growing gradually narrower from the bottom to the point; and the feed-covers indented and marked with little tubercles. The roots are of a brownish colour on the outside, and of a yellowish within, which grows deeper in drying.

The roots of the sharp-pointed dock have a bitterish astringent taste; and no remarkable smell: the roots of the other common wild docks are nearly of the same quality, equally discover their astringent matter both to the taste and by striking an inky blackness with solution of chalybeate vitriol, and have been often substituted in our markets to those of the sharp-pointed kind; which last are generally, and, so far as can be judged from their taste, justly, accounted the most efficacious. They are supposed to have an aperient and laxative, as well as an astringent and corroborating virtue; approaching in this respect to rhubarb, but differing widely in degree, their stypticity being greater, and their purgative quality, if really they have any purgative quality at all, far less. They stand recommended in habitual costiveness, obstructions of the viscera, scorbutic and cutaneous maladies: in which last intention, fomentations, cataplasms, or unguents of the roots, have been commonly joined to their internal use: in many cases, the external application alone is said to be sufficient. Their active matter is taken up both by water and rectified spirit, and, on inspissating the tinctures, remains in the extracts; both the watery and spirituous extracts are considerably bitter and very austere. A decoction of half an ounce or an ounce of the fresh roots, or of a dram or two

of

LAPATHUM.

of the dry roots, is commonly directed for a dose.

2. Hydrolapathum *five Herba britannica Pharm. Edinb. Lapathum aquaticum folio cubitali C. B. Rumex aquaticus Linn.* Great wild water-dock: with very large leaves, two or three feet long; the feed-covers not indented. The roots are externally blackish, internally white with a faint reddish tinge, which, in drying, changes in some parts to a yellowish: the internal part of the fresh root, exposed to the air, or of the dry root moistened, soon changes superficially to a deep yellow or brown.

The roots of the water-dock strike a black colour with solution of chalybeate vitriol, like those of the preceding species, but have a much stronger and more acerb taste; which is diffused equally, so far as can be judged, through the whole substance of the root. They give out their active matter both to water and rectified spirit, and tinge both menstrua of a pale yellowish or reddish brown colour, though in chewing they render the saliva only milky.

The *herba britannica* of the ancients, celebrated as an antiscorbutic, and of which the knowledge was long lost, was proved by Muntingius, towards the end of last century, to be no other than this great water-dock. Muntingius endeavours to prove also, that its name *britannica* was not derived from that of our island, but from Teutonic words expressing its power of fastening loose teeth, or of curing the disease which makes them loose. Later experience has shewn, that it is a medicine of very considerable efficacy, both externally in lotions against putrid spungy gums and ulcerations,

rations, and as an internal antiscorbutic: Boerhaave assures us, that in these cases he has known many instances of its happy effects. It is supposed to be of service also in cutaneous defedations different from the true scurvy, in rheumatic pains, and in chronical disorders proceeding from obstructions of the viscera. It has been chiefly used in medicated wines and small ales, with the addition generally of some spicy materials, and sometimes of other antiscorbutic plants, as scurvygrass, buckbean, horse-radish, &c.

3. RHABARBARUM MONACHORUM *Pharm. Paris. Lapathum hortense latifolium C. B. Hippolapathum; Patientia. Rumex Patientia Linn.* Monks rhubarb, garden patience: with large, broad, acuminated leaves; reddish, branched stalks; the leaves that cover the seeds unindented, and a tubercle on one of them: the root is of a yellow colour, with red veins, approaching in appearance to rhubarb.

This root is supposed to possess the virtues of rhubarb, in an inferiour degree. It is obviously more astringent than rhubarb: but comes very far short of it in purgative virtue, though given, as usually directed, in double its dose; nauseating the stomach, without producing any considerable evacuation. It communicates a deep yellow tincture both to water and spirit.

LAVENDULA.

LAVENDER: a shrubby plant, with its leaves set in pairs, the stalks square when young, and round when grown woody; producing, on
the

LAVENDULA.

the tops of the branches, naked fpikes of blue, fometimes white, labiated flowers, of which the upper lip is erect and cloven, the lower divided into three roundifh fegments.

1. LAVENDULA *Pharm. Lond. & Edinb. Lavendula anguftifolia* C. B. *Pfeudonardus quæ lavendula vulgo* J. B. *Lavendula minor five fpica* Ger. & Park. *Lavendula Spica Linn.* Lavender: with oblong, very narrow, fomewhat hoary, undivided leaves; a native of dry gravelly foils in the fouthern parts of Europe, common in our gardens, and flowering in July.

The flowers of lavender have a fragrant fmell, to moft people agreeable, and a bitterifh, warm, fomewhat pungent tafte: the leaves are weaker and lefs grateful. They are often employed as a perfume; and medicinally, as mild ftimulants and corroborants, in vertigoes, palfies, tremors, and other debilities of the nervous fyftem, both internally and externally.

The flowers are fometimes taken in the form of conferve; into which they are reduced, by beating them, while frefh, with thrice their weight of double refined fugar. Their fragrance is lefs injured by beating or bruifing them than moft of the other odoriferous flowers, but is neverthelefs confiderably diminifhed: the flavour of the leaves is of a much lefs deftructible kind.

Water extracts by infufion nearly all the virtue both of the leaves and flowers. In diftillation with water, the leaves yield a very fmall portion of effential oil; the flowers a much larger, amounting, in their moft perfect ftate, when they are ready to fall off fpontaneoufly and the feeds begin to fhew themfelves, to about one ounce from fixty. The oil is of a bright yellowifh

Ol. essentiale flor. lavend. *Ph. Lond. & Ed.*

lowish colour, a very pungent taste, and possesses, if carefully distilled, the fragrance of the lavender in perfection: it is given internally from one drop to five, and employed in external applications for stimulating paralytic limbs and for destroying cutaneous infects. The decoction, remaining after the distillation of the oil, is disagreeably bitterish and somewhat saline.

Rectified spirit extracts the virtue of lavender more completely than water. The spirit elevates also in distillation a considerable part of the odoriferous matter of the leaves, and greatest part of that of the flowers; leaving in the inspissated extracts, a moderate pungency and bitterishness with very little smell. A spirit prepared by pouring a gallon of proof spirit on a pound and a half of the fresh-gathered flowers, and drawing off five pints by the heat of a water bath; or by adding eight pounds of rectified spirit to two of the flowers, and drawing off seven pounds, is richly impregnated with the fragrance of the flowers. More compounded spirits of this kind, in which other aromatics are joined to the lavender, have been distinguished by the name of English or palsy drops: the college of London directs three pints of the simple spirit of lavender, and one pint of spirit of rosemary, to be digested on half an ounce of cinnamon, half an ounce of nutmegs, and one ounce of red saunders, as a colouring ingredient: the college of Edinburgh, to the same quantity of both spirits, orders one ounce of cinnamon, two drams of cloves, half an ounce of nutmegs, and three drams of red saunders. These preparations are taken internally, on sugar or in any convenient vehicle, from ten to an hundred drops, and used externally in embrocations, &c.

Spiritus lavend. *Ph. Lond.*

—simpl. *Ph. Ed.*

Tinct. lavend. comp. *Ph. Lond.*

Spirit. lavend. comp. *Ph. Ed.*

2. Lavendula

LAUROCERASUS.

2. LAVENDULA LATIFOLIA *C. B. Lavendula major seu spica* Pharm. Paris. *Pseudonardus quæ vulgo spica J. B. Lavendula major sive vulgaris* Park. Broad lavender: with longer, broader, and hoarier leaves, less numerous on the stalks and branches; and much larger spikes, though smaller flowers; common in the southern parts of Europe, but rare among us. The name spike is applied by foreign writers to this species, by some of ours to the first. Linnæus makes this only a variety of the former.

The broad-leaved lavender is stronger both in smell and taste than the narrow, and yields in distillation almost thrice as much essential oil, but the flavour both of the oil and of the plant itself, is much less grateful: the oil is likewise of a much darker colour, inclining to green. Watery and spirituous extracts, made from the two sorts of lavender, are very nearly alike; the difference seeming to reside only in the volatile parts.

LAUROCERASUS.

LAUROCERASUS Pharm. Paris. *Cerasus folio laurino* C. B. *Prunus Lauro-cerasus* Linn. LAUREL, CHERRY-BAY: an evergreen tree or shrub, with large, thick, oblong, glossy leaves, pointed at both ends, and slightly indented: towards the tops of the branches come forth pentapetalous flowers, in five-leaved cups, followed by clusters of berries, like cherries or damsons. It is cultivated in gardens, flowers in May, and ripens its fruit in August or September.

The leaves of the laurocerasus have a bitter taste, accompanied with a flavour resembling that of the kernels of certain fruits, as those of black cherries, apricots, bitter almonds, &c. Like those kernels, they communicate an agreeable flavour both to watery and spirituous liquors, by distillation and by infusion; and, like them also, they appear from some late trials to be poisonous. A distilled water, strongly impregnated with their flavour, given in the quantity of four ounces to a large mastiff dog, occasioned in a few minutes terrible convulsions, and within an hour put an end to his life: dogs have been killed also, in a few minutes, by smaller quantities, of the distilled water, of an infusion of the leaves in water, and of their expressed juice, taken into the stomach, or injected by the anus; and there are some instances of liquors flavoured with the distilled water being poisonous to human subjects. The dissections of dogs killed by this poison have shewn no other morbid appearances, or alterations, than such as may be reasonably supposed the immediate effect of the convulsions: when the distilled water, or the leaves in substance, were given in such small quantities as not to kill, and continued for some time, the pulse became quicker, and the blood more fluid, and of a more florid red colour *(a)*. It is said that infusions of the leaves (made probably very weak) are commonly used in Holland in disorders of the lungs *(b)*.

The kernel of the fruit is of the same nature with the leaves. The pulpy part discovers no ill quality to the palate, is coveted by birds, and appears to be innocent.

(a) See Dr. Langrish's *experiments on brutes*, and No. 418 and 420 of the *philosophical transactions*.

(b) Linnæi *Amænitat. academic.* iv. 40. i. 409.

LAURUS.

LAURUS.

LAURUS Pharm. *Lond. & Edinb.* Laurus vulgaris *C. B.* Laurus nobilis *Linn.* BAY: an evergreen tree or shrub, with oblong, stiff, smooth leaves, pointed at both ends, pale yellowish, monopetalous flowers, divided into four sections; and oblong dry berries, containing, under a thin black skin, a horny shell, within which are lodged two dark brownish seeds joined together. It is a native of the southern parts of Europe, and not uncommon in our gardens; it flowers in April and May, and ripens its berries in September. The shops have been commonly supplied with the berries from the Streights.

THE leaves of the bay have a light agreeable smell, and a weak aromatic roughish taste: in distillation with water they yield a small quantity of a very fragrant essential oil: with rectified spirit they afford a moderately warm pungent extract. The berries are stronger both in smell and taste than the leaves, and yield a larger quantity of essential oil: they discover likewise a degree of unctuosity in the mouth, give out to the press an almost insipid fluid oil, and on being boiled in water a thicker butyraceous one, of a yellowish green colour, impregnated with the flavour of the berry. *Oleum expressum baccarum lauri* Ph. Ed.

The leaves and berries of the bay are accounted stomachic, carminative, and uterine: in these intentions, infusions of the leaves are sometimes drank as tea; and the essential oil of the berries given, on sugar or dissolved by means of mucilages or in spirit of wine, from one to five or six drops. The principal use of these thefe fimples in the prefent practice is external: they are made ingredients in carminative glyfters, warm cataplafms, and uterine baths; and the butyraceous oil of the berries ferve as a bafis for fome nervine liniments, and mercurial and fulphureous unguents.

* Bergius relates that he has very frequently feen protracted intermittents cured by a mixture of two fcruples of powder of bay berries and fix grains of capficum feeds, divided into three dofes, one given at the firft acceffion of the rigor, another, the next day at the fame hour, and the third on the fucceeding day. *Mat. Med.* i. 144.

LAZULI LAPIS.

LAPIS CÆRULEUS; *Lapis cyanus; Cæruleum nativum.* LAPIS LAZULI: a compact ponderous foffil; lefs hard than flint; of a deep blue colour, variegated commonly with gold or filver coloured points or veins; retaining its colour in a moderately ftrong fire, in a very ftrong one calcining to a brown, and at length melting into a dufky coloured glafs with bluifh clouds; lofing its colour, and in great part diffolving, by digeftion in mineral acids; faid to be found in the mines of gold, filver, and copper, in the eaftern countries and in fome parts of Germany (a).

This ftone, levigated into an impalpable powder, and freed from the groffer parts by wafhing with water, has been given in dofes of half a dram and a dram, and faid to operate ftrongly by ftool and vomit. Some have re-

(a) See Cronftedt's *mineral fyftem*, and Marggraf's *chemical works*.

commended

LENTISCUS.

commended it in epilepfies and intermitting fevers: Dolæus tells us, that in this laſt diſorder, the above doſes, taken on the approach of a fit, with two or three fpoonfuls of brandy, were with him a fingular fecret. The ancients fuppoſed, that it evacuates chiefly what they called melancholic humours or aduſt black bile, probably, as Geoffroy fufpects, on account of its tinging the feces black; a property, from which it may rather be prefumed that the mineral participates of iron. The conftringing power, which is likewife afcribed to it, depends perhaps on this ingredient; but neither its real medical qualities, nor its chemical compofition, are as yet known. Its ferrugineous impregnation is apparent, from its yielding yellow martial flowers on fublimation with fal ammoniac, and from folutions of it in mineral acids affording a blue precipitate with the tincture of Pruffian blue deſcribed under the article *ferrum*. It has been generally fuppoſed to participate pretty largely of copper; but pure lapis lazuli gives no mark of copper; and thofe, who ſpeak of experiments difcovering that metal in it [a], have probably taken for lapis lazuli fome other blue ſtones, as the lapis armenus, which plainly contains copper, and which fome celebrated naturaliſts have ranked as a ſpecies of the lazuli. The lapis armenus may be readily diftinguiſhed, by its being lefs hard than the lazuli, foon lofing its blue colour in moderate fire, and raifing an effervefcence with acids, its bafis feeming to be a calcareous earth.

LENTISCUS.

LENTISCUS vulgaris C. B. *Lentiſcus verus ex infula chio cortice & foliis fuſcis* Commel. *Piſ-*

[a] Hoffman, *in notis ad Poterium, p.* 628.

tachia

tachia Lentiscus Linn. Lentisk or Mastich tree: an evergreen tree or shrub, with soft flexible branches hanging downwards, and small stiff leaves, pointed at both ends, set in pairs on a furrowed rib, which terminates in a soft prickle: some trees produce reddish imperfect flowers, in the bosoms of the leaves; others, clusters of black firm berries including a whitish kernel. It is a native of the southern parts of Europe, and bears the ordinary winters of our own climate: large plantations of it are cultivated in the island of Chio, on account of the resinous juice, called mastich, obtained from incisions made in the trunk. The wood is sometimes brought to us from Marseilles, in thick knotty pieces, covered with a brownish bark, internally of a whitish or pale yellowish colour.

This wood is accounted a mild balsamic restringent: infusions and decoctions of it are greatly commended, in the German ephemerides, against catarrhs, nauseæ, weakness of the stomach, and in general as a corroborant and an alterative or sweetener *(a)*. It may indeed be presumed, from its sensible qualities, to possess virtues of this kind, though in no very high degree. Its smell and taste are aromatic and resinous, but very weak: the small tough sprigs are stronger than the larger pieces, and the bark than the wood. It impregnates water with a red colour, and a light agreeable smell: to rectified spirit it gives a bright yellow tincture, and scarce any smell. On gently distilling off the menstrua from the filtered liquors, the remaining extracts prove resinous, subastringent,

(a) Wenck, *Acta nat. curiof. dec.* iii. *ann.* ix & x. *p.* 254.

and

LEPIDIUM.

and flightly pungent: the watery extract difcovers more of the flavour of the wood, and is in tafte rather ftronger, though much larger in quantity, than the fpirituous; the fpirit covering or fuppreffing the fmell, and not taking up enough of the gummy or mucilaginous matter to render the refin diffoluble in the mouth. According to Cartheufer's experiments, the watery extract amounts to one eighth the weight of the wood, the fpirituous to one twentieth or one fixteenth.

LEPIDIUM.

LEPIDIUM: a plant with undivided leaves, and fmall tetrapetalous white flowers on the tops of the ftalks and branches, followed by little fhort heart-fhaped fharp-pointed pods, which are divided longitudinally into two cells full of minute feeds.

1. LEPIDIUM *latifolium* C. B. & *Linn. Piperitis. Raphanus fylveftris Belgis & Gallis.* DITTANDER, PEPPERWORT, POOR-MANS-PEPPER: with oblong, broad, acuminated, ferrated leaves. It is perennial, and found wild in fome parts of England by the fides of rivers and in other moift fhady places.

2. IBERIS, *feu Cardamantica: Iberis latiore folio* C. B. *Lepidium gramineo folio five iberis Tourn. Lepidium Iberis Linn.* Sciatica creffes: with long narrow leaves, the lower on long pedicles and ferrated; the upper entire, without pedicles. It is annual; a native of the fouthern parts of Europe; and raifed in our gardens, as the preceding, for culinary ufe.

THESE herbs, when frefh, have a quick penetrating pungent tafte; which is in great part diffipated dissipated or destroyed by exsiccation, retained in the expressed juice, extracted by water and by rectified spirit, and elevated by both menstrua in distillation or evaporation. They are recommended as antiseptics, stomachics, attenuants, and aperients; and appear to be of the same general nature with the *cochleariæ, nasturtium,* and other acrid antiscorbutics. The second sort has been supposed particularly serviceable, externally, against the sciatica; whence its common English name.

LEVISTICUM.

LIGUSTICUM seu Levisticum, Pharm. Edinb. Ligusticum vulgare C. B. Angelica montana perennis paludapii folio Tourn. Ligusticum Levisticum Linn. Lovage: a tall umbelliferous plant, with large leaves divided and subdivided into sections like those of smallage: the umbels stand on short pedicles, with several little leaves at the origin of each of the primary ones, and a few at the ramifications; the seeds are of a pale brown colour, oblong, plano-convex, marked with five longitudinal ridges: the root thick, fleshy, juicy, branched, of a dark brownish colour on the outside, and a whitish or pale yellowish within. It is a native of the southern parts of Europe, and raised with us in gardens: it is perennial, flowers in June, and ripens its seeds in August.

All the parts of this plant are of the aromatic kind; of a strong flavour, somewhat like that of angelica, but less agreeable; supposed particularly useful in female disorders. The leaves, which have been generally made choice of in this intention, have the most unpleasant smell,

LICHEN.

smell, and suffer no great loss of it in keeping for some months; their taste is moderately warm, and acrid, and very durable in the mouth and throat. The root, whose smell is nearly of the same kind with that of the leaves, though more approaching to gratefulness, discovers to the taste a considerable sweetness joined to its mild aromatic warmth; an extract made from it by water retains little more than the sweet matter; the flavour exhaling in the inspissation, and impregnating the distilled fluid, from which, if the quantity of the root subjected to the operation be large, a small portion of essential oil separates: an extract made by rectified spirit retains the aromatic part as well as the sweet, and proves moderately warm, but much less so than the extract of angelica: towards the end of the inspissation of the spirituous tincture, a thin unctuous matter appears upon the surface, in taste highly aromatic, and which seems to be the part that gives activity to the rest of the mass. The seeds of the plant have little of the sweetness of the roots, but are rather of more warmth and pungency, and of a more agreeable flavour.

LICHEN.

LIVERWORT: a kind of imperfect vegetable production, consisting of spreading leaves, of a leathery crustaceous matter. A sort of flowers both male and female have been discovered in it, the latter producing innumerable seeds, like meal.

LICHEN *terrestris cinereus Raii. Lichen caninus Linn.* Ash-coloured ground liverwort: consisting of roundish pretty thick leaves, divided about

MATERIA MEDICA.

about the edges into obtuse segments, flat above, of a reticular texture underneath, fastened to the earth by small fibres; when in perfection, of an ash grey colour; by age turning darker coloured or reddish. It grows on commons and open heaths, spreads quickly on the ground, and is to be met with at all times of the year, but is supposed to be in its greatest vigour about the end of autumn.

This herb is said to be a warm diuretic. It is particularly celebrated as a preservative against the terrible consequences of the bite of a mad dog: an account of the remarkable efficacy, in this intention, of a powder composed of the dry leaves and black pepper, was communicated to the royal society by Mr. Dampier, and published in No. 237 of the Philosophical Transactions. This powder was afterwards inserted, in the year 1721, into the London pharmacopœia, at the desire of Dr. Mead, who had large experience of its good effects, and who declares, that he had never known it to fail, where it had been used, with the assistance of cold bathing, before the hydrophobia began. He directs a dram and a half of the powder to be taken in the morning fasting, in half a pint of cows milk warm, for four mornings successively: previously to these four doses, the patient is to be blooded nine or ten ounces; and after them, to be dipt in cold water every morning fasting for a month, and then dipt thrice a week for a fortnight longer (a). The powder was originally composed of equal parts of the lichen and pepper: but this quantity of

---

(a) *Mechanical account of poisons*, essay iii.

pepper

LICHEN.

pepper rendering the medicine too hot, only one part was afterwards ufed to two of the lichen. It is now expunged.

If cold bathing, bleeding, black pepper, and lichen, conjointly, be really of fufficient efficacy againft the poifon of the mad dog, it will not perhaps follow that any fhare of this efficacy belongs to the lichen: and indeed greater ftrefs has been laid in general on the cold bath, than on this or the other parts of the prefcription. The lichen does not promife to have any valuable medicinal power: to the organs of tafte or fmell it difcovers no activity: taken by itfelf, in double the quantity above prefcribed, it did not appear to have any fenfible operation. Digefted in rectified fpirit, it tinged the menftruum of a deep yellowifh green colour: on diftilling off the fpirit from the filtered tincture, the remaining grumous extract had very little tafte, and amounted only to twenty-fix grains from two ounces, or about one thirty-feventh of the weight of the lichen. A decoction of the herb in water was brownifh, and of a faint fmell, fomewhat like that of mufhrooms: the extract, obtained by infpiffating it, weighed one eighth of the lichen, and had fome tafte, but fo little, that it is hard to fay of what kind.

* LICHEN ISLANDICUS *Linn. Lichen Pharm. Edinb. Lichen terreftris, foliis Eryngii, Buxb. Cent.* II. *Lichenoides rigidum, Eryngii folia referens,* Raii & Dillen. Eryngo-leaved, or eatable Iceland, liverwort. This fpecies of lichen confifts of nearly erect leaves, ftiff when dry, but foft and pliant when moift, irregularly divided into broad diftant fegments, fmooth, and ciliated at the margins. It grows in the mountainous

mountainous parts of this country, and in various other parts of Europe.

The Iceland lichen infufed in water gives a bitterifh liquor, which is reddened by a mixture of martial vitriol. A decoction of it is very thick and vifcid; and on cooling concretes into a ftrong gelly. An ounce of the lichen boiled for a quarter of an hour in a pound of water, and afterwards ftrained, yielded feven ounces of a mucilage as thick as that procured by the folution of one part of gum-arabic in three of water.

The inhabitants of Iceland make great ufe of this lichen both as food and as phyfic. When frefh, according to Borrichius (a), it is employed as a purgative; but Olafsen (b) denies that it has any more than a very gently opening quality. It is ufually dried and ground into a meal, with which they make pottage and other preparations, adding either water or milk, and find it an agreeable and very nutritive article of food. It is beft firft to fteep it for a fufficient time in water, in order to extract the bitternefs.

The prepared lichen has been much ufed of late, particularly at Vienna, as a remedy for confumptive diforders. The celebrated Scopoli (c) has publifhed fome cafes of its fuccefsful exhibition in the phthifis; and other practitioners (d) have confirmed his account. It is ufed boiled in milk to a kind of pottage, of which the patient's diet is chiefly to confift. It is faid to be antifeptic, eafy of digeftion, and

(a) Bartholini Act. med. Hafn. 1671.
(b) Journey to Iceland.
(c) Annus 2. hiftorico-natural. p. 114.
(d) Bergius, mat. med. 858.

remarkably

LIGNUM ALOES.

remarkably nourishing. It is also recommended in other cases, where the stomach is so weak that common aliments are rejected. The Edinburgh college have received this lichen into their catalogue of simples.

LIGNUM ALOES.

LIGNUM ALOES, Xyloaloes, & Agallochum, Pharm. Parif. Lignum Calambac. Agallochum, Calambac, or Aloes wood: a wood brought from China, and the inner parts of Tartary, in small pieces, compact and ponderous, of a yellowish or rusty brown colour, with black or purplish veins, sometimes purple with ash-coloured veins, and sometimes all over blackish. Of its origin, we have no very satisfactory account: most of the writers, to whom we are indebted for information about the productions of those countries, report, that it is the internal part of certain trees; that a large tree affords only a very small quantity of this valuable part; and that there are several different sorts of it, of which the best is never brought to us, being sold in China itself for twice or thrice its weight of silver.

The best sort of agallochum wood brought into Europe, has a bitterish resinous taste, and a light aromatic smell. Set on fire, it seems to melt like wax, and emits, during the burning, an agreeable fragrance, which continues till the wood is wholly consumed. It is this fragrance in burning which makes the wood precious in the eastern countries for fumigations, and which affords the surest criterion of its genuineness and goodness. As this wood is apparently very resinous, rectified spirit takes up up more from it than watery menstrua: according to Cartheuser's experiments, an ounce yields with spirit three drams of extract, and with water only two. The watery decoction and extract are moderately bitter and subacrid. The spirituous make less impression on the organs of taste, being less dissoluble in the mouth, or less miscible with the saliva: the pure resin, obtained by precipitation with water from the somewhat inspissated spirituous tincture, as directed by the faculty of Paris, is still weaker in taste. Hoffman observes, that in distillation with water, it yields an essential oil, of a whitish colour, of a thick consistence, of great fragrance, but in small quantity, not exceeding half an ounce from one hundred and sixty ounces of the wood: this oil, in which the more valuable parts of the agallochum are concentrated, he recommends, dissolved in spirit of wine, as one of the best cordials and corroborants, in weaknesses of the stomach and depressions of strength *(a)*.

Resina ligni aloes Ph. Paris.

In our shops, we rarely meet with any agallochum that answers the above characters. In its place have been substituted woods of an inferiour kind, probably the *aspalathus, lignum aquilæ,* and *calambour* of authors; which are said to be woods of the nature of agallochum, but, when in their greatest perfection, far weaker.

LIGNUM CAMPECHENSE.

LIGNUM CAMPECHENSE *Pharm. Lond. & Edinb. Lignum campescanum & lignum indi-*

(a) Observ. physico-chym. lib. i. *obs.* 4. *Not. ad Poterium, p.* 487. *De medicament. balsamic.* § 15.

LIGNUM CAMPECHENSE.

cum Mont. exot. Lignum campechianum, species quædam brasil Sloan. Lignum sappan quibusdam.
CAMPEACHY WOOD or LOGWOOD: the wood of a prickly pod-bearing tree *(Hæmatoxylum Campechianum Linn.)* a native of Campeachy, in the bay of Honduras; from whence the wood is brought over in large compact hard logs of a red colour.

THIS wood, imported from America as a dying drug, has of late been introduced into medicine, and found to be a very useful reftringent and corroborant, in diarrhœas, dyfenteries, and other diforders from a laxity of the folids. It has a fweetifh fubaftringent tafte, and no remarkable fmell: extracts made from it, by water and fpirit, have a great degree of fweetnefs, mixed with a mild grateful aftringency. It gives a deep purplifh red tincture both to watery and fpirituous menftrua * *(a)*; and frequently tinges the ftools, and fometimes the urine, of the fame colour: of this the patient ought to be apprifed, that he may not be alarmed by judging the colour of the difcharge to be owing to blood.

Watery menftrua readily extract part of the virtue of this wood, but are very difficultly made to take up the whole. To promote the extraction, the wood fhould previoufly be reduced into fine powder, which is to be ftrongly boiled in the water, in the proportion, for example, of a pound to a gallon, till half the

---

* *(a)* Pure rain water acquires only a deep orange or mahogany colour from logwood; and rectified fpirit a fine yellowifh red. The purple hue feems to be communicated by fome extraneous faline matter, as the felenitic or aluminous falts in hard fpring water. A very fmall quantity of fixed alkali will alfo give it ftill more perceptibly.

liquor

liquor is wafted: the powder will ftill give a confiderable impregnation to the fame quantity of frefh water, and this repeatedly for four or five times or oftener: the extract obtained by infpiffating the decoctions, of a dark blackifh colour in the mafs, tinges water of a fine red, like that of the liquors before infpiffation, but does not totally diffolve: it is given in dofes of from ten grains to a fcruple and upwards. Rectified fpirit takes up more from the logwood than watery menftrua. Some digeft the powdered wood in four times its weight of fpirit, and aftewards boil it in water: the matters taken up by the two menftrua are then united into one extract, by infpiffating the watery decoction to the confiftence of honey, and then gradually ftirring in the infpiffated fpirituous tincture.

Extract. lign. campechenf. Ph. Lond.

Ph. Ed.

LIGNUM RHODIUM.

RHODIUM or ROSEWOOD: the wood or root of a tree of which we have no certain account; brought from the Canary iflands, in long crooked pieces, full of knots, externally of a whitifh colour, internally of a deep yellow, with a reddifh caft. The largeft, fmootheft, ftraighteft, heavieft, and deepeft coloured pieces fhould be chofen; and the fmall, thin, pale, light ones rejected.

This wood has a flightly bitterifh, fomewhat pungent, balfamic tafte, and a fragrant fmell, efpecially when fcraped or rubbed, refembling that of rofes. Digefted in rectified fpirit, it gives out pretty readily the whole of its active matter, and tinges the menftruum of a reddifh yellow colour: on committing to diftillation the filtered

LILIUM.

filtered tincture, the spirit brings over little or nothing of its flavour; the fine smell, as well as the balsamic pungency, of the rhodium, remaining nearly entire in the inspissated extract, which proves tenacious and adhesive like the turpentines. Infused in water, it gives out likewise great part of its smell and taste, together with a bright yellow colour: in evaporation, the water carries off the specific flavour of the wood, leaving in the extract only a slight pungency and bitterishness. Distilled with water, it gives over, somewhat difficultly and slowly, a highly odoriferous essential oil, at first of a gold colour, by age turning reddish, amounting, if the rhodium is of a good kind, to about one ounce from fifty: the distilled water is likewise agreeably impregnated with the fragrance of the rhodium, and resembles that of damask roses.

The essential oil is used as a perfume, for scenting pomatums, &c. and in this light only the rhodium wood is generally regarded. It promises, however, to be applicable to more important purposes, and bids fair to prove a valuable cordial and corroborant.

LILIUM.

LILIUM ALBUM Pharm. Edinb. *Lilium album flore erecto & vulgare* C. B. *Lilium candidum* Linn. WHITE LILY: a plant with a single straight round stalk, clothed with oblong, acuminated, thick, smooth, pale green, ribbed leaves, which have no pedicles; bearing on the top several elegant, naked, white, upright, hexapetalous, bell shaped flowers, which open successively, and are followed each by an oblong triangular capsule, divided into three cells full of brownish feeds: the root is a single bulb, composed of fleshy scales, with several fibres at the bottom. It is perennial, a native of Syria and Palestine, common in our gardens, and flowers in June.

The flowers of the white lily have a pleasant sweet smell, and a slightly mucilaginous taste. Their odorous matter is of a very volatile kind, being totally dissipated in drying, and totally carried off in evaporation by rectified spirit as well as water: both menstrua become agreeably impregnated with it by infusion or distillation, but no essential oil has been obtained, though many pounds of the flowers were submitted to the operation at once. The principal use of these flowers is for flavouring expressed oils; which, by insolation with fresh parcels of them continued about three days each time, are supposed to receive from them, along with their smell, an anodyne and nervine virtue. The distilled water has been sometimes employed as a cosmetic.

Oleum liliorum, susinum, &c.

The roots also have been used chiefly for external purposes; as an ingredient in emollient and suppurating cataplasms: they abound with a strong mucilage, and do not seem to have much active matter besides. Gerard indeed relates, that several persons were cured of dropsies, by the constant use, for a month or six weeks, of bread made of barley-meal with the juice of white lily roots: but there are examples of similar cures being obtained by the use of common dry bread; and probably in one case, as well as in the other, abstinence from liquids was the remedy.

LILIUM

LILIUM CONVALLIUM.

LILIUM CONVALLIUM album C. B. *Convallaria, Maianthemum. Convallaria maialis Linn.* LILY OF THE VALLEY, MAY LILY: a plant with two or three oblong, acuminated, ribbed leaves; in the bosoms of which arises a naked stalk bearing a number of small, naked, white, drooping, bell-shaped, monopetalous flowers, cut about the edges into six segments, and followed by red berries: the roots are long, slender, and white. It is perennial, grows wild in woods and shady places, and flowers in May.

THE flowers of this plant have a fragrant delightful smell, and a penetrating bitterish taste; both which they readily impart to watery and to spirituous menstrua. Their odorous matter, like that of the white lily, is very volatile; being totally dissipated in exsiccation, and elevated in distillation both by water and rectified spirit: there is no appearance of essential oil in either distillation; nor does the distilled spirit turn milky on the admixture of water, as those spirits do, which are impregnated with actual oil. The pungency and bitterness, on the other hand, reside in a fixt matter, which remains entire both in the watery and spirituous extracts, and which, in this concentrated state, approaches, as Cartheuser observes, to hepatic aloes.

It is principally from the volatile parts of these flowers, that medicinal virtues have been expected, in nervous and catarrhous disorders; but probably their fixt parts also have no small, perhaps the greatest, share in their efficacy. The flowers,

flowers, dried and powdered, and thus divested of their odoriferous principle, prove strongly sternutatory. Watery or spirituous extracts made from them, given in doses of a scruple or half a dram, act as gentle stimulating aperients and laxatives; and seem to partake of the purgative virtue, as well as of the bitterness, of aloes.

The roots have nothing of the fine smell which is admired in the flowers, but discover to the taste, a greater degree of penetrating bitterness. The bitter matter appears to be of the same kind in these as in the flowers; being equally extracted by water and spirit; remaining entire behind upon inspissating the tinctures or infusions; acting as a sternutatory when snuffed up the nose, and as a laxative or purgative when taken internally.

The leaves have the same kind of bitterness, in a lower degree, mixed with a considerable roughness, and a slight sweetishness.

LIMACES.

LIMACES terrestres sive Cochleæ terrestres.
THE SNAIL: an animal, lodged in a short thick turbinated shell, whose aperture is closed in the winter with a kind of cement. The large ash-coloured snail is said to be the species intended for medicinal use; but the smaller, dark-coloured, spotted, striped sort, more common in gardens, is taken indiscriminately, and appears to be not at all different in quality from the other.

THIS animal abounds with a viscid slimy juice, which it readily gives out, by boiling, to milk or water, so as to render them thick and glutinous. The decoctions in milk are apparently very nutritious and demulcent, and stand recommended in a thin acrimonious state of the humours, in consumptive cases, and emaciations.

LIMONES.

LIMON Pharm. Lond. Limonia mala Pharm. Edinb. LEMONS: the fruit of the *malus limonia fructu acido Pharm. Lond. malus limonia acida C. B. Citrus Limon Linn.* a tree resembling the orange; from which it differs chiefly in the leaves having no appendages at the bottom; and in the fruit having a nipple-like production at the end: it is a native of Asia, and cultivated in the warmer parts of Europe, from whence we are supplied with the fruit. There are many varieties of this tree in regard to the fruit: by Linnæus, the several citrons, as well as lemons, are reckoned varieties of one species, which is distinguished from those of the orange kind, only by the pedicles of the leaves being naked. The terms citron and lemon have been often confounded together; what is commonly called citron by the French (a) and Germans (b) being our lemon, and their lemon our citron.

THE yellow rind of lemons is a grateful aromatic, of common use in stomachic tinctures and infusions, and for rendering other medicines acceptable to the palate and stomach: its flavour is one of those which is best adapted for

(a) *Codex medicamentarius facultatis Parisienfis, p.* xxxviii. & lxx.

(b) Hoffman, *Differt. de citriis, Opera omnia, supplement.* ii. *par.* i. *p.* 720.

accompanying

accompanying medicines of the bitter kind. It is less hot than orange peel, and yields in distillation a less quantity of essential oil: the oil is extremely light, almost colourless, in smell nearly as agreeable as the fresh peel, and frequently employed as a perfume: it is generally brought to us from the southern parts of Europe, under the name of *essence* of lemons. The flavour of the lemon peel is more perishable in keeping than that of orange peel, yet does not rise so easily in distillation with spirit of wine: for a spirituous extract, prepared from the rind of lemons, retains the aromatic taste and smell of the peel in a much greater degree than an extract prepared in the same manner from that of oranges. After digestion in the spirit, lemon peel proves tough, that of oranges crisp.

Ol. stillat. cort. limon. Ph. Ed.

Essentia limonum Ph. Lond.

The juice of lemons differs from that of oranges only in being more acid. Six drams of it saturate about half a dram of fixt alkaline salt: this mixture, with the addition of a small quantity of some grateful aromatic water or tincture, as simple cinnamon water, is given in cases of nauseæ and reachings, and generally abates, in a little time, the severe vomitings that happen in fevers, when most other liquors and medicines are thrown up as soon as taken: it is used also as a saline aperient in icterical, hydropical, inflammatory and other disorders. A syrup made by dissolving fifty ounces of fine sugar in a quart † or two pounds and a half ‡ of the depurated juice, is mixed occasionally with draughts and juleps as a mild antiphlogistic, and sometimes used in gargarisms for inflammations of the mouth and tonsils. The London college also direct the inspissated juice to be kept.

Syrup. e succo limonum † Ph. Lond. ‡ Ph. Ed.

Succus spissat. limon. Ph. Lond.

LINARIA.

LINARIA.

LINARIA vulgaris lutea flore majore C. B. Ofyris, linaria, five urinaria Lobel. Antirrhinum linaria Linn. TOADFLAX: a plant with smooth round bluish stalks, and numerous, oblong, narrow, pointed leaves; greatly resembling the *esula minor* or pine spurge, so as scarcely to be otherwise distinguishable, before flowering, than by its wanting the milky juice with which the esula abounds: on the tops of the stalks and branches appear spikes of yellow, irregular, monopetalous, gaping flowers, with a long tail behind, followed by roundish bicapsular seed-vessels: it is perennial, grows wild about the sides of dry fields, and flowers in June and July.

THE leaves of this herb have a bitterish somewhat saline taste: and when rubbed betwixt the fingers, yield a faint smell, resembling that of elder. Taken internally, they are said to be powerfully resolvent, diuretic, and purgative: their principal use, however, has been external, in unguents and cataplasms, for painful swellings of the hemorrhoidal vessels; against which they have been said to be particularly effectual.

LINGUA CERVINA.

SCOLOPENDRIUM seu lingua cervina Pharm. Edinb. Lingua cervina officinarum C. B. Phyllitis Gerard. Asplenium Scolopendrium Linn. HARTS-TONGUE: a plant with long, uncut, narrow bright green leaves, set on long hairy pedicles, and nipt at the bottom: it has no stalks

ſtalks or manifeſt flowers; the ſeeds are a fine duſt, lying in large, rough, brown, tranſverſe ſtreaks on the backs of the leaves. It is perennial, and found green at all times of the year, in moiſt, ſhady, ſtony places.

The leaves of harts-tongue ſtand recommended as aperients and corroborants, in obſtructions of the hypochondriacal viſcera, laxities of the inteſtines, and ſome diſorders of the breaſt: they have been chiefly uſed in apozems and infuſions, along with maidenhair, ſpleenwort, and other plants of the ſame kind, with which they appear to agree in virtue. To the taſte they are ſlightly roughiſh and ſweetiſh: with ſolution of chalybeate vitriol, they ſtrike a blackiſh colour. When freſh, they yield, on being rubbed or bruiſed, a faint unpleaſant ſmell, which in drying is in great part diſſipated.

LINI SEMEN.

LINI SEMEN Pharm. Lond, & Edinb. Lini ſativi C. B. Lini uſitatiſſimi Linn. Linseed: reddiſh-brown, gloſſy, ſlippery, flat, pointed nearly oval ſeeds, of the common flax; an annual herb, cultivated in fields, on account of the mechanic uſes of its tough filamentous rind.

These ſeeds have an unctuous, mucilaginous, ſweetiſh taſte, and no remarkable ſmell. On expreſſion, they yield a large quantity of oil; which, when carefully drawn, without the application of heat, has no particular taſte or flavour, though in ſome properties it differs conſiderably from moſt of the other oils of this kind; not congealing in winter; not forming a ſolid ſoap

Oleum expreſſum feminum lini *Ph. Ed.*
Oleum e ſem. lini *Ph. Lond.*

LINI SEMEN.

soap with fixt alkaline salts *(a)*; acting more powerfully, as a menstruum, on sulphureous bodies, than any other expressed oil that has been tried. The seeds, boiled in water, yield a large proportion of a strong flavourless mucilage: to rectified spirit they give out little or nothing. Infusions of linseed, like other mucilaginous liquors, are used as emollients, incrassants, and obtunders of acrimony, in heat of urine, stranguries, thin defluxions on the lungs, and other like disorders: a spoonful of the seeds, unbruised, is sufficient for a quart of water, larger proportions rendering the liquor disagreeably slimy. The mucilage obtained by infpissating the infusions, or decoctions, is an excellent addition for reducing disgustful powders into the form of an electuary; occasioning the compound to pass the fauces freely, without sticking or discovering its taste in the mouth. The expressed oil is supposed to be more of a healing and balsamic nature than the other oils of this class; and has been particularly recommended in coughs, spitting of blood, colics, and constipations of the belly. The seeds in substance, or the matter remaining after the expression of the oil, are employed externally, in emollient and maturating cataplasms. In some places, these seeds, in times of scarcity, have supplied the place of grain, but appeared to be an unwholesome, as well as an unpalatable food: Tragus relates, that those who fed upon them in Zealand, had the hypochondres in a short time distended, and the face and other parts swelled; and that not a few died of these complaints.

(a) Geoffroy, *Memoires de l'acad. roy. des sciences de Paris, pour l'ann.* 1741.

LINUM CATHARTICUM.

LINUM CATHARTICUM Linn. Linum pratense floribus exiguis C. B. Chamælinum. Purging flax or Mill-mountain: a small plant, with little oblong smooth leaves, having one vein or rib running along the middle, joined in pairs close to the stalks, which are round, slender, reddish, divided towards the upper part into fine branches, bearing on the tops white pentapetalous uncut flowers, followed each, as in the common flax, by a roundish, ribbed, acuminated capsule, containing ten flattish slippery seeds in as many cells. It is annual, and grows wild on chalky hills, and in dry pasture grounds.

This herb is said to be an effectual and safe cathartic: an infusion of an handful of the fresh leaves in whey or white wine, or a dram of the leaves in substance with a little cream of tartar and aniseeds, are directed for a dose. Linnæus recommends an infusion of two drams of the dry leaves as a mild laxative. Their taste is bitterish and disagreeable.

LIQUIDAMBRA.

AMBRA LIQUIDA Pharm. Argent. Liquidamber: a resinous juice, of a yellow colour inclining to red, at first about the consistence of turpentine, by age hardening into a solid brittle resin; obtained from the tree that yields liquid storax, *styrax aceris folio Raii, Liquidambar styraciflua Linn.* growing in Virginia, Mexico, and other parts of America, and bearing the colds of our own climate.

This

LITHOSPERMUM. LOBELIA.

This juice has a moderately pungent, warm, balsamic taste; and a very fragrant smell, not unlike that of storax calamita heightened with a little ambergris. It was formerly in common use as a perfume, and might probably be applied to valuable medicinal purposes, but it is not at present much regarded, different artificial compositions having been often substituted to it in the shops.

LITHOSPERMUM.

LITHOSPERMUM MAJUS erectum C. B. Milium folis. Lithospermum officinale Linn. Gromwell: a rough plant, with stiff branched stalks, oblong acuminated leaves set alternately without pedicles; and whitish monopetalous flowers, scarcely longer than the cup, divided into five obtuse sections, followed by little roundish, hard, pearl-like seeds inclosed in the cup. It is perennial, grows wild in dry fields and by road-sides, and flowers in May and June.

The seeds of gromwell have been accounted notably diuretic; and recommended for cleansing the kidneys and urinary passages from viscid mucous matters, and promoting the expulsion of gravel. Their virtues do not appear to be very considerable; they have no smell, and their taste is little other than farinaceous. They have long been discarded from practice.

* LOBELIA.

LOBELIA Pharm. Edinb. Rapunculus galeatus virginianus flore majore violaceo Morison. Lobelia siphilitica Linn. Blue cardinal-flower: an herbaceous perennial plant, with an erect

erect stalk three or four feet high, and ovate-lanceolate subserrated leaves, bearing long spikes of labiated, irregular, blue flowers, each with five stamina having connate antheræ, succeeded by a bilocular capsule, containing many small seeds. The whole plant has a milky juice, and something of a rank smell. It grows in moist places in Virginia, and bears the winters of our climate.

The root of this plant consists of white fibres, a line in thickness, and about two inches in length. It resembles tobacco in its taste, which dwells long on the tongue, and is apt to excite vomiting. It was long a famous secret among the North American Indians for the cure of the venereal disease. The secret was purchased by Sir William Johnson, and has been made public in the writings of Bartram, Kalm, and others.

The dose and mode of administering this medicine are not exactly defined, but the following directions are given as the most accurate. A decoction is made of a handful of the roots in three measures of water. Of this, half a measure is taken in the morning fasting, and repeated in the evening; and the dose is gradually increased till its purgative effect becomes too violent, when the medicine is for a time to be intermitted, and then renewed, till a perfect cure is effected. One dose daily is sufficient during the latter part of the treatment; and the regimen during the whole process is to be equally strict with that observed in a course of mercurial salivation. From the third day, the ulcers, are to be well washed twice daily with the decoction; and it is said that when they are very deep and foul, the Indians sprinkle them with powder of the internal bark of the spruce-tree.

By

LUJULA.

By this method we are assured that inveterate venereal complaints are cured without the aid of mercury; and the Edinburgh college seem to give credit to the efficacy of the lobelia, by receiving it into their latest catalogue of simples.

LUJULA.

LUJULA Pharm. Lond. Trifolium acetosum vulgare C. B. Oxys alba Gerard. Alleluja, oxytriphyllum, & panis cuculi quorundam. Oxalis Acetosella Linn. WOOD SORREL: a plant, with the leaves and flowers issuing on separate pedicles from the root: the leaves are broad, shaped somewhat like a heart, and stand three together: the flowers are solitary, whitish, monopetalous, divided deeply into five segments, followed by angular capsules, which burst on being touched, and shed numerous small brownish seeds. It is perennial, grows wild in woods, and flowers in April.

THE leaves of the wood sorrel are useful saline antiseptics and antiphlogistics; similar, both in taste and in medicinal virtue, to those of the *acetosæ* or common sorrels, but somewhat more acid, and rather more grateful both to the palate and stomach. Beaten with thrice their weight of fine sugar, they form a grateful subacid conserve. Their expressed juice, depurated, is a very agreeable acid: duly inspissated, and set to shoot, it yields a crystalline acid salt of the same nature with that of the sorrels: the saline matter seems to amount to nearly one hundredth part of the weight of the fresh leaves.

Conserva lujulæ *Ph. Lond.*

LUM-

LUMBRICUS.

LUMBRICUS TERRESTRIS. *Vermes terreftres.* Earth worms: Thefe infects are fuppofed to have a diuretic and an antifpafmodic virtue. The faculty of Paris directs them to be prepared for medicinal ufe, by wafhing and drying them with a moderate heat. Moiftened with wine, or vinous fpirits, to prevent their putrefying, and fet in a cellar, they are almoft wholly refolved in a few days into a flimy liquor, which is faid, when mixed with alkaline falts, to yield cryftals of nitre.

LUPULUS.

LUPULUS Pharm. Parif. *Lupulus mas & femina* C. B. *Lupulus faliētarius* Ger. *Humulus Lupulus* Linn. Hop: a rough plant, with very long, angular, climbing hollow ftalks, and broad ferrated leaves, cut generally into three or five fharp-pointed fections, and fet in pairs at the joints: on the tops grow loofe fcaly heads, with fmall flat feeds among them. It is found wild in hedges and at the bottoms of hills, in England and other parts of Europe, but commonly cultivated in large plantations. It is perennial, and ripens in Auguft or September its leafy heads, which are cured by drying with a gentle heat on kilns made for that purpofe.

Hops have a very bitter tafte, lefs ungrateful than moft of the other ftrong bitters, accompanied with fome degree of warmth and aromatic flavour. They give out their virtue by maceration without heat, both to rectified and
proof

proof fpirit; and, by warm infufion, to water: to cold water they impart little, though macerated in it for many hours. The extracts obtained both by watery and fpirituous menftrua, particularly by the latter, are very elegant balfamic bitters, and promife to be applicable to valuable purpofes in medicine; though the hop is at prefent fcarcely regarded as a medicinal article, and fcarcely otherwife ufed than for the preferving of malt liquors; which by the fuperaddition of this balfamic, aperient, diuretic bitter, become lefs mucilaginous, more detergent, more difpofed to pafs off by urine, and in general more falubrious.

LYCOPERDON.

LYCOPERDON five Crepitus lupi. Fungus rotundus orbicularis C. B. *Bovifta officinarum* Dill. *Lycoperdon Bovifta Linn.* PUFFBALL: a round or egg-fhaped whitifh fungus, with a very fhort or fcarcely any pedicle, growing in dry pafture grounds; when young, covered with tubercles on the outfide, and pulpy within; by age becoming fmooth without, and changing internally into a very fine, light, brownifh duft.

THE dried fungous matter and the duft of lycoperdon have been long ufed among the common people, particularly in Germany, for reftraining the bleeding of wounds, and immoderate hemorrhoidal fluxes, and drying up running ulcers. In fome late trials, the duft has been found to produce the fame effect, in ftopping hemorrhages after amputation, as the celebrated agaric of the oak.

MACIS.

M A C I S.

MACIS Pharm. Lond. Macis officinarum C. B.
Mace: a pretty thick, tough, unctuous membrane, reticular or varioufly chapt, of a lively reddifh yellow colour approaching to that of faffron, enveloping the fhell of the fruit whofe kernel is the nutmeg. The mace, when frefh, is of a blood-red colour, and acquires its yellow hue in drying: it is dried in the fun, upon hurdles fixed above one another, and then, as is faid, fprinkled with fea-water, to prevent its crumbling in carriage.

Mace has a pleafant aromatic fmell, and a warm, bitterifh, moderately pungent tafte. It is in common ufe as a grateful fpice; and appears to be, in its general qualities, nearly fimilar to the nutmeg, both as the fubject of medicine and of pharmacy. The principal difference confifts in the mace being much warmer, more bitterifh, lefs unctuous, and fitting eafier on weak ftomachs; in its yielding by expreffion a more fluid oil, and in diftillation with water a more fubtile volatile one. What is called in the fhops expreffed oil of mace is prepared, not from this fpice, but from the nutmeg.

MAGNESIA.

MAGNESIA ALBA Pharm. Lond. & Edinb.
Magnesia: a fine white earth; foluble readily in all acids, the vitriolic as well as the others, into a bitter purgative liquor.

This earth has not hitherto been found naturally pure or in a feparate ftate: it was for feveral

veral years a celebrated secret in the hands of some particular persons abroad, till the preparation was made public by Lancisi in the year 1717 *(a)*, and afterwards by Hoffman in 1722 *(b)*. It was then extracted from the mother-lye, or the liquor which remains after the crystallization, of rough nitre; either by precipitation with a solution of fixt alkaline salt; or by evaporating the liquor, and calcining the dry residuum, so as to dissipate the acids by which the earth had been made dissoluble.

The magnesia, in this mother-lye, appears to have proceeded from the vegetable ashes, which are either made ingredients in the compositions from which nitre is obtained, or else added in the elixation of the nitre: for the ashes of different woods, burnt to perfect whiteness, and freed from their alkaline salt, were found to be, in part, of the same nature with the true magnesia *(c)*. But as quicklime also, in most of the German, French, and other European nitre-works, is commonly employed in large quantity, the earth obtained from the mother-lyes of those works is rather a calcareous earth than magnesia. What is now brought from abroad, under the name of magnesia, gives plain proofs of its calcareous nature, by its burning into quicklime, and forming a selenites with the vitriolic acid.

(a) Annot. in Mercati metallothec. vatican. Arm. ii. cap. x. p. 50.

(b) Observationes Physico-chymicæ, lib. ii. obs. 2.

(c) Of vegetable ashes, moderately or strongly calcined, only a part was found to dissolve in acids, and this part appeared to be perfect magnesia. It is probable that the remainder might be reduced to the same state by repeating the calcination.

The

The true magnesia is obtained in great purity, from a filtered folution of *fal catharticus amarus*, by adding a filtered folution of any alkaline falt. * This, by its fuperiour affinity with the vitriolic acid of the fal catharticus, precipitates its earthy bafis, which is the magnefia. The method of conducting the procefs is thus directed in the laft Edinburgh difpenfatory. Diffolve feparately equal weights of fal catharticus, and any pure fixed alkaline falt, in double their weight of water. Strain, and then mix them, and immediately add eight times the quantity of hot water. Let the liquor boil a while over the fire, and at the fame time agitate it. Then let it ftand till the heat be abated; and ftrain it through a linen cloth, on which the magnefia will be left. This is to be wafhed by affufions of pure water till it become perfectly taftelefs. The London pharmacopœia directs two pounds of each of the falts to be diffolved feparately in ten pints of water, and the liquors ftrained and mixed. The mixture is now to be boiled a little, and then ftrained through linen as in the former procefs.

* A method of preparing magnefia in the moft perfect and convenient manner, was publifhed by Mr. Henry in vol. ii. of the *Medical Tranfactions*. The fame writer, likewife, in a publication of *Experiments and Obfervations* on various fubjects, recommends the *calcination* of magnefia, as rendering it a fitter medicine in certain cafes. Magnefia is found by experiment to contain above half its weight of fixed air. The evolution of this in the ftomach may increafe flatulency, and caufe uneafinefs in weak bowels. By ftrong calcination the air is expelled from magnefia, while its purgative virtue remains unimpaired; nor does it acquire any

of

MAGNESIA.

of the acrimony or caufticity of lime. The London and Edinburgh colleges have now received this preparation, under the title of *magnefia ufta*.

The magnefia is recommended by Hoffman as an ufeful antacid, a fafe and inoffenfive laxative in dofes of a dram or two, and a diaphoretic and diuretic, when given in fmall dofes, as fifteen or twenty grains. Since this time, it has had a confiderable place in the practice of foreign phyficians, and has of late come into fome efteem among us, particularly in heart-burns, and for preventing or removing the many diforders which children are thrown into from a redundance of acid humours in the firft paffages. It is preferred, on account of its laxative quality, to the teftaceous and other abforbent earths, which, unlefs gentle purgatives are given occafionally to carry them off, are apt to lodge in the body, and occafion a coftivenefs very detrimental to infants. It muft be obferved, however, that it is not the magnefia itfelf which proves laxative, but the faline compound refulting from its coalition with acids: if there are no acid juices in the ftomach to diffolve it, it has no fenfible operation, and in fuch cafes it may be rendered purgative by drinking any kind of acidulous liquors after it. All the other known foluble earths yield with acids, not purgative, but more or lefs aftringent compounds.

It may be proper to obferve, that the name magnefia has been principally applied to a fubftance of a very different kind; a native mineral, found in iron mines, and in the lead mines of Mendip hills, in Somerfetfhire, ufually of a dark grey colour, fometimes bright and ftriated like antimony, fometimes dull, with only a few fmall ftriæ; remarkable for communicating, to a large

Magnefia, Manganefe, vitrariorum & minerologorum.

a large proportion of glafs in fufion, a purplifh or red tinge, which difappears on a continuance of the fire, at the fame time deftroying the effect of many other colouring matters, and rendering foul or coloured glafs clear: fuppofed to be an ore of iron, and recommended medicinally, when calcined by a ftrong fire, as an aftringent; but yielding no iron, or marks of iron, on any of the common trials by which that metal is diftinguifhed in ores; and in its nature and compofition as yet little known. Mr. Pott relates, that on being calcined with fulphur, and afterwards elixated with water, it yielded a large quantity of a white cryftalline falt, of a bitterifh aftringent tafte, followed by a kind of fweetnefs; and that the falt, after ftrong calcination, tafted like burnt alum, but more acid *(a)*; from whence it may be prefumed, that this mineral confifts in great part of an earth analogous to that of alum, which, in combination with acids, makes one of the ftrongeft ftyptics.

MAJORANA.

MAJORANA *Pharm. Lond. & Edinb.* Majorana vulgaris *C. B.* Origanum Majorana *Linn.* Sweet marjoram: a low plant, with flender, fquare, branched, woody ftalks; and little, oval, fomewhat downy leaves, fet in pairs: on the tops grow fcaly heads of fmall whitifh labiated flowers, whofe upper lip is erect and cloven, the lower divided into three fegments. It is fown annually in gardens, for culinary as well as medicinal ufes: the feeds, which rarely come to perfection in this country, are procured from the fouth of France, where the plant is faid to be indigenous.

(a) Mifcellanea Berolinenfia, tom. vi.

The

MAJORANA.

The leaves and tops of marjoram have a pleasant smell, and a moderately warm aromatic bitterish taste. Infusions of them in water, in colour brownish, smell pretty strongly, and taste weakly and unpleasantly of the herb: the blackish green tinctures, made in rectified spirit, have less smell, but a stronger and more agreeable taste. In distillation with water, an essential oil is obtained, amounting, as Hoffman observes, to about one ounce from sixty-four of the leaves slightly dried; when carefully drawn, of a pale yellow colour; by age, or too hasty fire in the distillation, contracting a reddish hue; of a very hot penetrating taste, and in smell not near so agreeable as the marjoram itself: the remaining decoction, thus divested of the volatile aromatic matter, is weakly, but unpleasantly bitterish and austere. Great part of the aromatic matter of the herb rises also in the inspissation of the spirituous tincture, and impregnates the distilled spirit: the remaining extract is stronger in taste than that made with water, its quantity being less, but has not much of the warmth or flavour of the marjoram.

This plant has been chiefly recommended in disorders of the head and nerves, in uterine obstructions and mucous discharges proceeding from what is called a cold cause (that is, from a laxity and debility of the solids, and a sluggish state of the juices) and in the humoural asthmas and catarrhs of old people. The powder of the leaves, their distilled water, and essential oil properly diluted, are agreeable errhines, and accounted particularly useful in pituitous obstructions of the nostrils, and disorders of the olfactory organs.

MALABATHRUM.

TAMALAPATRA Folium indum. Indian leaf: the leaf of the *cinamomum sive canella malavarica & javanensis* C. B. *Laurus Cassia Linn.* or casia-lignea tree, brought from the East Indies. It is of a firm texture; of an oblong oval figure, pointed at both ends; smooth and glossy on one side, which is the upper, and less so on the lower; of a yellowish green colour on the former, and a pale brownish on the latter; furnished with three ribs, running its whole length, very protuberant on the lower side, and two smaller ones which bound the edges.

These leaves have a remarkable affinity, in one respect, with the casia or bark of the tree, both the leaves themselves and their pedicles being, like it, extremely mucilaginous: chewed, they render the saliva slimy and glutinous: infused in water, they give out a large proportion of a strong tenacious mucilage. But of the aromatic flavour, which is strong in the bark, the leaves, as brought to us, have very little: they scarcely discover any warmth or pungency to the taste, and have little or no smell unless well rubbed, when they yield an agreeable, though weak, spicy odour. They are no otherwise made use of than as an ingredient in mithridate and theriaca; and are, when in their greatest perfection, far inferiour to the mace which our college directs as a succedaneum to them.

MALVA.

MALVA.

MALVA Pharm. Lond. & Edinb. *Malva silvestris folio sinuato* C. B. *Malva sylvestris* Linn. COMMON MALLOW: a plant with firm branched stalks, and roundish, notched leaves, set alternately on long pedicles: in their bosoms come forth bell-shaped monopetalous flowers, deeply divided into five heart-shaped sections, of a pale purplish or whitish colour, variegated with deeper streaks, followed by a number of capsules set in form of a flat disk, and containing each a kidney-shaped seed: the root is long, slender, and whitish. It is perennial, common in uncultivated grounds, and found in flower throughout the summer.

THE leaves and flowers of the mallow are in taste mucilaginous, and of no remarkable smell. The leaves were formerly of some esteem, as an emollient or laxative dietetic article, in dry constipated habits in the warmer climates: at present, infusions or decoctions of the leaves and flowers, and a conserve made by beating the fresh flowers with thrice their weight of fine sugar, are sometimes directed in dysuries, heat and sharpness of urine, and other like complaints; but the principal use of the herb is in emollient glysters, cataplasms, and fomentations. The roots have been recommended in disorders of the breast, and though now disregarded, may perhaps deserve some notice: they have a soft sweet taste, without any particular flavour, approaching in some degree to that of liquorice: an extract made from them by rectified spirit of wine is of great sweetness.

MANNA.

MANNA Pharm. Lond. & Edinb. Manna seu Ros calabrinus Pharm. Parif. MANNA: a sweet juice obtained from certain ash trees *(a)* in the southern parts of Europe, particularly in Calabria and Sicily, exuding from the leaves, branches, or trunk of the tree, and either naturally concreted, or exsiccated and purified by art.

* There are three ways in which manna is collected in Calabria. From the middle of June to the end of July, a very clear liquor exudes spontaneously from the trunk and branches of the tree, which by the sun's heat concretes into whitish masses, which are scraped off the next morning with wooden knives, and dried in the sun. This is called *Manna in the tear*. At the beginning of August, when this ceases to flow, the peasants make incisions in the bark, whence a juice flows, which concretes in larger masses and of a redder colour. This is the *fat* or *common Manna*. Besides these sorts, a third is procured by receiving the spontaneous exudations in June and July on straws or chips of wood fastened to the tree. This is the *cannulated* or *flaky Manna*, and is accounted the finest of the three.

Juices of the same nature are collected, in the eastern countries, from other trees and shrubs *(b)*: and similar exudations are sometimes found on different kinds of trees in Europe, as particularly on the larch in the Brian-

(a) Fraxinus rotundiore folio; & Fraxinus humilior minore & tenuiore folio *C. B.* Fraxinus Ornus *Linn.*

(b) Vide Clusii *exotic. lib.* i. *p.* 164. Rauwolf *itin. p.* 74. Teixeira *hist. Perf. p.* 29.

çonois

çamois in Dauphiny. How far the manna juices of different vegetables differ from one another, is not well known: but thus much is certain, that one and the same tree affords mannas very confiderably different, in their colour, in their taste, and in their difpofition to assume a folid concrete form; that is, in their purity, or the greater or less admixture of oily or refinous matter.

Manna brigantiaca.

The best sort of the officinal or Calabrian manna is in oblong pieces or flakes, moderately dry, friable, very light, of a whitish or pale yellow colour, and in some degree transparent: the inferiour kinds are moist, unctuous, and brown. Both sorts are said to be sometimes counterfeited by compofitions of sugar, honey, and purgative materials; compofitions of this kind, in a solid or dry form, may be diftinguished by their weight, compactness, and untranfparency: both the dry and moist compofitions may be diftinguished by their tafte, which is sensibly different from that of true manna, and with greater certainty by their habitude to menftrua.

This juice liquefies in a moift air, diffolves readily in water, and, by the assistance of heat, in rectified fpirit also; the impurities only being left by both menftrua. On infpissating the watery folution, the manna is recovered of a much darker colour than at first. From the faturated fpirituous folution, great part of it feparates as the liquor cools, concreting into a flaky mass, of a snowy whiteness, and a very grateful sweetness: the liquor, remaining after the feparation of this pure sweet part of the manna, leaves, on being infpissated, an unctuous, dark coloured, difagreeable matter, in greater or less quantity

quantity according as the manna made ufe of was lefs or more pure.

Manna, in dofes of an ounce and upwards, proves a gentle laxative: it operates in general with great mildnefs, fo as to be fafely given in inflammatory or acute diftempers, where the ftimulating purgatives have no place. It is particularly proper in ftomachic coughs, or thofe which have their origin in the ftomach; the manna, by its fweetnefs and unctuofity, contributing to obtund as well as to evacuate the offending humours: in this intention it is fometimes made into a linctus or lohoch, with equal quantities of oil of almonds and of fyrup of violets. In fome conftitutions, however, it acts unkindly, efpecially if given in confiderable quantity, occafioning flatulencies, gripes, and diftenfions of the belly; inconveniences which may be generally obviated by a fmall addition of fome grateful aromatic. It does not produce the full effect of a cathartic, unlefs taken in large dofes, as two ounces or more, and hence is rarely employed in this intention by itfelf: it may be commodioufly diffolved in the purging mineral waters, or acuated with the cathartic falts, or other purgatives: its efficacy is faid to be peculiarly promoted by cafia fiftularis, a mixture of the two purging more than both of them feparately. See *Cafia*.

MARGARITÆ.

PERLÆ. Uniones. PEARLS: fmall calculous concretions, of a bright femitranfparent whitenefs, found on the infide of the fhell of the *concha margaritifera* or mother-of-pearl fifh, as alfo of certain oyfters and other fhell-fifhes. The fineft pearls are brought from the Eaft and Weft

West Indies: the oriental, which are moſt eſteemed, have a more ſhining ſilver-like hue than the occidental, which laſt are generally ſomewhat milky: an inferiour ſort is ſometimes met with in the ſhell-fiſhes of our own ſeas, particularly on the coaſts of Scotland. The coarſe rough pearls, and the very ſmall ones which are unfit for ornamental uſes, called *rag pearl* and *ſeed pearl*, are thoſe generally employed in medicine.

It is ſaid, that counterfeit pearls are often brought from China, made of pellets of clay coated with the white matter of oyſter-ſhells. The clay may be diſtinguiſhed by its acquiring an additional hardneſs in the fire, and refiſting acids; whereas the true pearls calcine in the fire, and become quicklime, and readily diſſolve in acids; the vitriolic excepted, which precipitates them when previouſly diſſolved by other acids.

Theſe properties of the pearl, ſhew that it is an earth of the ſame kind with crabs-claws, oyſter-ſhells, and the other calcareous animal abſorbents. It has no other virtues than thoſe of the other ſubſtances of this claſs, and does not poſſeſs thoſe virtues in any greater degree than the common teſtacea.

MARRUBIUM.

MARRUBIUM Pharm. Edinb. Marrubium album Pharm. Lond. & C. B. Marrubium vulgare Linn. White Horehound: a hoary plant, with ſquare ſtalks, and roundiſh wrinkled indented leaves, ſet in pairs on long pedicles; in the boſoms of which come forth thick cluſters of whitiſh labiated flowers, in ſtriated cups, whoſe diviſions terminate in ſharp points or prickles.

prickles. It is perennial, grows wild in uncultivated grounds, and flowers in June.

The leaves of horehound have a moderately ſtrong ſmell, of the aromatic kind, but not agreeable, which by drying is improved, and in keeping for ſome months is in great part diſſipated: their taſte is very bitter, penetrating, diffuſive, and durable in the mouth. From theſe qualities, and their ſenſible operation, when taken in any conſiderable doſes, of looſening the body, it may be preſumed that this herb is a medicine of ſome efficacy, and has no ill claim to the corroborant and aperient virtues, for which it is recommended, in humoural aſthmas, and in menſtrual ſuppreſſions, cachexies, and other chronical diſorders proceeding from a viſcidity of the fluids and obſtructions of the viſcera: a dram of the dry leaves in powder, or two or three ounces of the expreſſed juice, or an infuſion of half a handful or a handful of the freſh leaves, are commonly directed for a doſe. The dry herb gives out its virtue both to watery and ſpirituous menſtrua, tinging the former of a browniſh, the latter of a green colour: on inſpiſſating the watery infuſion, the ſmell of the horehound wholly exhales, and the remaining extract proves a ſtrong and almoſt flavourleſs bitter: rectified ſpirit carries off likewiſe greateſt part of the flavour of the herb, leaving an extract in leſs quantity than that obtained by water, and of a more penetrating bitterneſs.

* The juices of horehound and plantain mixed are a remedy of great repute in America againſt the bite of the rattle-ſnake. They are adminiſtered by ſpoonfuls at ſhort intervals; while at the ſame time the wounded part is covered with
a cataplaſm

a cataplafm of the fame herbs bruifed. The good effects are faid to be fpeedy, and the recovery of the patient complete and certain.

MARUM.

MARUM SYRIACUM *Pharm. Lond.* *Majorana fyriaca vel cretica* C. B. *Marum cortufi* J. B. *Chamædrys maritima incana frutefcens foliis lanceolatis* Tourn. *Origanum Syriacum* Linn. MARUM, SYRIAN HERB-MASTICH: a low fhrubby plant, with fmall oval leaves, pointed at each end, fet in pairs, without pedicles, of a dilute green colour above, and hoary underneath: in their bofoms appear folitary, purple, labiated flowers, wanting the upper lip; the lower lip is divided into five fegments, the middlemoft of which is larger than the reft, and hollowed like a fpoon: each flower is followed by four roundifh feeds inclofed in the cup. It is faid to be a native of Syria, and of one of the Hieres iflands, on the coaft of Provence: in our climate it does not well bear fevere winters without fhelter.

The leaves of marum have a bitterifh, aromatic, very pungent tafte; and when rubbed a little, yield a quick piercing fmell, which provokes fneezing. They have been chiefly made ufe of as an ingredient in fternutatory powders, though, from their fenfible qualities, they promife to be applicable to more important purpofes, and to have no ill title to the ftimulating, attenuating, deobftruent, antifeptic virtues afcribed to them by Wedelius in a differtation on this plant: they feem particularly well adapted as an ingredient in the volatile oily aromatic fpirits with which their agreeable pungency in a great degree coincides.

The

The marum loses but little of its pungency on being dried, and in this respect it differs remarkably from many other acrid herbs, as those called antiscorbutic. It gives out its active matter partially to water, and completely to rectified spirit: the watery infusions, in colour yellow, though pretty strongly impregnated with the smell of the marum, have only a weak taste: the spirituous tinctures, in colour yellowish-green, are strongly impregnated with the taste, but have the smell in great measure covered by the menstruum. Distilled with water, it yields a highly pungent, subtile, volatile essential oil, similar to that of scurvygrass, but stronger, and of a less perishable pungency: the remaining decoction is little other than bitterish. Rectified spirit carries off likewise, in the inspissation of the spirituous tincture, a considerable share of the smell and pungency of the marum, but leaves much the greatest part concentrated in the extract; which, on being tasted, fills the mouth with a durable, penetrating, glowing warmth.

MARUM VULGARE.

SAMPSUCUS sive marum mastichen redolens C. B. Thymbra hispanica majoranæ folio Tourn. Clinopodium quibusdam, mastichina gallorum J. B. Thymus mastichina Linn. COMMON HERB-MASTICH: a low shrubby plant, with small oblong leaves, pointed at both ends, set in pairs, without pedicles: at the tops of the branches stand woolly heads, containing small white labiated flowers, whose upper lip is erect and cloven, the lower divided into three segments: each flower is followed by four seeds inclosed in the cup. It grows spontaneously on dry

dry gravelly grounds in Spain, and in the like foils it bears the ordinary winters of our own climate.

This plant is employed chiefly, like the foregoing, as an errhine. It is confiderably pungent, though far lefs fo than the *marum fyriacum*; and of a ftrong agreeable fmell, fomewhat refembling that of maftich.

MASTICHE.

MASTICHE Pharm. Lond. & Edinb. Mastich: concrete refin, obtained in the ifland Chio from the lentifk tree; brought over in fmall yellowifh tranfparent brittle grains or tears. From tranfverfe incifions made in the bark of the tree, about the beginning of Auguft, the refin exudes in drops, which running down, and concreting on the ground, are thence fwept up (a). The tree is raifed alfo in feveral parts of Europe; but no refin has been obferved to iffue from it in thefe climates: nor do all the trees of this fpecies, in the ifland Chio itfelf, afford this commodity.

This refin has a light agreeable fmell, efpecially when rubbed or heated: in chewing, it firft crumbles, foon after fticks together, and becomes foft and white like wax, without impreffing any confiderable tafte. It totally diffolves, except the earthy impurities, which are commonly in no great quantity, in rectified fpirit of wine, and then difcovers a degree of warmth and bitternefs, and a ftronger fmell than that of the refin in fubftance: the colour

(a) Tournefort, *Voyages to the Levant*, vol. i. p. 287.

of the solution is a pale yellow. Boiled in water, it impregnates the liquor with its smell, but gives out little or nothing of its substance; distilled with water, it yields a small proportion of a limpid essential oil, in smell very fragrant, and in taste moderately pungent. Rectified spirit brings over also, in distillation, the more volatile odorous matter of the mastich.

Mastich is recommended, in doses of from half a scruple to half a dram, as a mild corroborant and restringent, in old coughs, hemoptyses, diarrhœas, weakness of the stomach, &c. It is given either in substance, divided by other materials; or dissolved in spirit and mixed with syrups: or dissolved in water into an emulsion, by the intervention of gum-arabic or almonds: the decoctions of it in water, which some have directed, have little or nothing of the virtue of the mastich. It is said that this resin is commonly employed as a masticatory, in Chio and among the Turkish women, for sweetening the breath, and strengthening the gums and teeth; and that when thus used, by procuring a copious excretion of saliva, it proves serviceable in catarrhous disorders.

MATRICARIA.

MATRICARIA vulgaris seu sativa C. B. *Febrifuga Dorsten. Matricaria Parthenium Linn.*
FEVERFEW: a plant with firm branched stalks, and roughish leaves, each of which is composed of two or three pairs of indented oval segments set on a middle rib, with an odd one at the end, cut into three lobes: the flowers stand on the tops in the form of an umbel, consisting, each, of a number of short white petala, set round a yellow disk, which is followed by small
striated

striated seeds. It is biennial, or of longer duration; grows wild in hedges and uncultivated places, and flowers in June.

The leaves and flowers of feverfew have a strong, not agreeable smell, and a moderately bitter taste; both which they communicate, by warm infusion, to water and to rectified spirit. The watery infusions, inspissated, leave an extract of considerable bitterness, and which discovers also a saline matter, both to the taste, and in a more sensible manner, by throwing up to the surface small crystalline efflorescences in keeping: the peculiar flavour of the matricaria exhales in the evaporation, and impregnates the distilled water: on distilling large quantities of the herb, a yellowish strong-scented essential oil is found floating on the surface of the water. Rectified spirit carries off but little of the flavour of this plant in evaporation or distillation: the spirituous extract is far stronger in taste than that made with water, and more agreeable in smell than the herb itself. The quantity of spirituous extract, according to Cartheuser's experiments, is only about one sixth the weight of the dry leaves, whereas the watery extract amounts to near one half.

This herb is recommended as a warm, aperient, carminative bitter; and supposed to be particularly serviceable in female disorders. It appears, from the above analysis, to be a medicine of no inconsiderable virtue, in some degree similar to camomile.

MECHOACANNA.

BRYONIA MECHOACANNA alba C. B.
Convolvulus americanus mechoacan dictus Raii.
Jalappa

MATERIA MEDICA.

Jalappa alba. Rhabarbarum album quibusdam.
MECHOACAN: the root of an American convolvulus, *(Convolvulus Mechoacanna Linn.)* brought chiefly from a province in Mexico of the same name, in thin transverse slices, like jalap, but larger and whiter.

THIS root was first introduced, about the year 1524, and continued in esteem for a considerable time, as a mild cathartic, of very little taste or smell, not liable to offend the stomach, of slow operation, but effectual and safe: by degrees, it gave place to jalap, which has now, among us, almost wholly superseded its use. It seems to differ from jalap only in being weaker: the resins obtained from the two roots appear to be of the same qualities, but mechoacan scarcely yields one sixth part so much as jalap does, and hence requires to be given in much larger doses to produce the same effects. The dose of the mechoacan in substance is from one dram to two or more.

M E L.

MEL Pharm. Lond. HONEY: a sweet vegetable juice: collected by the bee from the flowers of different plants, and deposited in the cells of the combs; from which it is extracted, either by spontaneous percolation through a sieve in a warm place, or by expression. That which runs spontaneously is purer than the expressed; a quantity of the waxy and other impurities being forced out along with it by the pressure, especially when the combs are previously heated. The best sort of honey is of a thick consistence, a whitish colour, an agreeable smell, and a very pleasant taste: both the
colour

MEL.

colour and flavour are faid to differ in fome degree according to the plants which the bees collect it from.

Honey, expofed to a gentle heat, as that of a water bath, becomes thin, and throws up to the furface its waxy impurities, together with the meal or flower fometimes fraudulently mingled with it, which may thus be feparated by defpumation, fo as to leave the honey pure. On continuing the heat, there rifes a confiderable quantity of aqueous fluid, impregnated with the fine fmell of the honey: the infpiffated refiduum, like the honey at firft, diffolves both in water and in rectified fpirit, and promotes the union of oily and refinous fubftances with watery liquors. By treating the infpiffated mafs with moift clay, as practifed by the fugar-bakers for purifying fugar from its unctuous treacly matter, the unctuous parts of honey may in like manner be feparated, and its pure fweet matter obtained in the form of a folid, faline, white concrete.

This juice is an ufeful fweet, for medicinal as well as domeftic purpofes; more aperient and detergent than the fimpler fweet prepared from the fugar cane; particularly ferviceable for promoting expectoration in diforders of the breaft, and as an ingredient in cooling and detergent gargarifms. For thefe, and other fimilar intentions, it is fometimes mixed with vinegar, in the proportion of about two pounds to a pint, and the mixture boiled down to the confiftence of a fyrup; fometimes impregnated with the virtues of different vegetables, by boiling it in like manner with their juices or infufions, till the watery parts of the juice or infufion have exhaled and left the active matter incorporated

Oxymel fimplex *Ph.Lond.*

incorporated with the honey. It excellently covers the tafte of purging falts and waters. The boiling of honey, though it diffipates great part of its odorous matter, and thus proves in fome cafes injurious to it, is in fome cafes alfo of advantage: there are particular conftitutions with which honey remarkably difagrees, and in which even very fmall quantities occafion gripes, purging, and great diforder: by boiling, it lofes of that quality by which it produces thefe effects.

* The Edinburgh college feem at prefent of opinion that honey has no qualities which render it in any cafe preferable to fugar; fince they have entirely expunged it, and all preparations in which it entered, from their laft pharmacopœia (a).

MELILOTUS.

MELILOTUS officinarum germaniæ C. B. Lotus filveftris. Trifolium odoratum. Trifolium Melilotus officinalis Linn. MELILOT: a plant with fmooth oval ftriated leaves, ftanding three together, on flender pedicles; and round, ftriated, branched ftalks, terminated by long fpikes of papilionaceous flowers drooping downwards, which are followed by fhort thick wrinkled pods, containing, each, one or two roundifh feeds. It is annual or biennial, and found in flower, in hedges and corn fields, greateft part of the fummer.

MELILOT has been faid to be refolvent, emollient, anodyne, and to participate of the virtues of camomile. In its fenfible qualities, it differs

(a) It is preferved, perhaps from inattention, in the Electuarium Thebaicum Pharm. Edinb.

very

very materially from that plant: its tafte is unpleafant, fubacrid, fubfaline, but not bitter: when frefh, it has fcarcely any fmell; in drying, it acquires a pretty ftrong one, of the aromatic kind, but not agreeable. Linnæus obferves, in the third volume of the *Amænitates Academicæ*, that diftilled water of melilot, of little fmell itfelf, remarkably heightens the fragrance of other fubftances. The principal ufe of this plant has been in glyfters, fomentations, and other external applications: it formerly gave name to one of the officinal plafters; which received from the melilot a green colour and an unpleafant fmell, without any addition to its efficacy.

MELISSA.

MELISSA Pharm. Lond. & Edinb. Meliſſa hortenſis C. B. *Melyſſophyllum, mellifolium, mellitis, citrago, citraria, cedronella, apiaſtrum. Meliſſa officinalis Linn.* BALM: a plant with fquare ftalks; and oblong, pointed, dark green, fomewhat hairy leaves, fet in pairs; in the bofoms of which come forth pale reddifh labiated flowers, ftanding feveral together on one pedicle, with the upper lip roundifh, erect, and cloven, and the lower divided into three fegments. It is perennial; a native of mountainous places in the fouthern parts of Europe; and flowers in our gardens in June.

This plant, formerly celebrated for cephalic, cordial, ftomachic, uterine, and other virtues, is now juftly ranked among the milder corroborants. It has a pleafant fmell, fomewhat of the lemon kind, and a weak aromatic tafte; of both which it lofes a confiderable part on being dried;

a flight roughifhnefs, which the frefh herb is accompanied with, becoming at the fame time more fenfible. Infufions of the leaves in water, in colour greenifh or reddifh brown according to the degree of faturation, fmell agreeably of the herb, but difcover no great tafte, though, on being infpiffated, they leave a confiderable quantity of bitterifh and fomewhat auftere extract: the infufions are fometimes drank as tea in chronical diforders proceeding from debility and relaxation, and fometimes acidulated with lemon juice for a diluent in acute difeafes. On diftilling the frefh herb with water, it impregnates the firft runnings pretty ftrongly with its grateful flavour: when large quantities are fubjected to the operation at once, there feparates, and rifes to the furface of the aqueous fluid, a fmall portion of effential oil, in colour yellowifh, of a very fragrant fmell, apparently of great medicinal activity, commended by Hoffman as an excellent corroborant of the nervous fyftem. Tinctures of the newly-dried leaves made in rectified fpirit, in colour blackifh green, difcover lefs of the balm fmell than the watery infufions, but have its tafte in a greater degree: infpiffated, they leave an extract in fomewhat lefs quantity than that obtained by water, in tafte ftronger, and which retains a confiderable fhare of the fpecific fmell and flavour of the balm, but is lefs agreeable than the herb in fubftance.

MENTHA.

MINT: a perennial herb; with fquare ftalks; ferrated leaves fet in pairs; and fpikes of monopetalous flowers, each of which is cut into four fections, and followed by four feeds inclofed in the cup.

1. MENTHA

MENTHA.

1. MENTHA SATIVA *Pharm. Lond. & Edinb.* *Mentha angustifolia spicata* C. B. *Mentha viridis Linn.* Mint, hartmint, spearmint: with oblong, narrow, pointed leaves, joined close to the stalk; and small purplish flowers standing in long spikes on the tops. It is a native of the warmer climates, common in our gardens, and flowers in June and July.

This herb has a strong agreeable aromatic smell, and a bitterish, roughish, moderately warm taste. It is in general used as a restringent stomachic and carminative: in vomitings and weakness of the stomach, there are, perhaps, few simples of equal efficacy. Some report that it prevents the coagulation of milk, and hence recommend it to be used along with milk diets, and even in cataplasms and fomentations for resolving coagulated milk in the breasts: upon experiment, the curd of milk, digested in a strong infusion of mint, could not be perceived to be any otherwise affected than by common water, but milk, in which mint leaves were set to macerate, did not coagulate near so soon as an equal quantity of the same milk kept by itself.

The leaves are sometimes taken in substance, beaten with thrice their weight of fine sugar into a conserve. Moderately bruised, they yield upon expression about two thirds their weight of a turbid, brown-coloured, somewhat mucilaginous juice; which is commonly supposed to retain the full virtues of the mint, but which, though participating of the bitterness and sub-astringency of the herb, is found to have little or nothing of the peculiar aromatic flavour in which the principal virtue of this plant resides. The leaves lose in drying about three fourths of their weight, without suffering much loss of their smell or taste; nor is the smell soon dissipated by moderate warmth, or impaired in keeping.

Cold water, by maceration for six or eight hours on the dry herb, and warm water in a shorter time, become richly impregnated with its flavour: if the maceration be long continued, the grosser parts of the mint are extracted, and the liquor proves less grateful: on boiling the mint in water till the aromatic matter is dissipated, the remaining dark-brown liquor is found nearly similar to the recent juice; unpleasant, bitterish, subastringent, and mucilaginous. By distillation, a pound and a half of the dry leaves communicate a strong impregnation to a gallon of water: † the distilled water proves rather more elegant if drawn from the fresh plant in the proportion of ten pints from three pounds ‡ than from the dry plant, though the latter is frequently made use of as being procurable at all times of the year. Along with the aqueous fluid, an essential oil distils, of a pale yellowish colour, changing by age to a reddish, and at length to a dark red, in quantity near an ounce from ten pounds of the fresh herb in flower, smelling and tasting strongly of the mint, but somewhat less agreeable than the herb itself.

Dry mint, digested in rectified spirit, either in the cold, or with a gentle warmth, gives out readily its peculiar taste and smell, without imparting the grosser and more ungrateful matter, though the digestion be long continued. The tincture appears by day-light of a fine dark green, by candle-light of a bright red colour: a tincture extracted from the remaining mint by fresh spirit appears in both lights green:

Aq. menthæ sativ. Ph. Lond. †
Ph. Ed. ‡
Ol. menthæ sativ. essentiale Ph. Lond. & Ed.

the

MENTHA.

the colour of both tinctures changes, in keeping, to a brown. On gently infpiffating the filtered tinctures, little or nothing of their flavour rifes with the fpirit: the remaining extract poffeffes the concentrated virtues of about ten times its weight of the dry herb; and differs from the products obtained by diftillation with water, in this, that the bitternefs and fubaftringency of the mint, which are there feparated from the aromatic part, are here united with it.

Proof fpirit extracts the fmell and tafte of mint, but not its green colour. The tincture is brown, like the watery infufions; and, like them alfo, it becomes ungrateful, if the digeftion is long continued. On gentle diftillation, the more fpirituous portion, which rifes at firft, difcovers little flavour of the mint; but as foon as the watery part begins to diftil, the virtues of the mint come over plentifully with it. Hence the officinal fpirituous water, prepared by drawing off a gallon of proof fpirit from a pound and a half of the dried leaves, proves ftrongly impregnated with the mint. *Spiritus menthæ fativæ Ph. Lond.*

After mint has been repeatedly infufed in water, rectified fpirit ftill extracts from it a green tincture, and a fenfible flavour of mint: on the other hand, fuch as has been firft digefted in fpirit, gives out afterwards to water a brown colour, and a kind of naufeous mucilaginous tafte very different from that which diftinguifhes mint. The fpirituous tinctures mingle with watery liquors without precipitation or turbidnefs; but fpirituous liquors impregnated with its pure volatile parts by diftillation, turn milky on the admixture of water.

2. Mentha aquatica five Mentaſtrum *Pharm. Pariſ.* Mentha aquatica five fiſymbrium *J. B.* Mentha rotundifolia paluſtris five aquatica major *C. B.* Mentha aquatica *Linn.* Watermint: with ſomewhat oval leaves ſet on pedicles, and long ſtamina ſtanding out from the flowers.

3. Mentastrum hirsutum: *Auricularia officinarum* Dale: *Mentha paluſtris folio oblongo C. B.* Hairy water-mint: with long hairy leaves having no pedicles; and broad ſpikes of flowers.

Both theſe plants grow wild in moiſt meadows, marſhes, and on the brinks of rivers, and flower towards the end of ſummer. They are leſs agreeable in ſmell than the ſpearmint, and in taſte bitterer and more pungent: the ſecond ſort approaches in ſome degree to the flavour of pennyroyal. They yield a much ſmaller proportion of eſſential oil: from twenty pounds of the water-mint were obtained ſcarcely three drams. With regard to their virtues, they appear to partake of thoſe of ſpearmint; to which they are obviouſly far inferiour as ſtomachics. The hairy water-mint is ſuppoſed to be the *auricularia, planta zeylanica,* or earwort, celebrated by Marloe for the cure of deafneſs: though probably not more effectual againſt that complaint, than the other water-mint againſt nephritic ones, in which it is ſaid to have been formerly an empyrical ſecret *(a)*.

4. Mentha piperitis *Pharm. Lond. & Edinb.* Mentha ſpicis brevioribus & habitioribus, foliis

---

(a) In the firſt volume of Linnæus's *Amænitates Academicæ,* the auricularia is ſaid to be not at all of the mint kind, but a ſtellated plant, akin to galium.

MENTHA.

menthæ fufcæ, fapore fervido piperis Raii fynopf. Mentha piperita Linn. Pepper-mint: with acuminated leaves on very fhort pedicles; and the flowers fet in fhort thick fpikes or heads. It is a native of this kingdom; and, fo far as is known, of this kingdom only: it is much lefs common, however, than the other wild mints; but having been of late received in general practice as a medicine, it is now raifed plentifully in gardens; and does not appear, like many of the other plants that grow naturally in watery places, to lofe any thing of its virtue with this change of foil.

This fpecies has a more penetrating fmell than any of the other mints; and a much ftronger and warmer tafte, pungent and glowing like pepper, and finking as it were into the tongue. It is a medicine of great importance in flatulent colics, hyfteric depreffions, and other like complaints; exerting its activity as foon as taken into the ftomach, and diffufing a glowing warmth through the whole fyftem; yet not liable to heat the conftitution near fo much as might be expected from the great warmth and pungency of its tafte.

By maceration or infufion, it readily and ftrongly impregnates both water and fpirit with its virtue; tinging the former of a brownifh colour, and the latter of a deeper green than the other mints. In diftillation with water, it yields a confiderable quantity of effential oil, of a pale greenifh yellow colour, growing darker coloured by age, very light, fubtile, poffeffing in a high degree the fpecific fmell and penetrating pungency of the pepper-mint*(a): the decoction

Ol. effentiale menthæ piperitidis *Ph. Lond. & Ed.*

(a) Some particles of a true camphor were procured from dried pepper-mint by cohobation. *Gaubii Adverfar.*

remaining

remaining after the separation of this active principle, is only bitterish and subastringent, like those of the other mints. Rectified spirit, drawn off with a gentle heat from the tincture made in that menstruum, brings over little of the virtue of the herb, nearly all its pungency and warmth remaining concentrated in the extract, the quantity of which amounts to about one fourth of the dried leaves. A simple and a spirituous distilled water, drawn in the same proportions as those of spearmint, and the essential oil, are kept in the shops.

Aq. menthæ piperit. Ph. Lond. & Ed.
Aq. menthæ piperit. spirituosa Ph. Ed.
Spir. menthæ piper. Ph. Lond.

MERCURIALIS.

HERB-MERCURY: a plant with oblong, acuminated, indented leaves, standing in pairs: in their bosoms come forth, either spikes of imperfect flowers, set in three-leaved cups, falling off without any seeds; or little rough balls, joined two together, including each a single seed.

1. MERCURIALIS *testiculata sive mas, & mercurialis spicata sive femina dioscoridis & plinii* C. B. *Mercurialis annua Linn.* French mercury: with smooth glossy leaves, and branched stalks. The flowering plants, called female, and those which produce seeds, called male, are both annual, and grow wild together in shady uncultivated grounds.

The leaves of this plant have no remarkable smell, and very little taste: when freed by exsiccation from the aqueous moisture, with which they abound, their prevailing principle appears to be of the mucilaginous kind, with a small admixture of saline matter. They are ranked among the emollient oleraceous herbs, and said to

MEUM.

to gently loofen the belly: their principal ufe has been in glyfters.

2. CYNOCRAMBE: *Mercurialis montana tefti-culata & fpicata* C. B. *Mercurialis perennis Linn.* Dogs mercury, male and female: with rough leaves and unbranched ftalks. It is perennial, and grows wild in woods and hedges.

This fpecies has been faid by fome to be fimilar in quality to the foregoing, and to be more acceptable to the palate as an oleraceous herb: it has lately however been found to be poffeffed of noxious qualities, acting as a virulent narcotic. An inftance is related in N°. 203 of the Philofophical Tranfactions, of its ill effects on a family, who eat at fupper the herb boiled and fried: the children, who were moft affected by it, vomited, purged, and fell faft afleep: two flept about twenty-four hours, then vomited and purged again, and recovered: the other could not be waked for four days, and then opened her eyes and expired.

MEUM.

MEUM ATHAMANTICUM. *Meum foliis anethi* C. B. *Meu & athamanta & radix urfina quibufdam. Æthufa Meum Linn.* SPIGNEL, BAULDMONY: an umbelliferous plant, with bufhy leaves divided into flender fegments, like thofe of fennel, but finer; producing large, oblong, ftriated feeds, flat on one fide, and convex on the other: the root is long, varioufly branched, with generally a number of hairs or filaments at top, which are the remains of the ftalks of former years, of a brownifh colour on the outfide, pale or whitifh within, when dried of a fungous texture. It is perennial, grows wild

wild in meadows in some of the mountainous parts of England, and flowers in June.

The root of spignel, recommended as a carminative, stomachic, and for attenuating viscid humours, appears to be nearly of the same nature with that of lovage; differing, in its smell being rather more agreeable, somewhat like that of parsneps, but stronger, and in its taste being less sweet and more warm or acrid. The difference betwixt the two roots is most considerable in the extracts made from them by water; the extract of spignel root being unpleasantly bitterish, with little or nothing of the sweetness of that of lovage roots. The spirituous extract of spignel, more aromatic than that of the lovage, is moderately warm, bitterish, and pungent.

MILLEFOLIUM.

MILLEFOLIUM Pharm. Edinb. Millefolium vulgare album, & millefolium purpureum C. B. Achillea, Myriophyllon, Chiliophyllon, Militaris herba, Stratiotes, Carpentaria, Lumbus veneris, & Supercilium veneris. Achillæa Millefolium Linn. Milfoil or Yarrow: a plant with rough stiff leaves, divided into small segments, set in pairs along a middle rib like feathers: the little flowers stand thick together in form of an umbel on the top of the stiff stalk, and consist each of several whitish or pale purplish petala set round a kind of loose disk of the same colour, followed by small crooked seeds. It is perennial, grows plentifully by the sides of fields and on sandy commons, and is found in flower greatest part of the summer.

The

MILLEFOLIUM.

The leaves and flowers of milfoil are greatly recommended by some of the German physicians *(a)* as mild corroborants, vulneraries, and antispasmodics, in diarrhœas, hemorrhagies, hypochondriacal and other disorders. They promise, by their sensible qualities, to be of no inconsiderable activity. They have an agreeable though weak aromatic smell, and a bitterish, roughish, somewhat pungent taste. The leaves are chiefly directed for medicinal use, as having the greatest bitterishness and austerity: the flowers have the strongest and most subtile smell, are remarkably acrid, and promise to be of most efficacy, if the plant has really any such efficacy as an anodyne or antispasmodic. Dr. Grew observes, that the young roots have a glowing warm taste, approaching to that of contrayerva, and thinks they might in some measure supply its place; but adds, that they lose much of their virtue in being dried *(b)*, from whence it may be presumed that their active matter is of another kind.

The virtue of the leaves and flowers is extracted both by watery and spirituous menstrua: the astringency most perfectly by the former, their aromatic warmth and pungency by the latter, and both of them equally by a mixture of the two. The flowers, distilled with water, yield a penetrating essential oil, possessing the flavour of the milfoil in perfection, though rather less agreeable than the flowers themselves, in consistence somewhat thick and tenacious, in colour remarkably variable, sometimes of a

---

(a) Stahl, *Dissert. de therapeia passionis hypochondriacæ.* Hoffman, *De præstantia remed. domest.* § 18.

(b) *Idea of philos. hist. of plants*, § 29. *Of diversities of tastes,* chap. v. § 2.

greenish

greenifh yellow, fometimes of a deep green, fometimes of a bluifh green, and fometimes of a fine blue: thefe differences feem to depend in great meafure on the foil in which the plant is produced; the flowers gathered from moift rich grounds yielding generally a blue oil; whereas thofe, which are collected from dry commons, afford only, fo far as I have obferved, a green one with a greater or lefs admixture of yellow: the decoction remaining after the feparation of this volatile principle, leaves, on being infpiffated, a dark brownifh mafs, ungratefully auftere, bitterifh, and fomewhat faline. On infpiffating the yellowifh tincture made in rectified fpirit, fcarcely any thing of the flavour of milfoil exhales or diftils with the menftruum: the remaining deep yellow extract is more agreeable in fmell than the flowers themfelves, of a moderately warm penetrating tafte, fomewhat like that of camphor, but much milder, accompanied with a flight bitterifhnefs and fubaftringency.

MILLEPEDÆ.

MILLEPEDÆ *Pharm. Lond. & Edinb.* Centipedes & onifci quibufdam. Onifcus Afellus Linn. MILLEPEDES, WOOD-LICE: an oblong infect, with fourteen feet, and its body compofed of fourteen rings, rolling itfelf up into a round ball on being touched; found in cellars, and under ftones and logs of wood in cold moift places; rarely met with in the warmer climates. Two forts are commonly ufed indifcriminately; one large, of a dufky bluifh-black or livid colour; the other fmaller, flatter, thinner, of a pale brownifh grey, and differing alfo from the former in the laft divifion of the body being not annular,

MILLEPEDÆ.

annular, but pointed, and in the tail being forked. The firſt ſpecies is ſaid to be the true officinal ſort, though ſome have preferred the ſecond: but there does not ſeem to be any material difference between them in quality.

MILLEPEDES have a faint diſagreeable ſmell, and a ſomewhat brackiſh, ſweetiſh, unpleaſant taſte. They are celebrated as reſolvents, aperients, and diuretics; in jaundices, aſthmas, ſcrophulous and other diſorders; but that their virtues are ſo great as they are generally ſuppoſed to be, may be juſtly queſtioned, at leaſt when given in the cuſtomary doſes. I have known two hundred taken every day for ſome time together, without producing any remarkable effect; in large doſes, indeed, it is probable that their activity may be conſiderable; as they are ſaid to have ſometimes produced an univerſal heat and thirſt with a pain in the region of the pubes *(a)*, and ſometimes a ſcalding of urine *(b)*.

Theſe inſects may be commodiouſly ſwallowed entire, as they ſpontaneouſly contract themſelves, on being touched, in the form of a pill. In the ſhops they are commonly reduced into a powder; for which purpoſe they are prepared, by incloſing them in a thin canvas cloth, and ſuſpending them over hot ſpirit of wine in a cloſe veſſel, till they are killed by the ſteam and rendered friable. Of the extraction of their active matter by menſtrua, no direct experiments have been made: it is rather by expreſſion, than on the principle of extraction or diſſolution, that

Millepedæ præparatæ Ph. Lond. & Ed.

(a) Frid. Hoffman, *De mat. med. regn. animal. cap.* 18. *Opera omnia, ſuppl.* ii. *par.* iii. *p.* 157.

(b) Fuller, *Pharmacopœia extemporanea,* ſub. *Expreſſ. milleped. ſimp.*

their

their virtues are commonly endeavoured to be obtained in a liquid form; though some liquors are generally added previously to the expression, partly to improve their virtue for particular intentions, partly to preserve the animal juice from corruption, and partly to render it more completely separable. The college of Edinburgh directs two ounces of live millepedes to be slightly bruised, and digested for a night in a pint of rhenish wine, after which the liquor is to be pressed out through a strainer.

<small>Vinum millepedatum *Ph. Ed.*</small>

MOLDAVICA.

MOLDAVICA seu Melissa turcica: an Melissa americana trifolia odore gravi Tourn. inst.? Camphorosma Morison. hist. ox.? Dracocephalum canarense Linn. Turkey or rather Canary balm, commonly called Balm-of-gilead: a plant with square stalks, and acuminated leaves, slightly and obtusely indented, set generally three on one pedicle: of each three, the end one is largest, and the other two are nipt at the bottom on the upper side, or do not reach so far down their middle ribs on that side as on the other: the pedicles stand in pairs at the joints, with similar sets of smaller leaves in their bosoms. On the tops come forth thick spikes, or heads, of pretty large, reddish, labiated flowers; whereof both the upper and lower lip are cut into two parts, and the cup into five. It is perennial, a native of the Canary islands, and scarcely bears the winters of our climate without shelter.

This or some of the other species of the Turkey balm, of which there are several, is greatly commended by Hoffman, for strengthening the tone of the stomach, and the nervous system:

syſtem: in this country, it has not yet been, though it ſeems to have a good claim to be, received among the medicinal plants: infuſions of it may be drank as tea, and are very grateful. The leaves and flowery tops have a fragrant ſmell, ſomewhat reſembling that of balm, but far ſtronger, and approaching to that of the fine balſam from which the plant received its name. Their taſte is likewiſe agreeable, but ſo covered with the aromatic flavour, that its particular ſpecies is not eaſily determined: when the herb is infuſed in water, and the aromatic part diſſipated by inſpiſſating the filtered infuſion, the remaining extract impreſſes on the palate a moderately ſtrong, though only momentary, pungency and bitterneſs. In diſtillation with water, it yields a fragrant eſſential oil. *Ol. ſyriæ Germanis quibuſdam.*

MOSCHUS.

MOSCHUS Pharm. Lond. & Edinb. Musk: an odoriferous, grumous ſubſtance: found in a little bag, ſituated near the umbilical region of an oriental quadruped, which is ſaid by ſome to bear the greateſt reſemblance to the goat, by others to the ſtag kind. The beſt muſk is brought from Tonquin in China, an inferiour ſort from Agria and Bengal, and a ſtill worſe from Ruſſia.

Fine muſk comes over in round thin bladders, generally about the ſize of pigeons eggs, covered with ſhort brown hairs, well filled, and without any aperture or any appearance of their having been opened. The muſk itſelf is dry, with a kind of unctuoſity; of a dark reddiſh brown or ruſty blackiſh colour; in ſmall round grains, with very few hard black clots; perfectly free from any ſandy or other viſible foreign matter.

Chewed,

Chewed, and rubbed with a knife on paper, it looks bright, yellowish, smooth and free from grittiness. Laid on a red-hot iron, it catches flame, and burns almost intirely away, leaving only an exceeding small quantity of light greyish ashes: if any earthy substances have been mixed with the musk, the quantity of the residuum will discover them.

This concrete has a bitterish subacrid taste; and a fragrant smell, agreeable at a distance, but so strong as to be disagreeable when smelt near to, unless weakened by a large admixture of other substances. A small quantity, macerated for a few days in rectified spirit of wine, imparts a deep colour, and a strong impregnation to the spirit: this tincture, of itself, discovers but little smell, the spirit covering or suppressing the smell; but on dilution it manifests the full fragrance of the musk, a drop or two communicating to a quart of wine or watery liquors a rich musky scent. The quantity of liquor which may thus be flavoured by a certain known proportion of musk, appears to be the best criterion of the genuineness and goodness of this commodity; a commodity, which is not only said to vary in goodness according to the season of its being taken from the animal *(a)*, but which is oftentimes so artfully sophisticated, that the abuses cannot be discovered by any external characters, or by any other known means than the degree of its specific smell and taste, which the above experiment affords the most commodious method of measuring. The rectified spirit takes up completely the active matter of the musk; watery liquors extract it

(a) Strahlenberg, *Descript. Ruff. Siber. &c.* p. 340.

only

only in part. The shops endeavour to procure an union of its virtues with water by the intervention of sugar and gum arabic: forty grains of musk and a dram of fine sugar are rubbed together, and then a dram of powdered gum arabic is added; to this are poured by degrees six ounces by measure of rose water. But the most elegant of all the liquid preparations of this drug, is the tincture in rectified spirit, which may be occasionally diluted with any watery liquors, like the other spirituous tinctures. This is directed in the Edinburgh pharmacopœia in the proportion of two drams of musk to a pound of spirit. By distillation, water becomes strongly impregnated with the scent of the musk, and seems to elevate all its odoriferous matter; while rectified spirit, on the contrary, brings over little or nothing of it.

Mistura moschata Ph. Lond.

Tinct. mosch. Ph. Ed.

Musk, a medicine of great esteem in the eastern countries, has lately come into general use among us also, in some nervous disorders: though liable, by its strong impression on the organs of smell, to offend and disorder hysterical persons and constitutions of great sensibility, yet, when taken internally, it is found to abate symptoms of that kind which its smell produces, and to be one of the principal medicines of the antispasmodic class. Dr. Wall relates, that two persons labouring under a subsultus tendinum, extreme anxiety, and want of sleep, occasioned by the bite of a mad dog, were perfectly relieved by two doses of musk of sixteen grains each: that convulsive hiccups, attended with the worst symptoms, were removed by a dose or two of ten grains: but in some cases, where this medicine could not, on account of strong convulsions, be administered by the mouth, it proved of service when injected as a glyster: that he never met with any perfon, how averfe fo ever to perfumes, but could take it in the form of a bolus without inconvenience: that under the quantity of fix grains, he never found much effect from it, but that when given to ten grains and upwards, it produces a mild diaphorefis, without heating or giving any uneafinefs, but on the contrary, abating pain and raifing the fpirits; and that after the fweat has begun, a refrefhing fleep generally fucceeds *(a)*. This medicine is now received in general practice, in different convulfive diforders; and its dofe has been increafed, with advantage, to a fcruple, and half a dram, every four or fix hours. It has been tried alfo in fome maniacal cafes; in which it feemed to procure a temporary relief.

M O X A.

MOXA five lanugo artemifiæ japonicæ Pharm. Parif. Moxa: a foft lanuginous fubftance, prepared in Japan, from the young leaves of a fpecies of mugwort, by beating them, when thoroughly dried, and rubbing them betwixt the hands, till only the fine fibres are left. A like fubftance is faid, in the German ephemerides, to have been obtained, by treating the leaves of our common mugwort in the fame manner.

Moxa is celebrated in the eaftern countries, for preventing and curing many diforders, by being burnt on the fkin: a little cone of moxa, laid on the part previoufly moiftened, and fet on fire at top, burns down with a temperate glowing heat, and produces a dark coloured

(a) Philofoph. Tranfact. No. 474.

fpot,

MYROBALANI.

spot, the exulceration of which is promoted by applying a little garlic, and the ulcer either healed up when the eschar separates, or kept running for a length of time, as different circumstances may require. A fungous substance, found in fissures of old birch trees, is said to be in common use among the Laplanders for the same purposes *(a)*; and some have used cotton, impregnated with a solution of nitre, and afterwards dried, which answers the end as effectually as the moxa of the Japonese *(b)*. It is obvious, that all these applications are no other than means of producing an exulceration of the skin, and its consequence a drain of humours.

MYROBALANI.

MYROBALANI Pharm. Parif. MYROBALANS: dried fruits, of the plum kind, brought from the East Indies. Five sorts, produced by different trees, have been distinguished in the shops.

1. - MYROBALANI BELLIRICÆ: *Myrobalani rotundæ belliricæ, arabibus belleregi,* &c. C. B. Belliric myrobalans: of a yellowish grey colour, and an irregularly roundish or oblong figure, about an inch in length, and three quarters of an inch thick.

2. MYROBALANI CITRINÆ: *Myrobalani teretes citrini bilem purgantes* C. B. Yellow myrobalans: somewhat longer than the preceding; with generally five large longitudinal ridges,

(a) Linnæus. *Flora lapponica, p.* 264.
(b) Hagendorn, *Cynofbatologia, p.* 74.

and as many smaller between them; somewhat pointed at both ends.

3. MYROBALANI CHEBULÆ: *Myrobalani maximi angulosi pituitam purgantes, arabibus quebolia, &c. C. B.* Chebule myrobalans: resembling the yellow in figure and ridges, but larger, of a darker colour inclining to brown or blackish, and with a thicker pulp.

4. MYROBALANI EMBLICÆ, *arabibus embelgi, &c. C. B. Myrobalani emblicæ in segmentis nucleum habentes angulosæ J. B.* Emblic myrobalans: of a dark blackish grey colour, roundish, about half an inch thick, with six hexagonal faces opening from one another; the fruit of the *Phyllanthus Emblica* of Linnæus.

5. MYROBALANI INDICÆ: *Myrobalani nigræ octangulares C. B. Myrobalani indicæ nigræ sine nucleis J. B.* Indian or black myrobalans: of a deep black colour, oblong, octangular, differing from all the others, in having no stone, or only the rudiments of one; from whence they are supposed to have been gathered before maturity.

ALL the myrobalans have an unpleasant, bitterish, very austere taste; and strike an inky blackness with solution of chalybeate vitriol. They are said to have a gently purgative, as well as an astringent and corroborating virtue; and are directed to be given, in substance from half a dram to four drams, and in infusion or slight decoction from four to twelve drams. It is said also, that the fruit in substance acts barely as a styptic, without exerting its purgative quality; that this last is discovered only in the infusions infufions *(a)*, and that by boiling it is diffipated or deftroyed *(b)*. A difference of this kind, between the fruit and its infufions, might be eafily conceived, if the aftringency of the myrobalans was not extracted by watery liquors, but the contrary of this was found on trial to be true; the infufions, decoctions, and the decoctions infpiffated to the confiftence of an extract, being ftrongly ftyptic. In this country, they have long been entire ftrangers to practice, and are now difcarded, by the colleges both of London and of Edinburgh, from their catalogue of officinals.

MYRRHA.

MYRRHA Pharm. Lond. & Edinb. MYRRH: a gummy-refinous concrete juice, of an oriental tree, of which we have no certain account. * Mr. Bruce informs us that it grows fpontaneoufly in the eaftern part of Arabia felix, and in that part of Abyffinia which the Greeks named Troglodytria; and that the Abyffinian myrrh, which is the leaft plentiful, is the beft. The beft kind is that which flows from deep incifions of the larger branches, hardening upon the tree. It continues to diftil every year from the fame wound; but the myrrh is of an inferiour quality after the firft, being mixed with foreign impurities, and the decayed juices of the tree. The worft is that which comes from near the root, or the old trunks *(c)*. It comes over in glebes or drops, of various colours and magnitudes: the beft fort is fomewhat tranfparent,

(a) Geoffroy, *Tract. de materia medica*, tom. ii. *p.* 332.
(b) Benancius, *Declaratio fraudum & errorum apud pharmacopæos, è mufeo Bartholini*, *p.* 68.
(c) *Philof. Tranf.* vol. lxv. *p.* 408.

friable,

friable, in some degree unctuous to the touch, of an uniform brownish or reddish yellow colour, often streaked internally with whitish semicircular or irregular veins; of a moderately strong, not disagreeable smell; and a lightly pungent, very bitter taste, accompanied with an aromatic flavour, but not sufficient to prevent its being nauseous to the palate.

There are sometimes found among it hard shining pieces, of a pale yellowish colour, resembling gum-arabic, of no taste or smell: sometimes masses of bdellium, darker coloured, more opake, internally softer than the myrrh, and differing from it both in smell and taste: sometimes an unctuous gummy-resin, of a moderately strong somewhat ungrateful smell, and a bitterish very durable taste, obviously different both from those of bdellium and myrrh: sometimes likewise, as Cartheuser observes, hard compact dark coloured tears, less unctuous than myrrh, of an offensive smell, and a most ungrateful bitterness, so as, when kept for some time in the mouth, to provoke reaching, though so resinous, that little of them is dissolved by the saliva. Great care is therefore requisite in the choice of this drug.

This bitter aromatic gummy-resin is a warm corroborant, deobstruent, and antiseptic. It is given from a few grains to a scruple and upwards, in uterine obstructions, cachexies, putrid fevers, &c. and often employed also as an external antiseptic and vulnerary. * In doses of half a dram, Dr. Cullen remarks that it heated the stomach, produced sweat, and agreed with the balsams in affecting the urinary passages. It has lately come more into use as a tonic in

hectical

MYRRHA.

hectical cafes, and is laid to prove lefs heating than moft other medicines of that clafs.

Myrrh diffolves almoft totally in boiling water, but as the liquor cools, a portion of refinous matter fubfides. The ftrained folution is of a dark yellowifh colour, fomewhat turbid, fmells and taftes ftrongly of the myrrh, and retains both its tafte, and a confiderable fhare of its fcent, on being infpiffated with a gentle heat to the confiftence of an extract. By diftillation with a boiling heat, the whole of its flavour arifes, partly impregnating the diftilled water, partly collected and concentrated in the form of an effential oil; which is in fmell extremely fragrant, and rather more agreeable than the myrrh in fubftance, in tafte remarkably mild, fo ponderous as to fink in the aqueous fluid, whereas the oils of moft, perhaps of all, of the other gummy-refins fwim: the quantity of oil, according to Hoffman's experiments, is about two drams from fixteen ounces, and when the myrrh is of a very good kind, near three drams.

Rectified fpirit diffolves lefs of this concrete than water, but extracts more perfectly that part in which its bitternefs, flavour, and virtues, refide: the refinous matter, which water leaves undiffolved, is very bitter; but the gummy matter, which fpirit leaves undiffolved, is infipid, the fpirituous folution containing all the active parts of the myrrh. Tinctures of myrrh, made by digefting three ounces of the concrete in two pounds and a half of rectified or a pint and a half of proof ‡ fpirit, with half a pint of rectified, are kept in the fhops, and given fometimes internally from fifteen drops to a tea-fpoonful, but oftener ufed among us externally, for cleanfing ulcers and promoting

Tinct. myrrh.
† *Ph. Ed.*
‡ *Ph. Lond.*

the

the exfoliation of carious bones: both tinctures are of a reddish yellow colour. In distillation, rectified spirit brings over little or nothing of the flavour of the myrrh: the extract, obtained by infpiffating the tincture, is a fragrant, bitter, very tenacious refin, amounting to one third or more of the weight of the myrrh employed.

MYRTUS.

MYRTUS communis italica C. B. *Myrtus communis* Linn. MYRTLE: an evergreen shrub; with oblong leaves, pointed at both ends; in the bofoms of which spring folitary white pentapetalous flowers, followed by black oblong umbilicated berries full of white crooked feeds. It is a native of the fouthern parts of Europe, from whence the shops have been ufually fupplied with the berries, called *myrtilli,* which rarely come to perfection in our climate; nor does the shrub bear our fevere winters without shelter.

THE berries of the myrtle, recommended in alvine and uterine fluxes and other diforders from relaxation and debility, appear to be among the milder reftringents or corroborants: they have a roughish not unpleafant tafte, accompanied with a degree of fweetifhnefs and aromatic flavour. The leaves have likewife a manifeft aftringency, and yield, when rubbed, a pretty ftrong aromatic fmell, agreeable to moft people.

MYRTUS BRABANTICA.

MYRTHUS BRABANTICA Pharm. Parif. *Rhus myrtifolia belgica* C. B. Gale, *frutex odoratus*

NAPUS.

odoratus septentrionalium, elæagnus Cordo, chamælæagnus Dodonæo J. B. *Myrica Gale Linn.* GAULE, SWEET WILLOW, DUTCH MYRTLE: a small shrub, much branched; with oblong, smooth, whitish green leaves, somewhat pointed or converging at each end; among which arise pedicles bearing flowery tufts, and separate pedicles bearing scaly cones which include the seeds, one little seed being lodged in each scale. It grows wild in waste watery places in several parts of England: in the isle of Ely it is said to be very plentiful. It flowers in May or June, ripens its seeds in August, and loses its leaves in winter.

THE leaves, flowers, and seeds of this plant, have a strong fragrant smell, and a bitter taste. They are said to be used among the common people, for destroying moths, and cutaneous insects, being accounted an enemy to insects of every kind; internally, in infusions, as a stomachic and vermifuge; and, as a substitute to hops, for preserving malt liquors, which they render more inebriating, and of consequence less salubrious *(a)*: it is said that this quality is destroyed by boiling *(b)*.

NAPUS.

NAVEW: a plant of the turnep kind, with oblong roots growing slenderer from the top to the extremity. Two sorts of it, ranked among the articles of the materia medica, are supposed by Linnæus to be only varieties, and are therefore joined into one species, under the name

(a) Ray, *Historia plantarum,* tom. ii. *p.* 1707.
(b) Linnæi, *Amœn. Academic.* iii. 96.

of *braffica (napus) radice caulefcente fufiformi*. They are both biennial.

1. Napus, *dulcis officinarum. Napus fativa* C. B. Garden or fweet navew, or French turnep: cultivated for the culinary ufe of its roots, which are warmer and more grateful than thofe of the common turnep, and are faid to afford likewife, in their decoctions, a liquor beneficial in diforders of the breaft. The feeds, in figure roundifh and in colour reddifh, are the part principally directed for medicinal purpofes: they have a moderately pungent tafte, fomewhat approaching to that of muftard feed, of the virtues of which they appear to partake: with muftard feed they agree alfo in their pharmaceutic properties, their pungent matter being taken up completely by water, and only partially by rectified fpirit, and being diffipated in the infpiffation of the watery infufion, only an unpleafant bitterifhnefs remaining in the extract. As the navew feeds, nearly fimilar in kind to thofe of muftard, are apparently much inferiour in degree, the college of Edinburgh has difcarded them, and that of London retains them only as an ingredient in theriaca.

2. Bunias *Pharm. Parif. Napus filveftris* C. B. Wild navew, or rape: growing on dry banks and among corn: with leaves fomewhat different from thofe of the preceding, being more like thofe of cabbages than of turneps; the root fmaller, and of a ftronger unpleafant tafte; and the feeds alfo rather more pungent, on which account they are preferred by the faculty of Paris. The feeds of both kinds yield upon expreffion a large quantity of oil: the oil called rape-oil is extracted from the feeds of the wild

wild fort, which is cultivated in abundance, for that ufe, in fome parts of England; the cake, remaining after the expreffion of the oil, retains, like that of muftard, the acrimony of the feeds.

NARDUS CELTICA.

NARDUS CELTICA diofcoridis C. B. Spica celtica & faliunca quibufdam. Valeriana celtica Tourn. & Linn. CELTIC NARD: a fmall fpecies of valerian, with uncut, oblong, obtufe, fomewhat oval leaves. It is a native of the Alps, from whence the fhops have been generally fupplied with the dried roots, confifting of a number of blackifh fibres, with the lower parts of the ftalks adhering; which laft are covered with thin yellow fcales, the remains of the withered leaves.

THIS root has been recommended as a ftomachic, carminative, and diuretic: at prefent, it is fcarcely otherwife made ufe of, in this country, than as an ingredient in mithridate and theriaca, though its fenfible qualities promife fome confiderable medicinal powers. It has a moderately ftrong fmell, of which it is extremely retentive [a], and a warm bitterifh fubacrid tafte, fomewhat refembling thofe of common wild valerian: an extract made from it by rectified fpirit has a ftrong penetrating tafte, and retains in good meafure the particular flavour, as well as the bitternefs and pungency of the root.

[a] Linnæus obferves that this plant, in a dry herbal, has retained its fragrance above a century. *Amœnitat. Academic. vol.* iii. *p.* 71.

NARDUS

NARDUS INDICA.

NARDUS INDICA *quæ spica, spica nardi, & spica indica officinarum* C. B. INDIAN NARD, or SPIKENARD: the bushy top of the root, or the remains of the withered stalks and ribs of the leaves, of an Indian grassy-leaved plant of which we have no particular description. The nard, as brought to us, is a congeries of small tough reddish brown fibres; cohering close together, but not interwoven, so as to form a bunch or spike about the size of a finger: sometimes two or three bunches issue from one head, and sometimes bits of leaves and stalks in substance are found among them.

The Indian nard, now kept in the shops chiefly as an ingredient in the mithridate and theriaca, was formerly employed in the same intentions as the Celtic, and is said to be used among the orientals as a spice. It is moderately warm and pungent, accompanied with a flavour not disagreeable.

NASTURTIUM AQUATICUM.

NASTURTIUM AQUATICUM *Pharm. Lond. & Edinb. Nasturtium aquaticum supinum* C. B. *Sisymbrium aquaticum Tourn. Cressio quibusdam. Sisymbrium Nasturtium aquaticum Linn.* WATER CRESSES: a juicy plant, with brownish, oblong, obtuse leaves, set nearly in pairs, without pedicles, on a middle rib, which is terminated by an odd one larger and longer-pointed than the rest: the stalks are hollow, pretty thick, channelled, and crooked: on the tops grow tufts of small tetrapetalous white flowers, followed

NASTURTIUM AQUATICUM.

followed by oblong pods, which burfting throw out a number of roundifh feeds. It grows in rivulets and the clearer ftanding waters, and flowers in June:- the leaves remain green all the winter, but are in greateft perfection in the fpring.

The leaves of the water crefles have a moderately pungent tafte; and, when rubbed betwixt the fingers, emit a quick penetrating fmell, like that of muftard feed, but much weaker. Their pungent matter is taken up both by watery and fpirituous menftrua, and accompanies the aqueous juice which iffues copioufly upon expreffion: it is very volatile, fo as to arife, in great part, in diftillation with rectified fpirit as well as with water, and almoft totally to exhale in drying the leaves, or infpiffating by the gentleft heat, to the confiftence of an extract, either the expreffed juice, or the watery, or fpirituous tinctures: both the infpiffated juice and the watery extract difcover to the tafte a faline impregnation, and in keeping throw up cryftalline efflorefcences to the furface. On diftilling with water confiderable quantities of the herb, a fmall proportion of a fubtile, volatile, very pungent effential oil is obtained.

This herb is one of the milder acrid, aperient, antifcorbutics; of the fame general virtues with the *cochlearia*, but confiderably lefs pungent, and in great meafure free from the peculiar flavour which accompanies that plant. Hoffman has a great opinion of it, and recommends it as of fingular efficacy for ftrengthening the vifcera, opening obftructions of the glands, promoting the fluid fecretions, and purifying the blood and humours: for thefe purpofes the herb may be ufed as a dietetic article, or the expreffed juice

taken

taken in doses of from one to four ounces twice or thrice a day.

NASTURTIUM HORTENSE.

NASTURTIUM HORTENSE *vulgatum* C. B. *Lepidium sativum Linn.* GARDEN CRESSES: a low plant, with variously cut winged leaves, bearing on the top of the round stalk and branches tufts of tetrapetalous white flowers, which are followed by roundish capsules, flatted on one side, full of reddish round seeds. It is annual, and raised in gardens.

THE garden cress is an useful dietetic herb in scorbutic habits, viscidities of the juices, obstructions of the viscera, and for promoting digestion; nearly of the same quality with water cress, but somewhat milder. The seeds are considerably more pungent than the leaves, and agree in their general qualities with those of mustard.

NATRON.

NATRON, *Anatron, Soude blanche, Pharm. Paris. Nitrum antiquorum. Aphronitrum. Baurach.* NATRON, or MINERAL FIXT ALKALINE SALT: This salt is contained in great abundance in the waters of the ocean, and makes the basis of the neutral salt so plentifully extracted from them for alimentary uses. It is likewise discoverable in sundry mineral springs, even of those which do not participate of sea salt. The celebrated Seltzer waters, in the archbishoprick of Treves, appear to be no other than a dilute solution of this salt mixed with a little earthy matter: twelve ounces of the water, according to

Hoffman's

NATRON.

Hoffman's analyfis, yield a fcruple of the pure alkali. In fome of the eaftern countries it is found in confiderable quantities on the furface of the earth, fometimes pure, but more commonly blended with various heterogeneous matters, from which it is extracted by means of water. I have been favoured by Dr. Heberden with a fample of this falt in a very pure ftate, which was taken up on the Pic of Teneriffe, and with which fome parts of that mountain are covered. An account of this falt, as found foffil in a cryftalline ftate, in the country of Tripoli, is contained in the *Phil. Tranf.* vol. lxi. part ii. The alkali called *foda*, or *barilla*, prepared by incinerating the maritime plant kali or glafswort, contains a falt of the fame kind. * This is received into the London catalogue of fimples, and a purified falt is ordered to be prepared from it, by repeated folution in water, colature, and cryftallization.

* Barilla *Ph. Lond.* Natron præpar. *Ph. Lond.*

The mineral alkali agrees in its general qualities with the common lixivial falts of vegetables. The differences which have been obferved are, that it is milder and lefs acrid in tafte: that it melts eafier in the fire, and requires more water for its folution: that when diffolved in water it concretes, on evaporation, into cryftalline maffes: that when expofed to a moift air, though it grows fomewhat moift on the furface, it does not run into a liquid form: that in a dry air, the cryftals lofe the water neceffary for their cryftalline form, and fall by degrees into a white powder: that the neutral falt, refulting from its coalition with the vitriolic acid, *fal glauberi*, is very eafily diffoluble in water and fufible in the fire: that with the nitrous acid it forms cubical cryftals, *nitrum cubicum*; with the marine, perfect fea falt; with

Vol. II.   K   tartar,

tartar, a falt which eafily cryftallizes, *fal rupellenfe*. Made cauftic by lime, it proves greatly inferiour to the vegetable alkali in diffolving the urinary calculus *(a)*.

This falt appears to poffefs the fame general virtues with the vegetable alkalies; but as it does not liquefy in the air, it is better adapted for an ingredient in powders; and as it is lefs acrimonious, it may be prefumed to be lefs difpofed to ftimulate the firft paffages: fome of the chemifts have taken great pains, in the preparation of the common alkalies, to preferve in them a part of the oil of the plant, fo as to reduce them to fuch a degree of mildnefs, as this alkali, with much greater uniformity and certainty, poffeffes in its pure ftate.

N E P E T A.

MENTHA CATARIA vulgaris, & major. C. B. *Cataria & Herba felis quibufdam. Nepeta Cataria Linn.* Nep, or Catmint, fo called from its being often deftroyed by cats: a hoary plant; with fquare ftalks; heart-fhaped, acuminated, ferrated leaves, fet in pairs on long pedicles; and whitifh labiated flowers ftanding in fpikes on the tops of the branches: the upper lip of the flower is divided into two, the lower into three fections. It is fometimes found wild in hedges and on dry banks, and flowers in June.

The leaves of catmint have a moderately pungent aromatic tafte, and a ftrong fmell, not ill refembling that of a mixture of fpearmint and pennyroyal; of the virtues of which herbs, in weak-

(a) Med. Tranf. i. 124.

neffes

NEPHRITICUM LIGNUM.

nefies of the ftomach, and more particularly in uterine diforders, they appear alfo to participate. Their active matter is extracted both by water and rectified fpirit, moft perfectly by the latter: the watery tinctures are of a greenifh yellow or brownifh colour, the fpirituous of a deep green. In diftillation with water, they yield a yellowifh effential oil, fmelling ftrongly of the catmint, but rather lefs agreeable than the herb itfelf: the remaining decoction is ungratefully bitterifh and fubaftringent. Rectified fpirit elevates likewife a part of the fmell and aromatic warmth, but leaves the greateft fhare behind concentrated in the extract, which proves more grateful than the leaves in fubftance, having more of the mint and lefs of the pennyroyal flavour.

NEPHRITICUM LIGNUM.

LIGNUM peregrinum aquam cæruleam reddens C. B. NEPHRITIC WOOD: an American wood, brought to us in large compact ponderous pieces, without knots: the outer part is of a whitifh or pale yellowifh colour, the medullary fubftance of a dark brownifh or reddifh. It is the product of the *Guilandina Moringa* of Linnæus. This wood, macerated in water for half an hour or an hour, imparts a deep tincture, appearing, when placed betwixt the eye and the light, of a golden colour, in other fituations of a fine blue: a property in which it agrees with the bark of the afh tree, and differs from all other known woods. Pieces of a different kind of wood, are often mixed with it, which give only a yellow tincture to water.

NEPHRITIC WOOD has a flightly bitterifh fomewhat pungent tafte; and in rafping or fcraping emits

emits a faint smell of the aromatic kind. The blue watery tincture has neither smell nor taste: but a strong infusion, which appears not blue, but of a dark brownish colour, is manifestly bitter, and smells pretty agreeably; inspissated, it leaves a blackish brown extract, in which the bitterness is more considerable, and accompanied with a slight astringency. A saturated tincture made in rectified spirit, is of a blackish red colour; the extract, obtained by inspissating it, is a tenacious resin, larger in quantity and weaker in taste than the watery extract. According to Cartheuser, the spirituous extract amounts to about one fifth, the watery only to one twelfth the weight of the wood. Both menstrua seem to extract the whole of the active matter; for if the wood remaining after the action of the one, be digested or boiled in the other, and the liquors inspissated, the extracts thus obtained have neither smell nor taste.

This wood stands greatly recommended in difficulties of urine, nephritic complaints, and all disorders of the kidneys and urinary passages; and is said to have this peculiar advantage, that it does not, like the warmer diuretics, heat or offend the parts: the blue aqueous tincture is directed to be used as common drink, and fresh water to be poured on the remaining wood so long as it communicates any blueness. For my own part, I have never known its being given medicinally, nor is it received in practice: Geoffroy says he has seen some instances of its being used without success; and indeed, whatever may be the virtues of strong infusions or extracts of the wood, the exceedingly dilute blue tincture cannot be expected to have much efficacy.

NICO-

NICOTIANA.

TOBACCO: a plant with alternate leaves, and monopetalous tubulous flowers divided into five sections: the flower is followed by an oval capsule, which opening longitudinally, sheds numerous small seeds.

1. NICOTIANA *Pharm. Lond.* Nicotiana major latifolia. C. B. Nicotiana Tabacum Linn. Tobacco: with large, sharp-pointed, pale green, soft leaves, about two feet in length, joined immediately to the stalk without pedicles. It was brought into Europe by M. Nicot, from the island Tobago in America, about the year 1560, and is now cultivated for medicinal use in our gardens. It is perennial, as is said, in America; and annual with us.

THE leaves of tobacco have a strong disagreeable smell, and a very acrid burning taste. They give out their acrid matter both to water and spirit, most perfectly to the latter: the aqueous infusions are of a yellow or brown colour, the spirituous of a deep green. They yield nothing considerable in distillation with either menstruum: nevertheless their acrimony is greatly abated in the inspissation of the tinctures, the watery extract being less pungent than the leaves themselves, and the spirituous not much more so. The several sorts of tobacco brought from abroad, are stronger in taste than that of our own growth, and the extracts made from them much more fiery, but in less quantity.

Tobacco taken internally, even in a small dose, or decoctions of it used as a glyster, prove virulently cathartic and emetic, occasioning extreme anxiety, vertigoes, stupors and disorders of the senses: some have nevertheless ventured upon it both as an evacuant, and in minuter quantities as an aperient and alterant, in epilepsies and other obstinate chronical disorders; a practice which, though in some cases it may have been successful, appears much too hazardous to be followed, particularly in the more irritable, hot, dry, bilious constitutions. By long boiling in water, its deleterious power is abated, and at length destroyed: an extract made by long coction is recommended by Stahl and other German physicians, as the most effectual and safe aperient, detergent, expectorant, diuretic, &c. but the medicine must necessarily be precarious and uncertain in strength, and has never come into use among us.

* In the year 1785, Dr. Fowler published " Medical Reports of the Effects of Tobacco principally with regard to its diuretic Quality in the Cure of Dropsies and Dysuries." In these he represents it as a safe and effectual remedy, properly administered, proving a pretty certain diuretic, and generally an anodyne. Its operation is commonly attended with vertigo; and frequently with nausea. It often acts, in a full dose, as a laxative. The mode of exhibition which he generally used, was a watery infusion of an ounce to a pint, given by drops, from six to a hundred, twice a day.

The smoke of tobacco, received by the anus, is said to be of singular efficacy in obstinate constipations of the belly. Hoffman observes, that horses have often been relieved by this remedy, but in human subjects it has been rarely tried; and says he has known some of the common people, who laboured under excruciating pains of the intestines, freed in an instant from all pain by by swallowing the smoke. * Both the decoction and the smoke have not unfrequently been injected in cases of incarcerated herniæ, and often with success. The smoke thus applied is recommended as one of the principal means for the revival of persons apparently dead from drowning or other sudden causes; but some suspect the narcotic powers of tobacco, as unfavourable in these cases.

Tobacco is sometimes employed externally in unguents and lotions, for cleansing foul ulcers, destroying cutaneous insects, and other like purposes: it appears to be destructive to almost all kinds of insects, to those produced on vegetables as well as on animals. Beaten into a mash with vinegar or brandy, it has sometimes proved a serviceable application for hard tumours of the hypochondres[a]. Some caution however is requisite even in these external uses of tobacco, particularly in solutions of continuity: there are instances of its being thus transmitted into the blood, so as to produce virulent effects. Of the common uses of the leaves brought from America, prepared in different forms, both the advantages and inconveniences are too well known to require being mentioned here.

2. NICOTIANA MINOR *C. B. Priapeia quibusdam nicotiana minor J. B. Tobacco anglicum Park. Hyoscyamus luteus Ger. Nicotiana rustica Linn.* English tobacco: with short, somewhat oval leaves, set on pedicles. It is annual, originally a native of America, but now propagates itself plentifully in England and other parts of Europe.

[a] *Edinburgh medical essays, vol.* ii. *art.* 5.

The leaves of this species are said by some to be of the same quality with those of henbane; by others, to be similar to the preceding, but weaker, which, in point of taste, they manifestly are. They have been sometimes substituted, in our markets, to the true tobacco; from which they are readily distinguishable by their smallness, their oval shape, and their being furnished with pedicles.

NITRUM.

NITRUM Pharm. Lond. & Edinb. NITRE, or SALTPETRE: a neutral salt, formed by the coalition of the common vegetable fixt alkaline salt with a peculiar acid: of a sharp penetrating cooling taste: soluble in eight times its weight of very cold water, in less than thrice its weight of water temperately warm, and, as is said, in one third its weight of boiling water: concreting from its saturated solutions, on evaporation of a part of the fluid or a gradual diminution of the heat that kept it dissolved, into colourless transparent crystals, which in figure are hexagonal prisms terminated by pyramids of the same number of sides: melting thin as water in a moderate heat: when heated to ignition, deflagrating, on the contact of any inflammable substance, with a bright flame and a considerable hissing noise; and leaving, after the detonation, its fixed alkaline salt, the acid being destroyed in the act of accension.

Nitrum fixum.

The origin of nitre, or rather of the acid which makes the characteristic part of nitre, is unknown. Thus much only is known with certainty, that common waters, both atmospherical and subterraneous, often contain a little of this acid in combination with earthy or other bodies,

bodies, so as to yield, by cryftallization, on fupplying the vegetable fixt alkali, a perfect nitre: and that when animal and vegetable fubftances, mixed with porous abforbent earths, have lain expofed to the air till they are thoroughly rotted, they are found in like manner to contain a fmall portion of nitre or of nitrous acid, fo as to give out a little nitre to water, either without addition, or on being fupplied with the proper alkaline bafis. On this foundation, fome nitre is prepared in different parts of Europe: but the greateft quantities are the produce of the Eaft Indies; the means by which it is there fo plentifully obtained, or whether it is a natural or artificial production, have not yet, fo far as I can learn, been revealed.

Nitre, as brought into the fhops, has generally a greater or lefs admixture of fea falt; from which it is purified, by diffolving it in boiling water, and, after duly evaporating the filtered folution, fetting it in a cold place to cryftallize. The more impure brown nitre requires repeated diffolution and cryftallization: to promote the purification, it is commonly diffolved in lime-water, or the folution fuffered to percolate through quicklime or a mixture of quicklime and wood afhes. It is obfervable that nitrous folutions differ from thofe of moft other falts in contracting no pellicle in evaporation: if a folution of rough nitre, containing fea falt, be boiled down till a pellicle appears, or till a part of the falt begins to concrete and fall to the bottom, all that thus feparates is faid to be fea falt, boiling water keeping far lefs of this falt diffolved than it does of nitre: but if the liquor be now poured off, though it fhould ftill retain a quantity of the fea falt, only the nitre will

Nitrum purificatum Ph. Lond.

will cryftallize in cooling, fea falt continuing diffolved in nearly as little water when cold, as was fufficient to keep it diffolved when boiling.

This falt is one of the principal medicines of the antiphlogiftic clafs; of general ufe in diforders accompanied with inflammatory fymptoms whether chronical or acute, and as a corrector of the inflammation or irritation produced by ftimulating drugs. Hoffman thinks it has an advantage above the refrigerants of the acid kind, in not being liable to coagulate the animal juices; folutions of it mingling with or diffolving recent thick blood, and in fome degree preferving it from coagulation as well as corruption; at the fame time changing its colour, when dark or blackifh, to a crimfon, an effect which it produces alfo, in a lefs degree, upon the flefhy parts of dead animals *(a)*. It retards likewife the coagulation of milk, but feems, from Stahl's account, to increafe the confiftence of thin ferous humours; for he obferves, that when ufed in gargarifms for inflammations of the fauces in acute fevers, it thickens the falival fluid into a mucus, which keeps the parts moift for a confiderable time, whereas, when nitre is not added, a drynefs of the mouth prefently enfues *(b)*.

This medicine generally promotes urine, and often gives relief in ftranguries and heat of urine whether fimple or proceeding from a venereal

---

(a) Hoffman, *De falium mediorum virtute,* § 16. *De medicamentis felectioribus,* § 13. *De præftantiffima nitri virtute,* § 5.—We cannot, however, conclude much, from thefe kinds of experiments, in regard to the medical powers of nitre, or its effects on the animal fluids, whilft under the laws of the vital œconomy.

(b) De ufu nitri medico, Menfis martius, Opufc. p. 569.

taint.

taint. It sometimes loosens the belly, particularly in hot dispositions: in cold phlegmatic temperaments it rarely has this effect, though given in very large doses: the diarrhœas of acute diseases, and fluxes in other circumstances from an acrimony of the bile or inflammation of the intestines, have been frequently restrained by it. In high fevers, it often promotes a diaphoresis or sweat; in malignant fevers, where the pulse is low and the strength greatly depressed, it impedes that salutary excretion and the eruption of the exanthemata; in consequence of its general power of diminishing inflammation and heat. It seems to be prejudicial in disorders of the lungs, though some *(a)* have ventured to prescribe it in hæmoptyses.

The usual dose of nitre, among us, is from two or three grains to a scruple; though in many cases it may be given with great safety, and to better advantage, in larger quantities. It has been said, that nitre loses, in being melted, half its weight of watery moisture, and recovers this weight again on being dissolved and crystallized *(b)*; from whence it would follow, that one part of melted nitre is equivalent to two of the crystals: but there was probably some mistake in this experiment, for I have repeated it with different parcels of nitre, and never found the loss to be so much as one twentieth of its weight.

(a) Riverius, *Cent.* i. *obs.* 83. Stahl, *ubi supra, & Observ. chym. phys. med. curios.* p. 464. Tralles, *Virium terreis ascriptorum examen*, p. 246. Dickson, *Lond. med. obs.* iv. 106. This last writer ventures to assert, that he can depend upon an electuary of conserve of roses and nitre in the cure of an hæmoptoë almost equally with bark in an intermittent.

(b) Geoffroy, *Memoires de l'acad. des scienc. de Paris, pour l'ann.* 1717.

Nitre

MATERIA MEDICA.

Trochifci e nitro *Ph. Lond.*

Nitre may be commodioufly taken in the form of troches. The London college direct one part of the purified falt to be ground with three parts of fine fugar, and one and a half of gum tragacanth in powder, and the mixture made up with water. In this and all other folid forms it is accompanied, however, with one inconvenience; being liable, efpecially when the dofe is confiderable, to occafion a pain or uneafinefs at the ftomach, which can be prevented only by plentiful dilution. A liquid form is therefore, in general, the moft eligible; and may be eafily rendered grateful by a proper addition of fugar.

Sal prunellæ. Cryftallus mineralis.

The chemifts have thought to improve the virtue of nitre, by deflagration with a fmall portion of fulphur: they melt the nitre, in a crucible, and gradually fprinkle on it one twenty-fourth its weight of flowers of fulphur: when the deflagration is over, they pour out the melted falt into clean, dry, warm brafs moulds, fo as to form it into little cakes. In this procefs, a part of the acid of the nitre, and the inflammable principle of the fulphur, detonating together, are both deftroyed; while that part of the alkali of the nitre, which is thus forfaken by its acid, unites with the acid of the fulphur, which is the fame with that of vitriol, into a new neutral falt, the fame with vitriolated tartar; and the preparation is found to be no other than a mixture of unchanged nitre with a fmall portion of this vitriolated falt. If the nitre and fulphur be taken in equal quantities, the mixture injected by a little at a time into a red-hot crucible, and kept in till all detonation ceafes, nearly the whole of the nitre will thus be changed; and the remaining falt, purified by folution

Sal polychreft. *Ph. Ed.*

in

in water, proves almoſt wholly the fame with vitriolated tartar.

The fame falt is produced by pouring gradually on nitre the pure acid of vitriol or fulphur: this acid, uniting with the alkali, difengages the acid of the nitre, which begins to exhale, immediately on mixture, in yellow or red fumes, and may be collected by diſtillation in a glafs retort with a moderate fire. Two parts of nitre to one of vitriolic acid, is a proper proportion for difengaging all the acid of the nitre; the remaining falt is nearly a pure vitriolated tartar. If three parts of nitre be ufed to one of the vitriolic acid, a part of the nitre remains unchanged: on diſſolving the whole refiduum in hot water, and fetting the filtered folution to cryſtallize, the vitriolated falt ſhoots firſt, greateſt part of the nitre continuing diſſolved. *Acidum nitroſum vulgo Spiritus nitri glauberi Ph. Ed.* *Acidum nitroſum Ph. Lond.*

The nitrous fpirit is obtained alfo by diſtillation in a ſtrong fire with vitriol in fubſtance; the vitriol parting, when ſtrongly heated, with its own acid, which then acts upon the nitre and extricates its acid in the fame manner as when the pure vitriolic acid is ufed. The fpirit thus diſtilled, called aqua fortis, is more phlegmatic than the preceding, in proportion as the vitriol employed contains more phlegm than the oil of vitriol: it is likewife liable to an admixture of the vitriolic acid, more or lefs of which is generally forced over. The proportion ufually directed is three parts of nitre, three of green vitriol uncalcined, and one and a half of the fame vitriol calcined. The ingredients are well mixed together, the diſtillation performed in an earthen retort or an iron pot fitted with an earthen head and a receiver, and continued fo long as any red vapours arife. *The colleges of London and Edinburgh have now difcarded this *Aqua fortis.*

Acidum ni-
trofum tenue
Ph. Ed.
—dilutum
Ph. Lond.

this kind of preparation, and direct a weaker nitrous acid to be made by mixing equal parts of the strong acid and pure water.

The nitrous spirit, usually distilled from rough nitre, contains often an admixture of the marine acid as well as of the vitriolic. The first is discovered, and separated, by dropping in a little solution of silver, the latter by a solution of chalk or any other calcareous earth, made in the pure nitrous acid; the silver absorbing the marine acid, and the chalk the vitriolic, and forming with those acids, respectively, indissoluble concretes, which immediately render the liquor milky, and on standing settle to the bottom. The solutions are to be cautiously and slowly dropt in, so long only as they continue to produce a milkiness: in case of an excess in their quantity, if the spirit is required perfectly pure, it is to be rectified by redistillation.

By the property on which the above method of purification depends, the nitrous spirit may be readily distinguished from the other two mineral acids. By the red or yellowish red colour of its fumes; by its forming with one fourth its weight of sal ammoniac, or with sea salt or its acid, a menstruum that perfectly dissolves gold; by its deflagrating on the contact of any inflammable matter, when heated to ignition, whatever other body it be previously combined with; it may with certainty be distinguished both from those and from every other known species of acid.

This acid has been sometimes given as a diuretic, from two or three to fifty drops, diluted largely with water; but its principal use is in combination with other bodies.

Combined

NITRUM.

Combined with vegetable fixt alkalies, it re- produces common nitre. With the mineral fixt alkali, or *soda*, it compofes a fpecies of nitre in fome refpects different from the common, cryftallizing not into a prifmatic but a cubical figure; with volatile alkalies, a fubtile pungent falt remarkable for its folubility in fpirit of wine: *(a)* of thefe two compounds, the medicinal qualities are little known, though they fhould feem to be well deferving of inquiry.

_{Nitrum cu-
bicum.}

_{Nitrum flam-
mans, vola-
tile, five am-
moniacale.}

(a) NITRUM FLAMMANS, VOLATILE, AMMONIA- CALE. This falt diflolves readily in water, and becomes pappy or fluid in a moift air: by flow evaporation in gentle warmth it fhoots into large cryftals, much refembling thofe of common nitre. It diflolves in fix times its weight or lefs, of rectified fpirit of wine. In a heat equal to that of boiling water, it melts and looks like oil, without fuffering any lofs of its fubftance; on increafing the heat a very little beyond that degree, it begins to exhale, and in a little time is wholly diffipated: the fumes, caught in proper veffels, condenfe not into a concrete falt, but a fluid fpirit; in which, however perfect the neutralization was at firft, the acid appears now to prevail. The falt thrown into a red-hot crucible, without addition of any inflammable matter, emits bright flames, without detonation or noife: the flafhes continue to play on the furface till the whole quantity of the falt is diffipated.

This falt is in tafte fimilar to common nitre, but fomewhat fharper or more penetrating. Taken in dofes of from ten to twenty-five grains, it fenfibly promotes urine; and if the patient is kept warm, perfpiration or fweat: It is recommended by Kurella, preferably to the other neutral faline medicines, in inflammatory cafes, in exanthematous fevers, and as an attenuant and refolvent in obftructions of the vifcera. He gives it either in powder, mixed with abforbents neutralized by lemon juice, or diffolved in well dulcified fpirits of vitriol or nitre, in which laft form he finds it in fome cafes to anfwer beft: from fifteen to twenty-five drops of the faturated folution are given for a dofe in any agreeable warm liquor. He recommends it likewife externally againft inflammations, eryfipelafes, and gouty pains, diffolved in fpirit of wine, either by itfelf, or with the addition of camphor and opium. *M. S. of Dr. Lewis.*

The

The acid, in the moft concentrated ftate in which it is commonly met with, faturates about five fixths its weight of vegetable fixt alkali *(a)*.

Nitrum calcareum verum.

Solutions of calcareous earths in this acid are in tafte bitterifh and very pungent. They are difficultly made to affume a cryftalline appearance; and when evaporated and exficcated by heat, the dry falt deliquiates again in the air. This falt has not hitherto been employed medicinally, nor is it as yet much known. It is a common ingredient in waters, which when its quantity is confiderable, it renders hard and indifpofed to putrefy, apparently impeding putrefaction in a much greater degree than an equal quantity of fea falt. Alkaline falts, fixt or volatile, added to the folutions, precipitate the earthy bafis; and uniting with the acid in its ftead, compofe therewith, according to the fpecies of alkali employed, the common, cubical, or ammoniacal nitre mentioned in the preceding paragraph.

The nitrous fpirit diffolves zinc, iron, copper, bifmuth, lead, mercury, and filver, the moft readily of all the acids: tin it diffolves imperfectly: regulus of antimony it only corrodes: fee the refpective metals.

The concentrated acid, combined with a due proportion of rectified fpirit of wine, lofes its acidity; the coalition of the two producing a new compound, of a gratefully pungent tafte and colour, and which is given from a few drops to a tea-fpoonful or more as mildly aperient, diuretic, antiphlogiftic, in fome degree anodyne and antifpafmodic. On mixing the two fpirits together, a great heat, ebullition, and noxious

(a) Homberg, *Memoires l'acad. roy. des fcienc. de Paris, pour l'ann.* 1699.

red

NITRUM.

red vapours arife: this conflict is lefs violent when, cautioufly and by little and little, the acid fpirit is added to the vinous, than when the vinous is added to the acid. It is prudent alfo to place the bottle containing the fpirit of wine, in a veffel of cold water. One part of the ftrong acid fpirit is commonly taken to three of the fpirit of wine‡, or half a pound to a quart†: the mixture, after ftanding for fome time that the two liquors may in fome degree unite, is fet to diftil with a gentle fire, by which the union is completed, and the very volatile dulcified fpirit feparated from the more fixt acid that remains undulcified. The diftillation has been directed to be continued fo long as the fpirit that comes over raifes no effervefcence with fixt alkaline falts; it may be regulated more commodioufly by performing the procefs in a water bath‡, for all that rifes in this heat will be found to be a pure dulcified fpirit. *Spiritus ætheris nitrofi
† Ph. Lond.
Acidum nitri vinofum vulgo fpiritus nitri dulcis
‡ Ph. Ed.*

A fubtile ethereal fluid, fimilar in its general qualities to that defcribed under the head of vitriolic acid, is obtainable with the nitrous in a more compendious manner. If equal parts by meafure of fpirit of nitre and fpirit of wine, of moderate ftrength, be mixed together, the bottle clofely ftopt, and fet in a cool place, a large proportion of ether rifes to the furface in a few days: it may be purified from the adhering acid, by fhaking it with water in which fome fixt alkaline falt has been diffolved, and then drawing off the ether by diftillation. The medicinal qualities of this fubtile fluid are not as yet much known. *Nitrous ether.*

NUMMULARIA.

NUMMULARIA *Pharm. Parif.* *Nummularia major lutea* C. B. *Hirundinaria. Centimorbia. Lyfimachia Nummularia Linn.* Money-wort: a low creeping plant, with fquare ftalks, and fmooth little roundifh or heart-fhaped leaves fet in pairs at the joints upon fhort pedicles: in their bofoms appear yellow folitary monopetalous flowers, each divided into five oval fegments, and followed by a fmall round capfule full of minute feeds. It is perennial, grows wild in moift pafture grounds, and flowers from May to near the end of fummer.

This herb is accounted reftringent, antifcorbutic, and vulnerary. Boerhaave looks upon it as fimilar to a mixture of fcurvygrafs with forrel: it appears indeed to have fome degree both of pungency and acidity, but it is far weaker than thofe herbs, or than any mixture of the two.

NUX MOSCHATA.

NUX MOSCHATA *Pharm. Lond. & Edinb.* *Nux myriftica fructu rotundo* C. B. *Nucifta. Myriftica officinalis Linn.* Nutmeg: the aromatic kernel of a large nut, produced by a tree faid to refemble the pear tree, growing in the Eaft Indies. The outer part of the fruit is a foft flefhy fubftance like that of the walnut, which fpontaneoufly opens when ripe: under this lies a red membrane called mace, forming a kind of reticular covering, through the fiffures of which is feen the hard woody fhell that
includes

NUX MOSCHATA.

includes the nutmeg. Two forts of this kernel are diftinguifhed: one of an oblong figure, called male; the other roundifh, or of the fhape of an olive, called female: this laft is the officinal fpecies, being preferred to the other on account of its ftronger and more agreeable flavour, and its being, as is faid, lefs fubject to become carious. The nutmegs are cured, according to Rumphius, by dipping them in a fomewhat thick mixture of lime and water, that they may be every where coated with the lime, which contributes to their prefervation.

THE nutmeg is a moderately warm, grateful, unctuous fpice; fuppofed to be particularly ufeful in weaknefs of appetite, and the naufeæ and vomitings accompanying pregnancy, and in fluxes; but liable, when taken too freely, to fit very uneafy on the ftomach, and, as is faid, to affect the head. Roafted with a gentle heat, till it becomes eafily friable, it proves lefs fubject to thefe inconveniences, and is fuppofed likewife to be more ufeful in fluxes.

Nutmegs, diftilled with water, yield nearly one fixteenth (a) their weight of a limpid effential oil, very grateful, poffeffing the flavour of the fpice in perfection, and which is faid to have fome degree of an antifpafmodic or hypnotic (b) power: on the furface of the remaining decoction is found floating an unctuous concrete matter like tallow, of a white colour, nearly infipid, not eafily corruptible, and hence recommended as a bafis for odoriferous bal-

Ol. effent. nucis mofchatæ *Ph. Lond.*

(a) Hoffman, *Obfervationes phyfico-chymicæ*, lib. i. obf. 1.

(b) *Mifcell. nat. curiofor.* dec. iii. ann. ii. obf. 120. Bontius, *de medicina Indorum*, p. 20.

L 2 fams

fams *(a)*: the decoction, freed from this sebaceous matter, and infpiffated, leaves a weakly bitter fubaftringent extract. Rectified fpirit takes up, by maceration or digeftion, the whole fmell and tafte of the nutmegs, and receives from them a deep bright yellow colour: the fpirit, drawn off by diftillation from the filtered tincture, is very flightly impregnated with their flavour; greateft part of the fpecific fmell, as well as the aromatic warmth, bitterifhnefs and fubaftringency of the fpice remaining, concentrated in the extract. The effential oil, and an agreeable cordial water, lightly flavoured with the volatile parts of the nutmeg by drawing off a gallon or nine pounds of proof fpirit from two ounces of the fpice, are kept in the fhops. Both the oil, and the fpirituous tincture and extracts, agree better with weak ftomachs than the nutmegs in fubftance.

Spir. nucis mofch. *Ph. Lond.*
Aqua nucis mofch. *Ph. Ed.*

Nutmegs, heated, and ftrongly preffed, give out a fluid yellow oil, which concretes on growing cold into a febaceous confiftence. Rumphius informs us, that in the fpice iflands, when the nuts are broken, thofe kernels which appear damaged, carious, or unripe, are feparated for this ufe, and that feventeen pounds and a quarter of fuch kernels yield only one pound of oil, whereas, when the nutmeg is in perfection, it is faid to afford near one third its own weight.

Two kinds of febaceous matter, faid to be expreffed from the nutmeg, are diftinguifhed in the fhops by the name of oil of mace: the

---

* *(a)* After a fluid effential oil had been procured from nutmegs by diftillation, on repeating the procefs upon the refiduum, an oil of a butyraceous confiftence arofe, which poffeffed the tafte and odour of the nutmeg, and was perfectly foluble in alcohol. *Gaubii Adverfar.*

beft

best sort, brought from the East Indies in stone jars, is somewhat soft, of a yellow colour, and of a strong agreeable smell greatly resembling that of the nutmeg itself: the other comes from Holland in solid masses, generally flat and of a square figure, of a paler colour and much weaker smell. These oils are employed chiefly externally in stomach plasters, and in anodyne and nervine unguents and liniments. They appear to be a mixture of the gross sebaceous matter of the nutmeg with a little of the essential or aromatic oil; both which may be perfectly separated from one another by maceration or digestion in rectified spirit, or by distillation with water. The spirituous tincture, the distilled water, and the essential oil, are nearly similar to those drawn from the nutmeg itself, the pure white sebaceous substance being left behind.

Ol. nucis moſch. expressum, macis vulgo dictum Ph. Lond. & Ed.

NUX PISTACIA.

PISTACHIO NUT: an oblong, pointed nut, about the size and shape of a filberd; including a kernel of a pale greenish colour, covered with a yellowish or reddish skin. It is the produce of a large tree, with winged leaves, resembling those of the ash, *pistacia peregrina fructu racemoso sive terebinthus indica theophrasti C. B. Pistachia vera Linn.* which grows spontaneously in the eastern countries, and bears the cold of our own.

PISTACHIO NUTS have a pleasant sweetish unctuous taste, resembling that of sweet almonds: their principal difference from which consists in their having rather a greater degree of sweetness, accompanied with a light grateful flavour,

flavour, and in being more oily, and hence somewhat more emollient, and perhaps more nutritious. They have been ranked among the principal analeptics, and greatly esteemed by some in certain weaknesses and emaciations. They are taken chiefly in substance, their greenish hue rendering them unsightly in the form of an emulsion. They are very liable to grow rancid in keeping.

NUX VOMICA.

NUX VOMICA Pharm. Parif. Nux metella. Vomic nut: a flat roundish seed or kernel, about an inch broad and near a quarter of an inch thick, with a prominence in the middle on both sides, of a grey colour, covered with a kind of woolly matter, internally hard and tough like horn. It is the produce of a large tree growing in the East Indies, called by Plukenet *cucurbitifera malabarienfis, œnopliæ foliis rotundis, fructu orbiculari rubro cujus grana sunt nuces vomicæ officinarum*; by Linnæus, *Strychnos Nux Vomica.*

This seed discovers to the taste a considerable bitterness, but makes little or no impression on the organs of smell. It has been recommended in tertian and quartan fevers, in virulent gonorrhœas, and as an alexipharmac: Fallopius relates, that it was given with success in the plague; that in doses of from a scruple to half a dram, it procured a plentiful sweat; and that where this evacuation happened, the patient recovered(*a*). At present it is looked upon, and not without good foundation, as a deleterious drug; which, though like many

(*a*) *Tract. de tumoribus præternaturalibus, cap.* 27.

other deleterious substances; capable, in certain doses and in certain circumstances, of producing happy effects, has its salutary and pernicious operations so nearly and so indeterminably allied, that common prudence forbids its being ventured on. Hoffman tells us of a girl of ten years of age, to whom fifteen grains, given at twice, for the cure of an obstinate quartan, proved mortal (a). The principal symptoms it has been observed to produce, in human subjects and brutes (b), are, great anxieties, strong convulsions or epileptic fits, paralytic tremors and resolutions, a great increase of the motion of the heart and of respiration, and reachings and subversions of the stomach. Dissections of dogs killed by it have shewn no material injury of the grosser parts; from whence we may presume that it is the nervous system which it immediately offends. It is probable, that the active matter of this seed is of the same nature with that of bitter almonds, but more developed and in a more concentrated state.

* The nux vomica was lately used in Sweden in an epidemic dysentery, as it is said, with remarkably good effects. A scruple of the powder was given to adults once a day in barley water, proper evacuants having been premised. Bergius (c), however, asserts, that though the flux was suppressed for twelve hours by this medicine, it never failed to return. He also mentions a case in which the above dose caused convulsive stretchings and vertigo; and after the cure of the dysentery by other medicines, a pain in the stomach and epigastric region re-

(a) Philosophia corp. human. morbosi, P. ii. cap. viii. § 8.
(b) Vide Wepfer, De cicuta aquatica, cap. xiii. p. 194, & seq.
(c) Mat. Med. 145.

mained for a long time. In the isle of Ceylon the nux vomica is said to be used internally as a specific against the bite of a species of water snake.

THE wood or roots of the tree, or of other trees of the same genus, are sometimes brought from the East Indies under the name of *lignum colubrinum (Pharm. Parif.)* or snakewood, in pieces about the thickness of a man's arm, covered with a brownish or rusty coloured bark, internally of a yellowish colour with whitish streaks.

This wood, in rasping or scraping, emits a faint not disagreeable smell; and when chewed for some time discovers a very bitter taste. Cartheuser relates, that it gives a gold-coloured tincture both to water and spirit, and that the inspissated extracts are brownish; that the watery infusion has an agreeable smell like that of rhodium, the spirituous little or none; that the infusions and extracts made with both menstrua are very bitter; that the quantity of watery extract amounts to one sixth of the wood, and that of the spirituous to near one fourth; and that the wood remaining after the action of spirit, yields still, to water, a gold-coloured tincture, and one eighth its weight of a bitter subacrid extract: from whence water appears to be the proper menstruum of its active matter.

The lignum colubrinum has been recommended, in small doses, not exceeding half a dram, as an anthelmintic, and in obstinate quartans, jaundices, cachexies, and other chronical disorders: it is said to operate most commonly by sweat, sometimes by stool, and sometimes by vomit. It appears however to be possessed of the same ill qualities with the nux vomica itself,

NUX-VOMICA.

itself, though in a lower degree, having in sundry instances been productive of convulsions, tremors, stupors, and disorders of the senses.

The *faba indica* Pharm. Paris. *Faba sancti ignatii*, or *faba febrifuga*, is the produce of a tree of the same kind, growing in the East Indies and in the Philippine islands, called by Plukenet *cucurbitifera malabathri foliis scandens, catalongay & contara philippinis orientalibus dicta, cujus nuclei pepitas de besayas aut catbalogan & fabæ sancti ignatii ab hispanis, igasur & mananaog insulanis nuncupati*; by Linnæus, *Strychnos Ignatii*. The seeds of the gourd-like fruit, improperly called beans, are of a roundish figure, very irregular and uneven, about the size of a middling nutmeg, semitransparent, and of a hard horny texture.

These seeds have a very bitter taste, and no considerable smell: when fresh they are said to have somewhat of a musky scent. Neumann observes, that an extract made of them by rectified spirit impresses at first a very agreeable bitterness, somewhat like that of peach kernels, which going off leaves in the mouth a strong bitter; that an extract made with water is likewise bitter; that the watery extract is greenish and in quantity one half of the seeds, the spirituous yellowish and little more than one fifth; that the seeds remaining after the action of water scarcely gave out any thing to spirit, but that after spirit they yielded above one fourth of extract with water.

St. Ignatius's bean is said by father Camelli to be employed by the common people in the Philippine islands against all diseases. The effects attributed to it are similar to those of the two foregoing substances:' he observes, that it
generally

generally vomits, sometimes purges, and almost always produces in the Europeans, though not in the Indians, spasmodic motions; that the dose in substance, as an emetic, is ten or twelve grains, to be taken an hour after eating; and that in smaller doses it sometimes promotes a plentiful sweat *(a)*. Neumann says he has known intermitting fevers cured by drinking, on the approach of a paroxysm, an infusion of some grains of the seed made in carduus water *(b)*; and I have been informed, that two grains were found to have as much effect as a full dose of bark. This seed, nevertheless, as it apparently partakes of the qualities of the two preceding articles, seems much too hazardous for general use.

NYMPHÆA.

WATER-LILY: an aquatic plant, with thick firm roundish leaves, furnished with two obtuse ears at the pedicle, floating on the surface of the water: the flowers, which stand on separate pedicles, are large, composed of several petala with numerous stamina in the middle, followed by single capsules full of blackish shining seeds: the root is long, thick, internally white and fungous.

1. NYMPHÆA ALBA major *C. B. Leuconymphæa. Nenuphar. Nymphæa alba Linn.* White water-lily: with white flowers set in four-leaved cups: the seed vessels round, and the roots externally brownish or blackish.

2. NYMPHÆA LUTEA: *Nymphæa major lutea C. B. Nymphæa lutea Linn.* Yellow water-

(a) Philosophical transf. numb. 250.
(b) Chymia medica, &c. i. 717. *Chemical works, p.* 347.

lily: with yellow flowers set in large five-leaved cups, the seed vessels shaped like a pear, and the roots externally greenish.

BOTH these plants are found in rivers and large lakes; the yellow is most common: they are perennial, and flower usually in June. The roots and flowers have been employed, both internally and externally, as demulcent, antiinflammatory, and in some degree anodyne. Their virtues, however, do not appear to be very great, as they have no smell, at least when dried, and but little taste: extracts made from them both by water and spirit are weakly bitterish, subastringent, and subsaline. Lindestolpe informs us, that in some parts of Sweden, the roots, which are the strongest part, were in times of scarcity used as food, and did not prove unwholesome.

OCHRA.

OCHRA *five Minera ferri lutea vel rubra Pharm. Parif.* OCHRE: an argillaceous earth; less tenacious, when moistened, than the clays and the boles; impregnated with a calx of iron, and thereby tinged of a yellow or red colour. The dark red sort is called reddle or ruddle, *rubrica fabrilis* the yellow *sil*; *ochra plinio & latinis sil dicta Charleton.* Those which are naturally yellow become red by burning. Both kinds are dug in several parts of England.

THESE earths discover their argillaceous nature, by burning hard in the fire; and their ferrugineous impregnation, by digestion in aqua regis, which extracts the iron, leaving the earth nearly white. To the taste they seem somewhat astringent, in consequence, not of the metallic, but

but of the earthy part, for the iron is in such a state as not to be acted on by any fluid that exists in the bodies of animals: it may therefore be presumed, that they do not differ materially, in virtue, from the boles; except in being less viscid, and therefore of less efficacy for obtunding acrid humours: see *Bolus* and *Cimolia*. Among us they are rarely or never used medicinally under their own name; though sometimes applied in the shops to the counterfeiting of earths that are less common.

OCIMUM.

OCIMUM, *Basilicum, Herba regia.* BASIL: a plant, with square stalks; oval leaves set in pairs; and long spikes of labiated flowers, whose upper lip is divided into four parts, the lower entire: the cup also has two lips, one cut into four sections, the other into two.

1. OCIMUM *vulgatius* C. B. *Ocymum medium citratum* Ger. *Ocimum Basilicum Linn.* Common or citron basil: with most of the leaves indented, and the flower-cups edged with fine hairs.

2. OCIMUM CARYOPHYLLATUM: *Ocimum minimum* C. B. & *Linn.* Small or bush basil: with uncut leaves.

BOTH these plants are natives of the eastern countries, and sown annually in our gardens for culinary as well as medicinal uses. The seeds, which rarely come to perfection in this climate, especially those of the second sort, are brought from Italy and the south of France.

The

The leaves of basil are accounted mildly balsamic: infusions of them are sometimes drank as tea in catarrhous and uterine disorders, and the dry leaves in substance made an ingredient in cephalic and sternutatory powders. They are very juicy, of a weakly aromatic and very mucilaginous taste, and of a strong smell, which is somewhat disagreeable when the herbs are fresh, but is improved by drying: those of the first sort approach to the lemon scent, those of the second to that of cloves. Distilled with water, they yield a considerable quantity of essential oil, of a penetrating fragrance, commended by Hoffman as a nervine, similar, but greatly superiour, to oil of marjoram (a).

* OENANTHE.

THIS is the botanical name of a genus of plants of the umbelliferous class, of which there are three species natives of Great Britain. One of these only is known by its effects on the human body, the

OENANTHE *Chærephylli foliis* C. B. *Oenanthe crocata* Linn. Hemlock dropwort: this is a large umbelliferous plant, growing in ditches and other moist places; with pinnated leaves, resembling those of celery or chervil, and ribbed stalks. Its roots afford the easiest mark of distinction, which are white, thick, and short, and grow several together, forming a kind of bunch.

The hemlock dropwort has long been known as a most dangerous poison; the most virulent, perhaps, that this country produces. Its roots or leaves eaten by mistake, have frequently

(a) *Observationes physico-chymicæ*, lib. i. obs. 4.

proved

proved fatal, occasioning violent sickness and vomiting, rigors, convulsions, delirium, and other terrible affections of the nervous system. The head has been said to be affected even by being in the same room with a quantity of the plant. Like so many other deleterious vegetables, it, however, is capable of being rendered a powerful remedy. A case is published by Dr. Pulteney in the *Philof. Tranfact.* vol. lxii. in which this plant, used by mistake instead of the water parsnep, proved remarkably efficacious in removing an inveterate scorbutic complaint, which had resisted a variety of other remedies. The dose first given was a common spoonful of the juice of the root, which at the first exhibition produced very alarming effects. This was afterwards reduced to three tea-spoonfuls; which quantity was persisted in a considerable time, and then changed for a tea of the leaves. The medicine never proved purgative, but was diuretic. It always occasioned a degree of vertigo; accompanied, when the juice itself was taken, with nausea and sickness.

If this experiment be imitated, it is obvious that the greatest degree of caution will be necessary.

OLEA.

OLEA *sativa* C. B. *Olea europæa* Linn. OLIVE: an evergreen tree, with oblong, narrow, willow-like leaves, and monopetalous whitish flowers, cut into four sections, followed by clusters of oval black fruit, containing, under a fleshy pulp, a hard rough stone. It is a native of the southern parts of Europe, and bears the ordinary winters of our own climate.

OLIBANUM.

THE fruit of this tree *(oliva)* has a bitter, auftere, very difagreeable tafte: pickled, as brought from abroad, it proves lefs ungrateful, and is fuppofed to promote appetite and digeftion, and attenuate vifcid phlegm in the firft paffages: the Lucca olives, which are fmaller than the others, have the weakeft tafte; and the Spanifh, or larger, the ftrongeft: thofe brought from Provence, which are of a middling fize, are in general moft efteemed. But the principal confumption of olives is in the preparation of the common fallad oil *(oleum olivarum Pharm. Lond. & Edinb.)* which is obtained by grinding and preffing them when thoroughly ripe: the finer and purer oil iffues firft by gentle preffure; and inferiour forts, on heating the refiduum and preffing it more ftrongly. All thefe oils contain a portion of watery moifture; and of the mucilaginous fubftance of the fruit: to feparate thefe, and thus prevent the oil from growing rancid, fome fea falt is added, which not being diffoluble in the pure oil, imbibes the watery and mucilaginous parts, and finks with them to the bottom. As this oil grows thick in a moderate degree of cold, a part of the falt, thrown up by fhaking the veffel, is fometimes detained in it, fo as to render the tafte fenfibly faline. In virtue, it does not differ materially from the other flavourlefs expreffed oils: it is preferred to the others for dietetic ufes, and in plafters and unguents, but is more rarely employed as an internal medicine.

OLIBANUM.

OLIBANUM *Pharm. Lond. & Edinb.* OLIBANUM: a gummy refin brought from Turkey and the Eaft Indies, ufually in drops or tears, like

like those of mastich, but larger, of a pale yellowish colour, which by age becomes reddish. It is the product of a tree of the juniper kind growing in Arabia; the *juniperus lycia* of Linnæus.

This gummy-resin has a moderately strong, not very agreeable smell, and a bitterish somewhat pungent taste: in chewing, it sticks to the teeth, becomes white, and renders the saliva milky. Laid on a red-hot iron, it readily catches flame, and burns with a strong, diffusive, not unpleasant smell : it is supposed to have been the incense used by the ancients in their religious ceremonies, though it is not the substance now known by that name in the shops. On trituration with water, greatest part of it dissolves into a milky liquor, which on standing deposites a portion of resinous matter, and being now gently inspissated, leaves a yellow extract, which retains greatest part of the smell as well as the taste of the olibanum ; its odorous matter appearing to be of a less volatile kind than that of most other gummy-resins. Rectified spirit dissolves less than water, but takes up nearly all the active matter: the transparent yellowish solution, inspissated, yields a very tenacious resin, in which the active parts of the juice are so enveloped and locked up, that they are scarcely to be discovered, either by the smell or taste.

Olibanum is recommended in disorders of the head and breast, in hæmoptoës, and in alvine and uterine fluxes: the dose is from a scruple to a dram or more.

ONONIS.

ANONIS spinosa flore purpureo C. B. *Resta bovis. Aresta bovis. Remora aratri. Ononis spinosa nofa. Linn.* Rest-harrow: a plant with flexible branches terminating in sharp prickles; small oval indented leaves, standing generally three together, without pedicles; and purplish papilionaceous flowers, set in pairs, followed each by a short pod containing three unequal kidney-shaped seeds. It is perennial, grows wild in waste grounds and dry fields, and with its long tough spreading roots obstructs the plough or harrow.

The roots of rest-harrow have a faint unpleasant smell, and a sweetish, bitterish, somewhat nauseous taste. Their active matter is confined to the cortical part; which has been sometimes given in powder, in doses of a dram, and made an ingredient in apozems or decoctions, as an aperient and diuretic. Its virtue is extracted both by water and spirit.

OPIUM.

OPIUM Pharm. Lond. & Edinb. Opium: a concrete gummy-resinous juice, somewhat soft and tenacious, especially when much handled or warmed; of a dark reddish brown colour in the mass, and when reduced into powder yellow. It is brought from Egypt, Persia, and some other parts of Asia, in flat cakes or irregular masses, from four to about sixteen ounces in weight, covered with leaves to prevent their sticking together.

It is extracted from the heads of white poppies (see *Papaver*) which in those countries are cultivated in fields for this use. Kæmpfer reports, that the heads, when almost ripe, are wounded with a five-edged instrument, by which as many parallel incisions are made at once from top to bottom;

bottom; that the juice which exudes is next day fcraped off, and the other fides of the heads wounded in like manner; and that the juice is afterwards worked with a little water, till it acquires the confiftence, tenacity, and brightnefs of the fineft pitch. The beft opium was formerly called Thebaic opium, from its being prepared about Thebes in Egypt: no diftinction is now made in regard to the places of its production, though the epithet *thebaic* ferves to diftinguifh fome of its officinal preparations.

Opium has a faint difagreeable fmell, and a bitterifh, fomewhat hot, biting tafte. Watery tinctures of it ftrike a black colour with chalybeate folutions, and thus feem to difcover fome aftringency. Mixed with the ferum of blood, they thicken and render it whitifh; and on blood itfelf, newly drawn, they have nearly a like effect: Mr. Eller obferves, that on examining with a microfcope blood thickened by a vinous tincture of opium, the nature of its globules feemed to be deftroyed. But neither from thefe, nor any of the other known fenfible properties of this drug, can its furprizing operation in the human body be deduced.

Taken in proper dofes, it commonly procures fleep, and a temporary refpite from pain, or the action of any ftimulating power. The caufe of the pain it in many cafes confirms or augments; and in not a few, it fails even of giving palliating relief. The cafes in which it is proper or improper will be beft underftood from a view of its general effects; which, fo far as experience has hitherto difcovered them, are the following.

It renders the folids, while the operation of the opium continues, lefs fenfible of every kind

of

of irritation, whether proceeding from an internal caufe, or from acrimonious medicines, as cantharides, and the more active mercurials, of which it is the beft corrector—It relaxes the nerves; abating or removing cramps or fpafms, even thofe of the more violent kind; and increafing paralytic diforders and debilities of the nervous fyftem—It incraffates thin ferous humours in the fauces and adjacent parts; by which means, it proves frequently a fpeedy cure for fimple catarrhs and tickling coughs; but in phthifical and peripneumonic cafes, dangeroufly obftructs expectoration, unlefs this effect be provided againft by fuitable additions, as ammoniacum and fquills—It produces a fulnefs and diftenfion of the whole habit; and thus exafperates inflammations both internal and external, and all plethoric fymyptoms—It promotes perfpiration and fweat; but reftrains all other evacuations, unlefs when they proceed from a relaxation and infenfibility of the parts, as the colliquative diarrhœæ in the advanced ftage of hectic fevers—It promotes labour-pains and delivery *(a)* more effectually than the medicines of the ftimulating kind ufually recommended for that purpofe; partly perhaps by increafing plenitude, and partly by relaxing the folids or taking off fpafmodic ftrictures—And indeed all the preceding effects are perhaps confequences of one general power, being nearly allied to thofe which natural fleep produces *(b)*.

The operation of opium is generally accompanied with a flow but ftrong and full pulfe, and a flight rednefs, heat and itching of the fkin: it is followed by a weak and languid pulfe,

(a) Mead, *Monita & præcept. med.* p. 253.
(b) See Young's *treatife on opium.*

lowness of the spirits, some difficulty of breathing or a sense of tightness about the breast, a slight giddiness of the head, dryness of the mouth and fauces, and some degree of nausea. Given on a full stomach, it commonly occasions a nausea from the beginning, which continues till the opium is rejected along with the contents of the stomach. Where the evacuation of acrid humours, accumulated in the first passages, is suppressed by it, great sickness and uneasiness are generally complained of, till the salutary discharge either takes place again spontaneously or is promoted by art.

An over dose occasions either immoderate mirth or stupidity, a redness of the face, swelling of the lips, relaxation of the joints, vertigo, deep sleep with turbulent dreams and startings, convulsions, and cold sweats. Geoffroy observes, that those who recover, are generally relieved by a diarrhœa, or by a profuse sweat, which is accompanied with a violent itching. The proper remedies, besides emetics, blisters, and bleeding, are acids and neutral mixtures: Dr. Mead says he has given, with extraordinary success, repeated doses of a mixture of salt of wormwood with lemon juice.

A long continued use of opium is productive of great relaxation and debility, sluggishness, heaviness, loss of appetite, dropsies, tremors, acrimony of the humours, frequent stimulus to urine, and propensity to venery. On leaving it off, after habitual use, an extreme lowness of the spirits, languor, and anxiety, succeed; which are relieved by having again recourse to opium, and in some measure by spirituous or vinous liquors.

With regard to the dose, one grain is generally a sufficient, and sometimes too large a one: maniacal persons, and those who labour under violent spasms, require oftentimes two, three, or more grains; though even in these cases, it is generally more advisable to repeat the dose at proper intervals, than to enlarge it. By frequent use, much greater quantities may be borne: the Turks, who habituate themselves to opium as a succedaneum to spirituous liquors, are said to take commonly a dram at a time, and Garcias says that he knew one who every day took ten drams.

Opium appears to consist of about five parts in twelve of gummy matter, four of resinous matter, and three of earthy or other indissoluble impurities (a). From these last it has been purified, in the shops, by softening the opium with boiling water, in the proportion of a pint to a pound, into the consistence of a pulp, with care to prevent its burning; and whilst it remains quite hot, strongly pressing it from the fæces through a linen cloth: the strained opium is then inspissated in a water-bath, or other gentle heat, to its original consistence. When thus softened with a small quantity of water, the gummy and resinous parts pass the strainer together; whereas, if dissolved by a larger quantity, they would separate from one another.

* A more perfect method of purification is now directed by the London college, which consists in dissolving one pound of opium in twelve pints of proof spirit, straining the solution, and then distilling the spirit from it, till it be reduced to a due consistence. This preparation is ordered

Opium purificat. Ph. Lond.

(a) Alston, *Edinburgh medical essays,* vol. v. art. 12.

to be kept in a soft form, for making pills, and a hard one, for powdering.

It has been disputed, whether it is in the gummy or in the resinous parts of opium, that its activity resides. From the experiments of Hoffman *(a)* and Neumann *(b)*, it seems to be neither in the direct gum, nor in the direct resin, but in a certain subtile part of the resinous matter, somewhat analogous to essential oils, but of a much less volatile kind: they report, that on boiling the opium in water, there arises to the surface a frothy, viscid, unctuous, strong-scented substance, to the quantity of two or three drams from sixteen ounces: that this substance, in the dose of a few grains, has killed dogs that could bear above a dram of crude opium; that in distillation with water, though it does not rise itself, it gives over, at least in part, the active principle of which it is the matrix; impregnating the distilled liquor with its scent and its soporific power; as essential oils exhale their odoriferous principle in the air, without being dissipated themselves. What this subtile and highly active principle really is, in essential oils, in odorous vegetables that yield no oil, and in opium, is equally unknown.

Both water and rectified spirit extract, difficultly, by maceration or digestion, the active matter of opium, and receive from it a yellow or brownish tincture. The watery solution is found to contain great part of the resin along with the gum; and the spirituous, a smaller proportion of the gum along with the resin. Such part of the gum as is left by spirit, and such part of

(a) Diss. de opii correctione genuina & usu, Oper. supplement. ii. P. i. p. 645. Not. ad Poterium, p. 437.

(b) Chymia medica, vol. i. p. 996. Chemical works, p. 308.

the refin as is left by water, feem to be equally inert.

Tinctures of opium in water, wine, and proof fpirit, have the fame effects as the opium in fubftance; with this difference, that they exert themfelves fooner in the body, and are lefs difpofed to leave a naufea on the ftomach. Tinctures made in rectified fpirit are faid to act with greater power than the others: Geoffroy relates, from his own obfervation, that while the watery and vinous tinctures occafioned quiet fleep, the fpirituous brought on a phrenzy for a time. It is faid likewife, that alkaline falts diminifh the foporific virtue of the opium; that fixt alkalies render it diuretic, whilft volatile ones determine its action to the cutaneous pores; and that acids almoft entirely deftroy its force.

The officinal tinctures of opium are made in wine or proof fpirit. The college of London directs ten drams of ftrained opium, dried and powdered, to be macerated without heat for ten days in a pint of proof fpirit: the college of Edinburgh orders two ounces of crude opium to be digefted for four days in a pound and a half of fpirituous cinnamon water: a mixture of wine and proof fpirit has been fometimes made choice of, in order to prevent in fome meafure an inconvenience which both of them feparately, confidered as officinals, are liable to, being apt to throw off in long ftanding a part of the opium, which in wine falls to the bottom, and forms a cruft on the furface of fpirit. Of the firft of the above tinctures twenty drops, and of the latter twenty-five drops are reckoned to contain one grain of opium:* but as thefe quantities of the menftrua *Tinctur. opii Ph. Lond.* *Tinctura thebaica vulgo laudanum liquidum Ph. Ed.*

* (a) This calculation refers to the preceding editions of the difpenfatories. The London college have apparently much

menſtrua do not eaſily diſſolve all the active matter of ſo large a proportion of the opium, thoſe doſes are generally obſerved to have ſomewhat leſs effect than a grain of the drug in ſubſtance. As drops alſo, according to different circumſtances, vary in quantity, though in number the ſame, it were to be wiſhed that the ſhops were furniſhed with a ſolution of this drug, made in a quantity of menſtruum large enough not only for the complete extraction of the active parts, but to admit of the doſe being exactly determined by weight or meaſure.

In a ſolid form, independently of ſuch materials as may be ſubſervient to the other indications of cure, it is ſometimes mixed with ſponaceous or gummy ſubſtances which promote its diſſolution in the ſtomach, and ſometimes with reſinous ones, which render its diſſolution and operation more gradual and ſlow: to theſe is commonly ſuperadded ſome aromatic ingredient, to prevent its occaſioning a nauſea.

*The London and Edinburgh colleges have now, however, preſerved only a ſingle form each of opiate pills, in which the promotion of its ſolubility ſeems the only ſubject conſidered. The former unites two drams of hard ſtrained opium with one ounce of extract of liquorice. The latter directs the combination of one part of opium, four of extract of liquorice, three of Spaniſh ſoap, and two of powdered Jamaica pepper.

Pil. ex opio
Ph. Lond.

Pil. thebaicæ
vulgo pacificæ
Ph. Ed.

much diminiſhed the quantity of opium in their tinctures as it was formerly made with two ounces of ſtrained opium to the pint. But it is probable that they have found by experiment that the loſs of weight in drying the opium to powder is equivalent to the difference. By uſing this drug in powder, the difference of ſtrength reſulting from the unavoidable difference of conſiſtence in various parcels of the crude or ſtrained opium, is obviated.

Many

Many have endeavoured to correct certain ill qualities, which they suppose opium to be possessed of, by roasting it, by fermentation, by long continued digestions, or boiling, by repeated dissolutions and distillations. These kinds of processes, though recommended by several late writers, do not promise any singular advantage. That they weaken the opium is indeed very probable; but this intention is answered as effectually, and with far greater certainty, by diminishing the dose of the opium itself: for the ill effects, which opium produces in certain circumstances, do not depend on any distinct property or principle, and appear to be no other than the necessary consequences of the same power, by which in other circumstances it proves so beneficial: the only rational way of improving or correcting this valuable drug seems to be, by joining or interposing such medicines, as may counteract or remove those particular effects of it, which in particular cases may be injurious.

OPOBALSAMUM.

OPOBALSAMUM or BALSAM OF GILEAD: a resinous juice, obtained from an evergreen tree or shrub *(balsamum syriacum rutæ folio C. B.)* said to grow in Arabia. The best sort, which naturally exudes from the plant, is scarce known in Europe; and the inferiour kinds, said to be extracted by lightly boiling the branches and leaves in water, are very rarely seen among us.

THE true opobalsam, according to Prosper Alpinus, is at first turbid and white, of a very strong pungent smell, like that of turpentine; but much sweeter and more fragrant, and of a bitter,

bitter, acrid, aftringent tafte: on being kept for fome time, it becomes thin, limpid, light, of a greenifh hue, and then of a gold yellow, after which it grows thick like turpentine, and lofes much of its fragrance *(a)*. Some refemble the fmell of this balfam to that of citrons, others to that of a mixture of rofemary and fage flowers. I have fometimes met with a curious balfam of this laft kind of fmell, exceedingly fragrant, limpid, and thin: dropt on water, it fpread itfelf all over the furface, imparting to the liquor a confiderable fhare of its tafte and fmell: the groffer part, that remained on the top of the water, was fo tenacious, as to be eafily taken up at once with the point of a needle, which is reckoned, by Alpinus and others, as a characteriftic of the true balfam.

This precious balfam is of great efteem in the eaftern countries, both as a medicine, and as an odoriferous unguent and cofmetic. Its great fcarcity has prevented its coming into ufe among us: nor are its virtues, probably, superiour to thofe of fome of the refinous juices more common in the fhops; all thefe fubftances being in their general qualities alike, though differing in the degree of their gratefulnefs, pungency, and warmth.

OPOPANAX.

OPOPANAX Pharm. Lond. OPOPANAX: a concrete gummy-refinous juice, obtained from the roots of an umbelliferous plant, which grows fpontaneoufly in the warmer countries, and bears the colds of our own *(Paftinaca opopanax Linn.)* The juice is brought from Turkey and

(a) Vide Alpini *dialogum de balfamo.*

the East Indies, sometimes in little round drops or tears, more commonly in irregular lumps, of a reddish yellow colour on the outside with specks of white, internally of a paler colour and frequently variegated with large white pieces.

This gummy-resin has a strong disagreeable smell, and a bitter, acrid, somewhat nauseous taste. It readily mingles with water, by triture, into a milky liquor, which on standing deposites a portion of resinous matter and becomes yellowish: to rectified spirit it yields a gold-coloured tincture, which tastes and smells strongly of the opopanax. Water distilled from it is impregnated with its smell, but no essential oil is obtained on committing moderate quantities to the operation.

Opopanax is an useful attenuant and deobstruent, and in considerable doses loosens the belly. It is given from a scruple to a dram, in the same intentions as ammoniacum or galbanum; and joined in smaller doses as an auxiliary to those and the other deobstruent gums.

ORIGANUM.

ORIGANUM *Pharm. Lond. Origanum silvestre, cunila bubula plinii* C. B. *Agrioriganum sive onitis major Lob. Origanum anglicum Ger. Origanum vulgare Linn.* Origanum or Wild Marjoram: a plant with firm round stalks, and oval, acuminated, uncut, somewhat hairy leaves, set in pairs upon short pedicles: on the tops grow scaly heads of pale red labiated flowers, whose upper lip is entire and the lower cut into three segments, set in form of a convex umbel, intermixed with roundish purplish leaves: each flower is followed by four minute seeds inclosed in the cup. It is perennial, grows wild on dry chalky hills and gravelly grounds, in several parts of England, and flowers in June. The flowers, or rather flowery tops, of a somewhat different species, *origanum creticum*, were formerly brought from Candy, but have long given place to those of our own growth, which are nearly of the same quality.

The leaves and flowery tops of origanum have an agreeable aromatic smell, and a pungent taste, warmer than that of the garden marjoram, and much resembling thyme; with which they appear to agree in medicinal virtue. Infusions of them are sometimes drank as tea, in weakness of the stomach, disorders of the breast, for promoting perspiration and the fluid secretions in general: they are sometimes used also in nervine and antirheumatic baths; and the powder of the dried herb as an errhine. Distilled with water, they yield a moderate quantity of a very acrid penetrating essential oil, smelling strongly of the origanum, but less agreeable than the herb itself: this oil is applied on a little cotton for easing the pains of carious teeth; and sometimes diluted and rubbed on the nostrils, or snuffed up the nose, for attenuating and evacuating mucous humours.

OSTEOCOLLA.

OSTEOCOLLA, aliis ossifragus, osteites, ammosteus, osteolithos, holosteus, stelochites, Worm. mus. Osteocolla, or Bone-binder: a fossil substance, found in some parts of Germany, particularly in the marchè of Brandenburgh, and in other countries. It is met with in loose sandy grounds, spreading, from near the surface to

OSTREUM.

to a confiderable depth, into a number of ramifications like the roots of a tree : it is of a whitifh colour, foft while under the earth, friable when dry, rough on the furface, for the moft part either hollow within, or filled with folid wood, or with a powdery woody matter *(a)*.

This earth has been celebrated for promoting the coalition of fractured bones and the formation of a callus; a virtue to which it does not feem to have any claim. It is found to be compofed of two different earthy fubftances, which are nearly in equal proportions, and which may be feparated from one another, by wafhing the powdered ofteocolla with water: the finer matter, which wafhes over, appears from its burning into quicklime, and its properties in other experiments, to be mere calcareous earth, not different in quality from chalk: the groffer matter that remains is no other than fand.

OSTREUM.

The Oyster; a common, bivalvous, marine fhell fifh.

The fhell of the oyfter, levigated into a fubtile powder, is employed as an abforbent, in heart-burns and other like complaints arifing from acidities in the firft paffages: the hollow fhells are generally made choice of, as containing more, than the thinner flat ones, of the fine white earth, in proportion to the outer rough coat, which laft is found to be confiderably

Teftæ oftreorumprepar.
Ph. Lond.

(a) A more particular account of this foffil may be feen in *Neumann's chemical works*, *p.* 11. and the *Memoires de l'academie royale des fciences de Berlin, pour l'ann.* 1748.

impregnated

impregnated with fea falt. By calcination, they are converted into a ftrong quicklime, which imparts to water a greater degree of lithontriptic power than the mineral limes; fee *Calx viva*.

O V U M.

OVUM gallinaceum Pharm. Lond. HENS EGG.

EGGS are accounted very nutritious, but difficult of digeftion, efpecially if boiled hard. In medicine, the yolk has been employed as an intermedium for rendering refinous juices and balfams foluble in water: it anfwers this purpofe lefs effectually, and lefs elegantly than vegetable gums, the folutions obtained by means of the animal matter being apt on ftanding to become putrid or rancid. The yolk, exficcated by a gentle warmth, forms a friable concrete; the white, a firm femitranfparent one, in appearance refembling amber or gum-arabic, and foluble again in watery liquors. The boiled white, placed in a moift cellar, deliquiates fpontaneoufly, and gummy-refinous fubftances, included in it, diffolve along with it: preparations of this kind have been directed for medicinal ufes, but it does not appear that more of the gummy-refin is thus diffolved by the liquamen of the egg than by fimple water.

The fhells of eggs, freed, after boiling, from the inner fkin, and levigated into fine powder, are fometimes ufed as abforbents, and fuppofed, when combined with the acid humours in the firft paffages, to be lefs difpofed to bind the belly than moft of the other teftaceous powders,

PÆONIA.

PÆONIA.

PÆONIA.

PÆONIA folio nigricante splendido quæ mas, & pæonia femina flore pleno rubro majore C. B. *Pæonia officinalis Linn.* MALE and FEMALE PEONY or PIONY: a plant with large leaves, divided deeply into oblong segments, or rather composed of a number of these segments set on divided pedicles: on the tops of the branches grow large rose-like flowers, followed each by two or more horned pods, internally of a deep red colour, containing roundish shining red or black seeds. The male sort has dark green leaves, pale red single flowers, long thick roots, and the stalks and pedicles streaked with red: the female has longer, narrower, and paler leaves, deep red double flowers, and irregular roots composed of several tuberous pieces hanging by tough filaments from one head. They are both found wild in some parts of Europe, and cultivated with us in gardens: they are perennial, produce their flowers in May, and very soon shed them.

THE male peony has been generally preferred for medicinal use: but the female, which is the largest and most elegant, and for this reason the most common, is the species which the shops have been principally supplied with. In quality, there does not appear to be any material difference betwixt the two; and hence the college allow both sorts to be taken indiscriminately.

The roots and seeds of peony have, when fresh, a faint unpleasant smell, somewhat of the narcotic kind: and a mucilaginous subacrid taste, with a slight degree of bitterishness and astringency.

astringency. In drying, they lose their smell, and part of their taste. Extracts made from them by water are almost insipid as well as inodorous; but extracts made by rectified spirit are manifestly bitterish and considerably astringent.

The leaves are nearly inodorous. To the taste, the leaves themselves discover a moderate degree of roughness, and their pedicles of sweetness; both which are preserved in great measure in the watery, but more perfectly in the spirituous extracts.

The flowers have rather more smell than any of the other parts of the plant, and a rough sweetish taste, which they impart, together with their colour, both to water and spirit: the watery infusion leaves, on being inspissated, a blackish red, austere, sweetish, and somewhat bitterish extract: the spirituous tincture yields an extract of a beautiful bright red, of an agreeable though weak smell, a moderate astringency, and an almost saccharine sweetness.

The roots, flowers, and seeds, are looked upon as lightly anodyne and corroborant; to the latter, at least, of which virtues, they appear from the above experiments to have some claim. They have been principally recommended in spasmodic and epileptic complaints; in which, we are afraid, their effects are not very considerable.

PALMA.

PALM: a tall unbranched tree, with long reed-like leaves elegantly disposed on the top. Different species of it grow spontaneously in the eastern countries, and in the warmer parts of the West Indies.

PALMA.

The *palma major* C. B. *Phœnix dactylifera* Linn. is cultivated in some of the southern parts of Europe. Its fruit, the dates of the shops, is of an oblong shape, like an acorn, but generally larger; and consists of a thick fleshy substance including, and freely parting from, an oblong hard stone, which has a remarkable furrow running its whole length upon one side. The best dates come from Tunis: they should be chosen large, softish, not much wrinkled, of a reddish yellow colour on the outside, with a whitish membrane betwixt the flesh and the stone. They have an agreeable sweet taste, accompanied with a slight astringency; and hence stand recommended in tickling coughs and thin acrid defluxions on the lungs, and in alvine fluxes. Among the Egyptians and Africans, they are said to be a principal article of food, and when used too freely, to be difficult of digestion, occasion head-achs, sometimes gripes, and, in length of time, obstructions of the viscera, cachectic, and melancholic disorders.

The *palma oleosa* (*palma foliorum pediculis spinosis, fructu pruniformi luteo oleoso* Sloan. jam.) is a native of the coast of Guinea(*a*) and the Cape Verd islands, from whence it has been introduced into Jamaica and Barbadoes. From its fruit is extracted an oil, which, as brought to us, is about the consistence of an ointment, of a strong, not disagreeable smell, and scarcely any particular taste: by long keeping it loses its high colour, and becomes white, and in this state is to be rejected. The inhabitants of the Oleum expressum palmæ *Ph. Ed.*

---

*(*a*) According to Bergius, another species of the oil palm grows on the coast of Guinea and in Senegal, the *palma altissima non spinosa, fructu pruniformi minore, racemo sparso* Sloan. Jam. & Adanson.

Guinea coast are said to employ the palm oil for the same purposes as we do butter. With us, it is only used in some external applications, for pains and weakness of the nerves, cramps, sprains, and other like complaints. The common people sometimes apply it to chilblains; and, when used early, not without benefit. It is said to be peculiarly serviceable in hardness of the belly, both of adults and children. *(a)*

The medullary part of certain oriental palm trees *(palma indica caudice in annulos protuberantes distincto, fructu pruniformi, Raii. Sagus, seu palma farinaria Rumph. Amb. (b)* affords another article of food to the natives, and of the materia medica to us. The farinaceous medulla, freed from the filamentous matter with which it is enveloped, is beaten with water, and made into cakes, which are afterwards reduced into small grains, and dried. The cakes are said to be the bread used by the Indians in scarcity of rice: the grains are the sago or sagou of the shops. This substance, commonly recommended as a restorative in phthises and emaciations and for restraining fluxions, appears to be a light, moderately nutritious demulcent food; in which view it is by some directed *(c)* as a proper aliment for young children, in preference to the more tenacious and less digestible preparations of wheat flour. It dissolves in water into a viscid mucilage; is less acescent and flatulent than other farinæ; keeps longer in the grain, even for twenty years in a dry place, and also in its mucilaginous state a long time *(d)*.

(a) Bergii Mat. Med. 882.
* *(b)* The *Cycas circinalis Linn.* has been given as the sago plant, but, as Bergius supposes, erroneously.
(c) Albertus Seba, *Thesaur.* vol. i. p. 40. *Act. nat. curios.* vol. i. *Append.*   *(d)* Cullen, Mat. Med.

PAPA-

PAPAVER.

POPPY: a plant with oblong leaves and round ftalks, divided into a few branches, each of which is terminated by a large tetrapetalous flower, fet in a two-leaved cup that falls off as the flower opens: the flower itfelf likewife foon falls, leaving a fmooth roundifh head or capfule, covered with a radiated crown, and containing a number of fmooth roundifh feeds. It is annual, and flowers from June to near the end of fummer.

1. PAPAVER ALBUM *Pharm. Lond. & Edinb. Papaver hortenfe femine albo* C. B. *Papaver fomniferum Linn.* White poppy: with fmooth, flightly indented leaves; and whitifh flowers and feeds.

2. PAPAVER NIGRUM: *Papaver hortenfe nigro femine* C. B. Black poppy: a variety of the former, with fmooth, flightly indented leaves, purple flowers, and black feeds.

These plants are found wild in fome parts of Europe; and feveral varieties of them, in regard to the flowers, are produced by culture in our gardens. The heads, ftalk, and leaves, have an unpleafant fmell, and a bitterifh biting tafte, of the fame kind with thofe of opium. Their fmell and tafte is lodged in a milky juice; which abounds chiefly in the cortical part of the heads; which may be collected, in confiderable quantity, by flightly wounding them when almoft ripe; and which, on being expofed for a little time to a warm air, thickens into a tenacious dark-coloured mafs, fimilar to the opium brought from abroad, but ftronger in fmell and tafte. The juices thus obtained from the two forts of poppies, appear to be of the fame quality, the difference being only in the quantity afforded: the white poppy, which is the largeft, is the fort cultivated by the preparers of opium in the eaftern countries, and for medicinal ufes in this." *The following extract from Mr. Kerr's account of the culture of this plant, and the preparation of opium, in the province of Bahar in the Eaft Indies, may convey ufeful information.

"The feeds are fown in October or November. The plants are allowed to grow fix or eight inches diftant from each other, and are plentifully fupplied with water. When the young plants are fix or eight inches high, they are watered more fparingly. But the cultivator ftrews all over the areas a nutrient compoft of afhes, human excrements, cow-dung, and a large portion of nitrous earth, fcraped from the highways, and old mud-walls. When the plants are nigh flowering, they are watered profufely to increafe the juice.

"When the capfules are half grown, no more water is given, and they begin to collect the opium. At fun-fet they make two longitudinal double incifions upon each half-ripe capfule, paffing from below upwards, and taking care not to penetrate the internal cavity of the capfule. The incifions are repeated every evening, until each capfule has received fix or eight wounds; they are then allowed to ripen their feeds. The ripe capfules afford little or no juice. If the wound was made in the heat of the day, a cicatrix would be too foon formed. —The night-dews, by their moifture, favour the exftillation of the juice. Early in the morning

ing old women, boys, and girls, collect the juice, by fcraping it off the wounds with a fmall iron fcoop, and depofite the whole in an earthen pot, where it is worked by the hand in the open fun-fhine, until it becomes of a confiderable fpiffitude: it is then formed into cakes of a globular fhape, and about four pounds in weight, and laid into little earthen bafins to be further exficcated. Thefe cakes are covered over with the poppy or tobacco leaves, and dried until they are fit for fale. Opium is frequently adulterated with cow-dung, the extract of the poppy-plant procured by boiling, and various other fubftances which they keep in fecrecy."(a)

The collection of the pure milky juice of the poppy has not, among us, been as yet practifed in large, or with a view to the fupplying of the common demand of opium. Inftead of this troublefome procefs, we extract the narcotic matter by menftrua; the active parts of opium, as obferved under that article, being completely diffoluble both by water and rectified fpirit. A portion of the herbaceous inert fubftance of the plant is indeed, at the fame time, taken up, at leaft when water is made ufe of, fo as to render an enlargement of the dofe neceffary: but this addition to the bulk of a dofe of opium would be of no inconvenience, if the compound was always of the fame ftrength, or the narcotic and inert matter in the fame proportions to one another; a point which cannot be attained with fo much precifion as could be wifhed, but which may neverthelefs, by due care in the preparation, be adjufted as nearly as common practice in moft cafes requires.

(a) Lond. Med. Obf. and Inq. vol. v. p. 318.

MATERIA MEDICA.

3. The college of London directs the dried heads, cut and cleared from the seeds, to be boiled in water, in the proportion of three pounds and a half to eight gallons, in the heat of a brine bath, till it is reduced to three gallons: the liquor is then to be expressed, and boiled down to four pints, which is to be strained hot, first through a sieve, and then through a thin woollen cloth, and set by for twelve hours, that the dregs may subside. The liquor poured off clear is to be reduced to three pints, in which six pounds of sugar are to be dissolved. An ounce of this syrup is reckoned equivalent to about a grain of opium. *The Edinburgh college directs two pounds of poppy heads without the seeds to be macerated, for a night in thirty pounds of boiling water, the liquor then boiled down till only a third part remains, which is to be strongly expressed and strained, then boiled again to the half, strained, and made into a syrup with a sufficient quantity of sugar. They also allow this syrup to be made by dissolving one dram of the extract of white poppy heads in two pounds and a half of simple syrup.

Syr. Papav. alb. Ph. Lond.

Syrupus papaveris albi, seu de meconio, vulgo diacodion Ph. Ed.

Extract. capitum papaveris albi Ph. Ed.

A decoction of poppy heads in water, strongly pressed out, depurated by settling, then clarified with whites of eggs, and inspissated, yields an extract amounting to one fifth or one sixth the weight of the heads: it is said, that two grains of this preparation are equivalent to one grain of opium, and that the extract is not liable to produce a nausea or giddiness which generally follow the use of pure opium(a): but the consequential effects which opiates produce, in different subjects, and in different circumstances, are so variable, that the trials which have been

(a) Mr. Arnot, Edinburgh medical essays, vol. v. art. 11.

made

made of this preparation, however fuccefsful, do not appear fufficient for eftablifhing this fuperiority. Of tinctures or extracts made with fpirituous menftrua, no medicinal trials, fo far as I can learn, have as yet been made: in fmell and tafte they approach more to opium than any other preparation of the poppy I have feen.

Many have fuppofed the feeds of the poppy to be, like the other parts of the herb, narcotic*(a)*; mifled, perhaps, by analogical reafoning from other plants. Though the feeds of many plants are more efficacious than the veffel in which they are lodged; thofe of the poppy have nothing of the narcotic juice which is diffufed through their covering, through the ftalks, and more fparingly through the leaves. If emulfions of poppy feeds have been found ferviceable in coughs, catarrhs, heat of urine, and other like diforders; it is not to an anodyne, but an emollient quality, that this virtue is to be afcribed. The feeds in fubftance have a fweetifh unctuous farinaceous tafte, and yield upon expreffion a large quantity of infipid oil: both the feeds themfelves and the oil are faid to be in fome places common articles of food*(b)*.

3. PAPAVER ERRATICUM *Pharm. Lond.* *Papaver erraticum majus* C. B. *Papaver Rhœas* Linn. Wild or red poppy, or corn-rofe: with deep red flowers, dark-coloured feeds, hairy leaves and ftalks, and the leaves cut almoft, or quite, to the pedicle into indented fegments. It

(a) Hermann, *Cynofur. mat. med. edit. Boecler, p.* 436. Juncker, *Confpectus therapiæ generalis, p.* 279.

(b) Profper Alpinus, *De medicina Ægyptiorum, lib.* iv. *cap.* 1. Geoffroy, *Mat. med. tom.* ii. *p.* 715. Linnæi, *Amœnitat. Academic.* iii. 71.

is common in corn-fields; and is sometimes, like the others, made to vary its flowers by culture.

The heads of this species appear to contain the same kind of narcotic juice with those of the two preceding, but in so much smaller quantity that they are wholly neglected. The only part made use of is the flowers, which are supposed to be likewise impregnated in some degree with the same anodyne principle, and stand recommended in catarrhs, coughs, spitting of blood, and other disorders: they have a slight narcotic smell, and a very mucilaginous taste, accompanied with a sensible bitterishness. They are at present regarded rather on account of their colour, than for any great virtues expected from them: they yield upon expression a deep red juice, and impart the same colour to watery liquors, and a brighter though paler red to rectified spirit. A strong infusion of them is prepared in the shops, by pouring four pints and a half of boiling water upon four pounds of the fresh flowers, stirring them over the fire till the flowers are all immerged, and setting them by to steep for a night: without the application of fire so as to scald or shrink the flowers a little, they can scarcely be moistened with the water; if the heat is continued longer than this effect is produced, the liquor turns out quite slimy. This infusion, pressed out and depurated by settling, is reduced, by a proper addition of sugar, into a deep red syrup. The colouring matter of the red poppy differs from that of clove-gilly flowers, red roses, and other bright red flowers, in this; that on the admixture of alkaline liquors, it does not change, like them, to a green, but to a dark purple.

Syr. papav. errat. *Ph. Lond.*

PARA-

PARALYSIS.

PARALYSIS.

PARALYSIS: a plant with oblong wrinkled leaves, hairy on the upper fides of the ribs; and naked ftalks, bearing monopetalous flowers, each of which is divided about the edge into five, fegments, and fet in a loofe tubulous, ridged cup, which, after the flower has fallen, inclofes a hufk full of roundifh feeds. It is perennial, aud flowers early in the fpring.

1. VERBASCULUM *pratenfe odoratum* C. B. Primula *veris major* Gerard. *Primula veris officinalis* Linn. Cowflip, paigil, or peagle: with feveral. flowers fet together on one ftalk, of a deep yellow colour, drooping downwards. It grows wild in marfhes and moift meadows.

COWSLIP FLOWERS have a moderately ftrong pleafant fmell, and a fomewhat roughifh bitterifh tafte; both which they impart, together with a yellow tincture, to watery and to fpirituous menftrua. Vinous, liquors, impregnated with their flavour by maceration or fermentation, and ftrong infufions of them drank as tea, are fuppofed to be mildly corroborant, antifpafmodic, and anodyne. An infufion of three pounds of the frefh. flowers in five pints of boiling water is made in the fhops into a fyrup, of a fine yellow colour, and agreeably impregnated with the flavour of the cowflips.

2. PRIMULA VERIS *minor* Ger. *Verbafculum filveftre majus fingulari flore* C. B. *Primula veris acaulis* Linn. Primrofe: with pale yellow folitary flowers. It grows wild in woods and hedges.

THE

The flowers of this species are much weaker and less agreeable in smell than those of the preceding. The leaves and the roots seem to partake in some degree of the nature of those of asarum; acting as strong errhines or sternutatories, when snuffed up the nose, and as emetics (the roots at least) when taken internally. Gerard reports, as from the experience of a skilful practitioner, that " a dram and a half of the powder of the dried roots (taken up in autumn) purgeth by vomit very forcibly, but safely, in such manner as asarum doth."

PAREIRA.

PAREIRA BRAVA *Pharm. Lond. Pareyra, Ambutua, Butua, Overo brutua, Zan. hift. Pharm. Parif.* PAREIRA BRAVA: the root of an American climbing plant (*convolvulus brazilianus flore octopetalo monococcos Raii hift. Ciffampelos Pareira Linn.*) brought from Brazil, generally in crooked pieces of different sizes, some no bigger than the finger, others as large as a child's arm: the outside is brownish and variously wrinkled; the internal substance of a pale dull yellowish hue, and interwoven as it were with woody fibres, so that on a transverse section, there appears a number of concentric circles, crossed with striæ running from the centre to the circumference.

This root is extolled by the Brasilians and Portuguese in a variety of diseases, particularly in suppressions of urine and in nephritic and calculous complaints. Geoffroy is of opinion, that its virtue consists in dissolving and attenuating tenacious juices; and reports, that in sundry disorders arising from their viscidity, it

was

was found. remarkably beneficial: that in nephritic pains and fuppreffions of urine, he has often given it with happy fuccefs: that he has fometimes feen the patient freed from pain almoft in an inftant, and a plentiful difcharge of urine brought on: that in ulcers of the kidneys and bladder, where the urine was mucous and purulent, and could fcarcely be voided, or not without great uneafinefs, the fymptoms were foon relieved by pareira, and the ulcer at length healed by joining to it balfam of copaiba: that in an afthmatic cafe, where the patient was almoft fuffocated by thick phlegm, an infufion of pareira, after many other medicines had been tried in vain, brought on a copious expectoration, which proved a folution of the difeafe: that a perfon who, from an acute pain under the liver, had become in a few hours icterical, had the pain relieved, after bleeding, by the third cup of the decoction, and all the fymptoms removed by a continuance of it; and that the fame diforder frequently returning, fhe always found relief from the fame medicine: but that in another icterical cafe, where the liver was fwelled, it did no good. He cautions againft giving too large dofes, which might, he obferves, raife a heat, and perhaps an inflammation in the kidneys: of the root in fubftance he prefcribes from twelve grains to half a dram, and in decoction or infufion two or three drams; this quantity of the root, bruifed, he directs to be boiled in a pint and a half of water till only a pint remains, which is to be ftrained off, fweetened with a little fugar, divided into three portions, and drank as tea at intervals of half an hour.

The ufe of this root has not been in general accompanied with fo much fuccefs: but though, like

like many other medicines, it has not been found to anfwer the character at firft given of it, and has thence fallen into neglect, we may prefume, from its fenfible qualities, that it is not deftitute of medical virtue. It has no remarkable fmell; but to the tafte it manifefts a notable fweetnefs, of the liquorice kind, together with a confiderable bitternefs and a flight roughnefs covered by the fweet matter. It gives out great part both of the bitter and the fweet fubftance to watery and fpirituous menftrua: in evaporating the watery decoction, a confiderable quantity of refinous matter feparates, which does not mingle with the remaining extract or diffolve in water, but is readily taken up by fpirit; whence fpirit appears to be the moft perfect diffolvent of its active parts. Both the fpirituous tincture and extract are in tafte ftronger than the watery.

PARIETARIA.

PARIETARIA Pharm. Lond. & Edinb. Parietaria officinarum & dioscoridis C. B. Parietaria officinalis Linn. PELLITORY OF THE WALL: a plant with tender reddifh ftalks; rough, uncut, oblong leaves, pointed at both ends; and imperfect rough flowers, growing in clufters along the ftalks, followed each by a fmall fhining feed. It is perennial, common on old walls and among rubbifh, and flowers in May.

THE leaves of pellitory of the wall have been ufed in cataplafms for difcuffing inflammatory fwellings: decoctions of them, and their expreffed juice, have been given as emollient diuretics in nephritic cafes and ifchuries, and are

faid,

PASTINACA.

said, when long perfisted in, to be useful aperients or sweeteners in cutaneous defedations. The plant appears to be of no great activity, being rather oleraceous than medicinal: to the taste, the leaves in substance and their juice, are little other than herbaceous and watery.

PASTINACA.

PASTINACA: an umbelliferous plant, with naked umbels, yellow flowers, and flat seeds surrounded with a leafy margin: the leaves are oblong, and stand in pairs on a middle rib, without pedicles.

1. PASTINACA: *Pastinaca latifolia sativa C. B. Pastinaca sativa Linn.* Garden parsnep: with pale-coloured smooth indented leaves, and a large fleshy root.

2. ELAPHOBOSCUM: *Pastinaca silvestris latifolia C. B. Bancia & branca leonina quibusdam.* Wild parsnep: with dark green rough indented leaves and slender woody roots; common about the sides of fields; flowering, as the other, in June an July, and ripening its seeds in September. d The garden sort is supposed to be only a variety of this, and to owe its differences to culture.

THE roots of the garden parsnep, in taste considerably sweetish, are accounted a very nutritious aliment: they yield with rectified spirit a very sweet extract, and in distillation with water a small portion of essential oil possessing the specific flavour of the roots. It is said that by standing in the ground for some years, it
contracts

contracts pernicious qualities, so as to occasion disorders of the senses(*a*).

The seeds of the garden sort are somewhat aromatic; those of the wild a little more so; of considerable smell, but no great pungency or warmth. By infusion, they impregnate water moderately with their smell, but communicate very little taste: in distillation they give over a small quantity of a pale yellowish essential oil, in taste moderately pungent, and, smelling strongly of the seeds: five pounds of the seeds of the garden parsnep yielded little more than a dram. Rectified spirit takes up by digestion the whole of their active matter, and carries off little in the inspissation of the tincture: the extracts of both sorts have a moderate warmth and bitterishness, differing in degree as the seeds themselves. These seeds have been commended as diuretics, similar to those of daucus, but weaker, which, in their sensible qualities, they apparently are: Haller reports, that those of the wild species, made into pills, with extract of liquorice, were much used by Boerhaave against nephritic complaints and ulcerations of the bladder.

3. PANAX: *Panax heracleum Morison. Panax pastinacæ folio* C. B. *Sphondylio vel potius pastinacæ germanicæ affinis panax vel pseudocostus flore luteo* J. B. *Laserpitium Chironium Linn.* Hercules's allheal or wound wort: with uncut leaves, somewhat heart-shaped, but having one of the sides lower than the other: the middle ribs, bearing the several sets of leaves, stand in

---

(*a*) Ray, *Historia plantarum*, i. 420. Dan. Hoffman, *Acta acad. cæsar. nat. curiosor. vol.* vi. *anno* 1742. *obs.* 128. *p.* 426.

pairs

pairs along a larger rib. It is a native of the warmer climates, and bears the colds of our own.

Both the feeds and the roots of this species are confiderably warmer than thofe of the two preceding. The roots and ftalks have a ftrong fmell and tafte refembling thofe of opopanax; and Boerhaave relates, that on wounding the plant in fummer, he obtained a yellow juice, which, being infpiffated a little in the fun, agreed perfectly, in both refpects, with that exotic gummy-refin.

PENTAPHYLLUM.

PENTAPHYLLUM *Pharm. Lond.* Quinquefolium majus repens C. B. Potentilla reptans Linn. Cinquefoil or Five-leaved grass: a trailing plant, with oval ferrated leaves, fet five together on long pedicles, and pentapetalous yellow flowers ftanding folitary on like pedicles: the cup is divided into ten unequal fegments, the five innermoft of which form a covering to a button of feeds: the root is long and flender, dark coloured on the outfide, and reddifh within. It is perennial, grows wild on open clayie grounds, and flowers in June.

The roots of pentaphyllum are mild aftringents, and give out their aftringent matter both to water and fpirit. They have been ufed in diarrhœas and other fluxes, in intermitting fevers, fometimes as corroborants and antifeptics in low colliquative acute fevers, in gargarifms for ftrengthening the gums, &c. Their virtue is confined chiefly to the red cortical part, the whitifh woody fibre in the middle being nearly infipid.

PERSICA.

PERSICA.

PERSICA *molli carne vulgaris viridis & alba* C. B. *Amygdalus Persica Linn.* PEACH: a tree common in gardens; with oblong, narrow, pointed, serrated leaves; pale reddish flowers, composed of five broad petala with numerous stamina in the middle, set in five-leaved reddish cups, adhering to the branches without pedicles; and a fleshy fruit covered with downy matter and including a furrowed stone.

THE flowers of the peach tree have an agreeable but weak smell, and a bitterish taste: Boulduc observes, that when distilled without addition, by the heat of a water-bath, they yield one sixth their weight or more of à whitish liquor, which communicates, to a considerable quantity of other liquids, a flavour like that of the kernels of fruits. These flowers appear to be gently laxative: it is said, that an infusion in water of half an ounce of the fresh gathered flowers, or of a dram of them when dried, sweetened with sugar, proves, for children, an useful purgative and anthelmintic; and that the leaves, more unpalatable than the flowers, are somewhat more efficacious. The fruit is of the same quality with the other dulco-acid summer fruits: see *Fructus horæi.*

PERSICARIA.

ARSMART: an annual plant with oblong uncut leaves pointed at both ends, and imperfect flowers set in spikes on the tops of the stalks: the cup is thick and fleshy, divided into five oval segments, which, closing, form a cover to an angular glossy seed.

1. PERSICARIA

PERSICARIA.

1. PERSICARIA MITIS: *Perficaria mitis maculofa* C. B. *Pharm. Parif. Plumbago. Polygonum Perficaria Linn.* Spotted arfmart; fo called from moft of the leaves having a blackifh fpot in the middle. It grows wild in moift watery places, and flowers in July.

THIS plant is faid to be a good vulnerary and antifeptic; and decoctions of it in wine, to reftrain the progrefs of gangrenes *(a)*. It has a flightly acerb tafte inclining to acidity, and no remarkable fmell.

2. PERSICARIA URENS *five hydropiper* C. B. *Polygonum Hydropiper Linn.* Biting arfmart, lakeweed, water pepper: diftinguifhed from the former by the fpikes of flowers being flenderer, the leaves fhorter, narrower, and without any fpots; but more remarkably by its tafte. In our markets, a plant of a different genus, the fecond of the *ranunculi* hereafter defcribed, is fometimes fold for it.

THE leaves of this fpecies have an acrid burning tafte, and feem to be nearly of the fame nature with thofe of arum; their acrimony not rifing in diftillation, and being deftroyed in the procefs *(b)*. They are commended as antifeptic, aperient, diuretic; in fcurvies and cachexies, humoural afthmas, hypochondriacal and nephritic complaints, and in the wandering gout. The frefh leaves have been fometimes

(a) Tournefort, *Memoires de l'acad. des fcienc. de Parif. pour l'ann.* 1703.

(b) Rutty, *Synopfis of mineral waters, p.* 524. Dr. Cullen however fays, " its acrimony operates chiefly on the " kidneys. What is remarkable, it gives out its diuretic " virtue in diftillation to water." *Mat. Med.* 308.

applied

applied externally, in stimulating cataplasms, and for cleansing foul ulcers and consuming fungous flesh; in which last intention they are said to be used by the farriers.

PERUVIANUS CORTEX.

PERUVIANUS CORTEX Pharm. Lond. & Edinb. Peruvian bark: the bark of a middling-sized tree, growing in Peru, called by the Spaniards, from its efficacy against intermitting fevers, *palo de calenturas,* or the fever tree; by Linnæus, *Cinchona officinalis.* This virtue of the bark is said to have been discovered by the Indians about the year 1500, but not revealed to their European masters till 140 years after; when a signal cure having been performed by it on the Spanish viceroy's lady, the countess del Cinchon, it came into general use in those parts, and was distinguished by the appellations *pulvis comitissæ, cortex china china* or *chinchina, kina kina* or *kinkina,* and *quina quina* or *quinquina.* In 1649, a jesuit brought a large quantity of it into Italy, which was distributed by the fathers of that order, at a great price, in different parts of Europe: about the same time a quantity was purchased by cardinal de Lugo for the use of the poor at Rome. From these it received the names of *cortex* or *pulvis jesuiticus, pulvis patrum,* and *pulvis cardinalis de Lugo.*

This bark is brought to us in pieces of different sizes, some rolled up into short thick quills, and others flat: the outside is brownish, and generally covered in part with a whitish moss: the inside is of a yellowish, reddish, or rusty iron colour. The best sort breaks close and smooth, and proves friable betwixt the teeth:

teeth: the inferiour kinds appear when broken of a woody texture, and in chewing feparate into fibres. The former pulverifes more eafily than the latter, and looks, when powdered, of a light brownifh colour, refembling that of cinnamon, but fomewhat paler.

A bark was fome time ago brought from America under the name of the female Peruvian bark. This was found, from experience, to be lefs effectual as a medicine than the genuine fort, which it was frequently fubftituted to or mixed with in France, infomuch that its importation, as the editor of Geoffroy informs us, was prohibited by law. It is confiderably thicker, whiter on the outfide, redder within, and weaker in fmell and tafte than the true bark.

Peruvian bark has a flight fmell, approaching as it were to muftinefs, yet fo much of the aromatic kind as not to be difagreeable. Its tafte is confiderably bitter, aftringent, very durable in the mouth, and accompanied with fome degree of aromatic warmth, but not fufficient to prevent its being ungrateful.

The febrifuge virtue, for which alone this medicine was at firft recommended, has now been eftablifhed by the daily experience of about a century: and that, when judicioufly and feafonably adminiftered, it proves as fafe as it is effectual, is now alfo beyond difpute. An emetic, which is in moft cafes neceffary, being taken towards the approach of a paroxyfm, that its operation may be over before the fit comes on; the bark is begun at the end of the paroxyfm, or even in the time of the hot fit, and repeated, in dofes of half a dram or more, every third or fourth hour, during the intermiffion: after the fever has been removed, the medicine is is continued for a time, but more sparingly, to prevent a return. During the use of the bark, the pulse, which betwixt the paroxysms is generally weak and slow, becomes stronger and quicker, the appetite mends, the patient grows more cheerful, and perspiration increases: these may be looked upon as sure presages of its success. At first it frequently occasions a looseness, and this also is salutary; but if the purging runs on too long, as the fever rarely yields while this evacuation continues, it is usually checked by the addition of a little opium: if too great costiveness ensues, recourse is had to glysters. In gross impure habits, gentle purgatives are premised to the bark, or given for a time in conjunction with it: in agues of the inflammatory kind, or accompanied with great heat, a little nitre is joined or interposed: in lax spongy constitutions, and a thin watery state of the blood, the bark is assisted by bitters, snakeroot, camphor, and chalybeates: where obstructions of the abdominal viscera are apprehended, it is not ventured on without the addition of fixt alkaline salts, sal ammoniac, or other aperients. In all cases, moderate exercise, and the drinking of warm liquids, promote its effects. As the bark is hurtful in the inflammatory diathesis, it is not near so effectual in vernal, as in summer and autumnal intermittents *(a)*.

In remitting fevers, this medicine is less successful than in those which have perfect intermissions: in hectics, or wherever pus is formed, or juices are extravasated, it does harm. In the decline of long nervous fevers or after a remission, and in those of the low malignant kind where the blood is colliquated and the strength exhausted,

* *(a) Cull. Mat. Med.* 292.

exhaufted, it proves an excellent cordial, corroborant, and antifeptic.

Peruvian bark has likewife been found ferviceable in gangrenes and mortifications, and in foul obftinate ulcers and running fores of other kinds: in thefe cafes, taken in large and repeated dofes, it frequently brings on a laudable fuppuration, which degenerates on difcontinuing the ufe of the medicine, and again turns kindly upon refuming it. The like effects have been obferved from it in variolous cafes, where either the puftules did not duly fuppurate, or petechiæ fhewed a difpofition to a gangrene: by the ufe of bark, the empty veficles filled with matter, watery fanies changed into thick white pus, and the petechiæ became gradually paler and at length difappeared. The principal fymptom in this difeafe that contraindicates this valuable fuppurant and antifeptic, is great obftruction at the breaft or difficulty of breathing; which are always by this medicine increafed, infomuch that fmall dofes have in fome cafes endangered fuffocation.

In tumours of the glands, the Peruvian bark appears to promote, not fuppuration, but refolution. In the Medical Obfervations and Inquiries publifhed by a fociety of phyficians in London, there are feveral inftances of its being given with fuccefs in fcrophulous complaints. Dr. Fothergill obferves, that inveterate ophthalmiæ generally yield to it: that beginning glandular tumours are very frequently refolved and their farther progrefs ftopt by it: that fwelled lips, cutaneous blotches arifing from a like caufe, are healed, and the tendency to a ftrumous habit corrected: that it does not fucceed in all cafes, but that there are few in which a trial can be attended with much detriment: that he has never known it to avail where the bones were affected, or where the scrophulous tumour was so situated as to be attended with much pain, as in the joints or under the membranous covers of the muscles; for when it attacks these parts, the periosteum, and consequently the bone, seldom escape being injured; that here the bark, instead of lessening, adds to the fever which accompanies these circumstances, and if it does not increase the force of the mischief, seems at least to hasten its progress.

Peruvian bark has been applied likewise, in conjunction with other appropriated medicines, and often with good success, to the cure of periodic head-achs, hysterical, hypochondriacal, vertiginous and epileptic complaints, and other disorders that have regular intermissions. By its bitterness, astringency, and mild aromatic warmth, it strengthens the whole system, and proves a medicine of great utility in weakness of the stomach, uterine fluxes, and sundry chronical diseases proceeding from a laxity and debility of the fibres. To strengthen the solids appears indeed, in all cases, to be its primary operation; and its salutary virtues in different diseases, to be no other than consequential effects of this general power. In all the distempers where bark is known to take place, other astringent and bitter medicines, singly or combined, have likewise been of service, though not equally with this natural combination of them*(a).

THE virtues of this bark are very difficultly extracted by long coction in water, and part of

*(a) Dr. Percival found, that on mixing infusion of bark with putrid or ox gall, an instant coagulation ensued, and the foetor was increased. Hence he accounts for the disagreement of this medicine in the bilious fevers of the West Indies, Ess. Med. and Exper. vol. ii. p. 24.

PERUVIANUS CORTEX.

what the liquor is by heat enabled to take up begins to separate as soon as it is cold. This resinous part, which is rather melted out by the boiling heat than dissolved by the water as a menstruum, seems to contain chiefly the astringency of the drug: the bitter matter appears to be perfectly dissoluble, though more difficult to be got completely out. *After repeated infusion in cold water, till the liquor came off colourless and suffered no change from solution of vitriol, warm water extracted a considerable colour, and vitriol produced with this infusion an opake black: after warm water would extract no more, very hot water received a deeper colour than that of the strongest cold infusion of fresh bark; and this likewise struck a deep black with vitriol: boiling water had the same effect, after very hot water had ceased to act *(a)*.

On boiling a pound of finely powdered bark for an hour or two in five or six quarts of water, the decoction whilst hot looks clear and reddish, but in cooling becomes turbid and of a pale yellowish or wheyish hue: in this state it is found to partake, in a great degree, both of the bitterness and astringency of the bark, but in proportion as it deposites the matter that made it turbid, it loses more and more of its stypticity, the bitterness seeming to continue undiminished. The remaining bark, boiled in fresh water, exhibits the same appearance for two or three times successively; and when, at length, it ceases to render the water turbid, it imparts a bitterness without astringency *(b)*, retaining still some

(a) M. S. *of Dr. Lewis.*

(b) In the above experiments, I judged of the astringency only from the taste: solution of chalybeate vitriol, so useful on other occasions for discovering astringent matter

share of bitterness itself. The vapour which exhales in the first coction being caught in proper vessels, condenses into a limpid liquor which smells strongly of the bark; though no separable oil is obtained on submitting many pounds to the operation. The several decoctions, strained and inspissated together, yield an extract, rather less bitter, and much less styptic, than the bark in substance: this extract is kept in the shops in a soft and a hard form; the one of a proper consistence for making into pills; the other fit for being reduced into powder.

Extr. cort. peruv. molle & durum Ph. Lond.

As in vegetable decoctions or infusions, seemed here to fail; for having often mixed it, in different quantities, with even the first decoctions of bark, it produced, not a black, but a deep green. I have since observed, that when the vitriolic solution is used in very small proportion, it strikes a black with the turbid decoctions of bark, as with other astringents; and that even the green mixtures, resulting from a greater addition of the vitriol, on being largely diluted with water, become black or bluish like diluted ink. The resinous matter, which subsides on standing from the turbid decoctions, being dissolved in spirit of wine, gave likewise a black with vitriol. But when the bark had been boiled in fresh waters, till it no longer gave any turbidness to the liquor, the last transparent decoctions, though still pretty strong in taste, gave no blackness at all.

Some doubts having arisen with regard to this experiment, I have repeated it twice, and found the event both times the same as before. The last decoctions, on dropping in the chalybeate solution, contracted indeed a slight dusky hue, which in certain positions might be mistaken for a low degree of blackness; but the mixtures, held between the eye and the light, appeared only of a kind of olive yellowish or brownish colour, and, on standing for a little while, deposited, not a black, but an ochery precipitate; whereas the first infusions or decoctions, though so far diluted with water as scarcely to discover any taste, struck a bluish colour like that of diluted ink, and what little precipitate could be separated was black.

After the boiling of the bark in water had been repeated till the filtered liquor no longer made any change with solution

PERUVIANUS CORTEX.

As this drug gives out its virtue fo difficultly and imperfectly to boiling water, it has not been fufpected that cold water would have any confiderable action on it: I have neverthelefs found, that an infufion in cold water, though perfectly tranfparent, is rather ftronger in tafte than even the turbid decoction, though the latter has fomewhat more of a kind of fulnefs in the mouth *(a)*. It is by means of a gummy matter

tion of vitriol, the remaining bark gave no tincture at all to rectified fpirit.

But frefh bark, boiled in fucceffive portions of rectified fpirit, till it ceafed to impart any colour to the menftruum, gave ftill a deep tincture to boiling water; and this decoction, on the addition of folution of vitriol, exhibited nearly the fame appearances as the laft decoctions above-mentioned, only in a higher degree, the precipitate being much more copious, and its colour deeper.

Though by repeated boilings in water the bark may be fo exhaufted as to give out nothing to fpirit, but after the repeated action of fpirit ftill gives out fomething to water; yet fpirit appears, to be the moft active menftruum of its medicinal parts. For all, that fpirit can diffolve, is extracted by a far lefs quantity of fpirit than of water; and what fpirit leaves undiffolved is of little tafte. Equal quantities of bark being digefted for the fame length of time with equal quantities of water and rectified fpirit, with or without heat; the fpirituous tinctures proved always ftronger in tafte than the watery, and left on evaporation a larger proportion of extract * ||.

* || This affertion feems contrary to the refult of fome experiments by Dr. Percival, related in his firft vol. of *Eff. Med. and Exper.* p. 91, in which a dram of bark infufed feven days in three ounces each of rectified fpirit, proof fpirit, and water, loft in the firft, fix grains; in the fecond, eight and a quarter; and in the third, eight. Thefe accounts can be reconciled only upon the fuppofition that watery liquors do, indeed, extract more of the inert gummy matter of bark; but fpirituous, more of the active matter.

(a) I have endeavoured to compare the ftrength of the two preparations by characters that may be thought more fatisfactory than the tafte. A cold infufion and decoction were made with equal quantities of bark and water, and both

matter in vegetables, that the refinous parts become diffoluble in watery liquors; and it feems probable that, in boiling, part of the gummy principle of the bark is haftily diffolved and difunited from the refinous, whereas cold water, acting more gradually, extracts them both together. I have given the infufions in intermitting fevers as well as other diforders, with all the fuccefs that could have been expected from any preparation of this valuable medicine: the proportions commonly followed were, one ounce of the bark in fine powder, and eight or twelve of water, which were macerated without heat for twenty-four hours *(a)*, and

both liquors paffed through a filter: the infufion ran through faft; the decoction exceeding flowly, and continued turbid and opake after filtration. The two liquors, examined hydroftatically, were found very nearly of the fame fpecific gravity. Equal quantities of them being turned black with equal quantities of folution of vitriol, the quantity of water neceffary for diluting the blacknefs of the mixtures to an imperceptible degree, was very nearly the fame for both. Thefe experiments were often repeated, and feemed to prove, that the infufion and decoction are not confiderably different in the quantity of matter taken up from the bark, but that this matter is in the cold infufion tranfparently diffolved, whereas in the decoction great part of it is only diffufed through the liquor in an undiffolved ftate.—In the infufion itfelf, however, the folution does not appear to be very intimate. The tranfparent liquor becomes in a day or two turbid, and on ftanding for fome weeks (being now and then fhaken to prevent its growing mouldy) depofites fo much of the refinous part, that it is in tafte fimply bitter, and produces no blacknefs with vitriol. The refinous fediment gives to fpirit of wine a dark-coloured aftringent tincture, which ftrikes a black with vitriol like the tincture of bark itfelf.

(a) Since the above account was written, this preparation has been received in general practice, and found to anfwer the character here given of it. The time of maceration has been diminifhed to twelve hours, and fome late experiments

PERUVIANUS CORTEX.

and the clear liquor given in doses of two or three ounces.

* The London college seems now convinced that long coction of the bark is either unnecessary, or hurtful by dissipating some of the more volatile parts, and precipitating the resinous ones; for they have given a formula for a decoction of bark, in which one ounce of the powder is boiled in a pint and three ounces of water for ten minutes only, in a close vessel, and then strained off while hot.

Decoct. cort. peruv. Ph. Lond.

It is a common opinion, that bark in substance is more effectual than any preparation of it. Thus much is plain, that the infusions, as well as the decoctions, have not near so much effect as the quantity of bark they were made from, as the menstruum does not in either case completely extract its active matter: but their effects are evidently the same in kind, and the difference in degree may be compensated by an increase in the quantity.

The turbid decoctions, on the addition of any of the concentrated mineral acids, in the proportion of one drop to about a quarter of an ounce, become transparent, of a bright pale yellow colour, and of a rougher or more acerb

---

experiments shew, that it may be still further reduced, without any injury to the medicine. A mixture of one part of bark and eight of water being filtered after standing for one hour, the liquor appeared, from its taste, from its colour, from its specific gravity, and from the trial with solution of vitriol, to be very nearly, if not fully, as strong, as those which had stood 2, 4, 6, 8, 12, 24 hours. On doubling the quantity of bark, and shaking it with the water for only two or three minutes, the liquor proved rather stronger than any of the preceding; and being afterwards kept 24 hours on the same bark, it gained no sensible addition to its strength. So that a very strong infusion may be obtained in a very expeditious manner.

taste,

taste, but with the loss of their bitterness: the vegetable acids, added in proportionably larger quantity, render them likewise transparent and improve their roughness, without much diminishing their bitterness: all these mixtures deposite, on standing, a little powdery sediment. Alkalies, both fixt and volatile, occasion a more copious precipitation, and instead of making the turbid decoctions clear, make the clear turbid.

Rectified spirit of wine receives from bark a deep reddish brown colour, and takes up much more of its active matter than watery liquors *(a): by digesting the powder first in some rectified spirit, and then boiling it in water, nearly the whole of its virtue is pretty readily got out. On inspissating the filtered tincture, the spirit carries off nothing remarkable of its smell or taste: the remaining extract retains the peculiar flavour of the bark, as well as its astringency and bitterness, and proves a very elegant preparation, preferable to the pure resin obtained by precipitation from the tincture by water, as containing a part of the gummy matter, which is a medicinal principle of the bark as well as the resin. The spirituous tincture, and the decoction of the residuum, may be united into an extract, possessing this advantage in a greater degree, by inspissating them separately to the consistence of a syrup, then mixing them together, and continuing the evaporation with a gentle heat.

Extr. cort. peruv. Ph. Ed. Extr. cort. peruv. cum resina, Ph. Lond.

Proof spirit extracts less from bark than rectified spirit, but more than water. Four ounces of the powder, macerated for some days

* (a) See Dr. Percival's different opinion, at the note in page 201.

without

without heat, in a quart † or two pounds and a half ‡ of proof spirit, impart a considerable degree both of bitterness and astringency: on applying heat †, the taste becomes stronger, the colour darker, and the liquor somewhat turbid; from whence it may be concluded, that the resinous part is not by this menstruum completely dissolved.

† Tinct. cort. peruv. *Ph. Lond.*
‡ *Ph. Ed.*

Spirit of sal ammoniac made with fixt alkaline salt, by maceration with powdered bark in the above proportion, receives from it very little taste or colour. The spirit prepared with quicklime, and the dulcified spirit, extract in a few hours a very deep colour, and become strongly impregnated with its virtue. Though the spirit made with quicklime is held too acrimonious to be given internally by itself, it is not liable to that objection here; its pungency being sheathed by the substance which it dissolves.

Among the several substances which I have tried for covering the taste of bark, to some persons offensive, liquorice seemed to answer the best. Aromatics alone leave the taste of the bark very sensible in the mouth, but liquorice appeared to cover it effectually, whether in draughts or electuaries, with the bark in substance or its preparations: to this compound any proper aromatic material may be superadded, to give a grateful flavour. For liquid forms, an infusion of the liquorice, and for electuaries the extract should be used: for making up the electuaries, mucilages are more proper than syrups, as the former occasion the compound to pass down freely without sticking about the mouth and fauces.

* PERUVIANUS CORTEX RUBER: Red Peruvian bark. In the year 1779, a Spanish ship from Lima

Lima was taken by an Englifh frigate, and carried into 'Lifbon. Her cargo chiefly confifted of bark, part of which was afterwards brought to London, and purchafed by feveral druggifts. From its large coarfe appearance, it was fometime before practitioners could be prevailed on to ufe it. At length, it was tried in fome of the hofpitals, and found to be fo efficacious, that an opinion foon prevailed of its being of a much fuperiour quality to the beft common bark. Trials were multiplied throughout the kingdom, in a year when intermittents were remarkably frequent and obftinate; and its reputation increafed with every experiment. Chemical tefts were equally favourable to it, as they proved it to contain a much greater proportion of active matter, than the other forts. At length, Dr. Saunders, a phyfician in London, eminent for chemical knowledge, publifhed a treatife, in which various experiments on this bark were related, and atteftations of its great medical efficacy from feveral practitioners were annexed. From this pamphlet, together with the editor's own experiments, the following account is extracted.

The *red bark*, as it is called, is in much larger and thicker pieces than the common. Moft of the pieces are concave, though not rolled together, like the quilled bark. They break fhort, like the beft common bark; and appear evidently compofed of three layers. The outer is thin, rugged, frequently covered with a moffy fubftance, and of a reddifh brown colour. The middle is thicker, more compact, and of a darker colour: it is very brittle and refinous. The innermoft layer is more woody and fibrous, and of a brighter red. In powdering this bark, the middle layer, which feems to contain the greateft proportion of refinous matter,

ter, does not break so readily as the rest; a circumstance to be attended to, lest the most active part should be left out of the fine powder.

This red bark to the taste discovers all the peculiar flavour of the Peruvian bark, but much stronger than the common officinal sort. An infusion in cold water is intensely bitter; more so than the strongest decoction of common bark. Its astringency is in an equal degree greater than that of the infusion of common bark, as is shewn by the addition of martial vitriol. The spirituous tincture of the red bark is also proportionally stronger than that of the pale. The quantity of matter extracted by rectified spirit from the powder of the former, was to that from the latter, as 3 to 2 in one experiment, and as 229 to 130 in another. And yet, on infusing the two residuums of the first experiment in boiling water, that of the red bark gave a liquor considerably bitter, and which struck a black with martial vitriol; while that yielded by the other was nearly tasteless, and void of astringency.

With respect to medical properties, from numerous and repeated trials it appears, that the red bark possesses the same virtues with the common, but in a much higher degree. A single half ounce of this has radically cured an obstinate intermittent, where many ounces of the other kind had either had no effect, or merely a temporary one.

Upon the whole, there is the strongest reason to conclude, with Dr. Saunders, that the *red bark* is the true Peruvian bark, of the best quality, or in its highest perfection. It was probably the kind of bark first introduced into Europe, and which acquired so much reputation in the hands of Sydenham and Morton.

It

It is the sort still preferred by the Spaniards for their own use; and they are surprized at our preference of an inferiour kind. Whether it be, as Dr. Saunders first imagined, the bark of the *trunk* of *full grown* trees, the *branches* or *young* trees of which yield the pale bark; or whether the trees be different *species*, or, at least, *varieties*, does not seem accurately determined. The latter opinion is, perhaps, rendered the most probable, by an observation in the third edition of Dr. Saunder's pamphlet. He says, that " he has lately seen some exceeding good red bark imported by a Spanish merchant, a considerable part of which was as small as the quilled bark in common use, yet still preserved its redness in that form, approaching, however, to the colour of cinnamon. It was extremely resinous, and gave evident proofs of its being the *quill* of the larger red bark which was in the same chest." This idea seems to be confirmed by some curious remarks on the natural history of the cinchona, communicated by Dr. Simmons from the papers of the late M. Jussieu, and subjoined to the same edition.

This writer makes several different species of bark, which may, however, be reduced to two. The first includes the *red*, the *yellow*, and the *knotty* barks, all of which have very smooth leaves, purplish flowers, with a bark that is bitter to the taste, and more or less coloured. Of these, the *red* is held in the highest estimation, and was that first imported into Europe, but is now become exceeding scarce, so that its place has been supplied by the yellow and knotty kinds. The second species includes the *white* barks, of which there are four varieties. All these have broad hairy leaves, and red,

very

PERUVIANUS CORTEX.

very odoriferous flowers, furnished with hairs on their infide. In two of thefe varieties the inner layers of bark are of a reddifh hue. Thefe have a flightly bitter tafte, and fomewhat of a febrifuge quality, which, however, they foon lofe. The bark of the other two is quite white and infipid.

There have been lately difcovered in the province of Santa-Fe, four degrees and a half *north* of the equator, two kinds of cinchona, one of which appears to be the fame with the *red* bark of Peru; the other, one of the *white* fpecies. This is a fortunate difcovery, as it points out a new ftore of this moft valuable medicine, when the ancient ones fhall be exhaufted. We fhall fee in the next article, that our own fettlements are not unprovided with a plant of the fame genus, and fimilar virtues.

* CINCHONA CARRIBÆA *Linn. Cinchona Jamaicenfis* D^{ris}. *Wright, Phil. Tranf.* vol. lxxvii. part ii. This is a fpecies of the Jefuit's bark, produced in Jamaica and the Carribee iflands, of which an accurate defcription, with an account of its virtues, has been publifhed by Dr. Wright in the volume of Philofophical Tranfactions above referred to; and fome additions are made to this, in a letter from the fame phyfician to Dr. Duncan. *Med. Comment.* vol. v. p. 398.

This tree, called in Jamaica the *fea-fide beech*, grows to the height of from twenty to forty or fifty feet. The outer bark of the large trees is white, furrowed, and very thick. This is inert, and may be knocked off from the inner. This latter is of a dark brown colour. Its flavour is at firft fweet, with a mixture of the tafte of horferadifh and of the eaftern aromatics; but when

when swallowed, it has that very bitterness and astringency which characterize the Peruvian bark. It yields its virtues both to cold and warm water; and a decoction of half an ounce of it boiled in a quart of water to the consumption of a pint, proved as strong as a decoction of an ounce and a half of the true bark. With the addition of orange peel it makes an elegant and grateful bitter tincture.

Its medicinal powers have been frequently tried by Dr. Wright, and it was found very efficacious in the dangerous remittent fevers of the West Indies, and also in nervous fevers. It has been administered in London in an intermittent, and effected a cure as completely as the Peruvian bark. From these accounts, we may hope that it will prove an useful and efficacious substitute for the cinchona of Peru, if ever the supplies of this medicine should fail.

PETASITES.

PETASITES major & vulgaris C. B. Galerita & tussilago major quibusdam. Tussilago Petasites Linn. Butterbur or Pestilentwort: a perennial plant, found wild by the sides of ditches and in meadows; producing early in the spring a thick naked roundish stalk, with a spike of small naked purplish flosculous flowers on the top: the flowers and stalks soon wither, and are succeeded about May, by very large, roundish or somewhat heart-shaped leaves, standing on long pedicles, somewhat hollowed in the middle so as to resemble a bonnet *(petasos)*: the root is long, thick, of a dark brownish or black colour on the outside, and white within.

THE

PETROLEUM.

The roots of butterbur are recommended as aperient and alexipharmac; and promife, though now difregarded in practice, to be of confiderable activity. They have a ftrong fmell, and a bitterifh acrid tafte, of the aromatic kind, but not agreeable, very durable and diffufive, fcarcely to be concealed, as Fuller obferves, by a large admixture of other fubftances. Their virtue appears to refide in a refinous matter; which is diftinguifhable by the eye in the dried root, and which is readily extracted by fpirit of wine.

PETROLEUM.

PETROLEUM, Pharm. Lond. *Oleum petræ, Oleum terræ.* Rock oil: a fluid bitumen or mineral oil; exuding from the clefts of rocks or from the earth, or found floating on the furface of waters, in different parts of Europe, and more plentifully in the warmer countries; fimilar, in its general properties, to the oils extracted by diftillation from pitcoal, amber, and other folid bituminous bodies. The more fluid petrolea have been diftinguifhed by the name of *naphtha;* and the thicker by thofe of *piffafphaltum* and *piffelæum.*

1. Petroleum album. White petroleum: nearly colourlefs; almoft as clear, fluid, and tranfparent, as water; of a ftrong penetrating fmell, not difagreeable, fomewhat refembling that of rectified oil of amber. The principal, or only, part of Europe, in which it is found, is the dutchy of Modena in Italy.

2. Petroleum flavum *feu italicum Pharm. Parif.* Yellow petroleum: of a clear yellow colour;

colour; somewhat lefs fluid than the former; in smell rather lefs penetrating, lefs agreeable, and more nearly allied to that of oil of amber. This alfo is found chiefly in the dutchy of Modena, and does not appear to differ very materially from the white fort.

3. PETROLEUM RUBRUM *feu gabianum, five Oleum gabianum Pharm. Parif.* Red petroleum: of a blackifh red colour; of a thicker confiftence, and a lefs penetrating and more difagreeable fmell, than either of the foregoing; found in Italy, and about the village Gabian in Languedoc.

There are many variations of thefe oils in regard to colour, fluidity, fubtility, and the pungency of their fmell and tafte: the moft fluid are in general the moft fubtile and pungent. Among us, the finer kinds are rarely to be met with; and even the inferiour forts are rarely unfophifticated.

Fine petroleum catches fire on the approach of a flaming body, even without the contact of its fubftance with the flame; and burns entirely away. The hafty affufion of concentrated mineral acids, which raifes a violent ebullition with diftilled vegetable oils, and generally fets them on fire, makes no great conflict with petroleum: its confiftence becomes thicker by this admixture, and its fmell more fragrant. By diftillation, it lofes much of its natural fcent, and becomes fomewhat more pellucid than at firft; a fmall quantity of a brownifh or yellowifh matter, fimilar to amber *(a)*, remaining behind. Dropt on water, it fpreads to a

(a) Borrichius, *Acta medica & philofoph. Hafnienfia,* tom. i. *obf.* 57.

great

great diftance, forming a various-coloured film on the furface. It floats' alfo on rectified fpirit of wine, and appears to be indiffoluble in this menftruum; but unites with the effential oils of vegetables *(a)*.

The finer petrolea, more agreeable than oil of amber, and more mild than that of turpentine, partake of the virtues of both. They have been fometimes taken internally in nervous complaints and as a diuretic; but ufed chiefly as an external ftimulant, againft rheumatic pains, palfies, chilblains, &c. In thefe intentions, fome mineral oils, procurable among ourfelves, are ufed by the common people, and often with benefit: the empyrical medicine, called Britifh oil, is of the fame nature with the petrolea; the genuine fort being extracted by diftillation from a hard bitumen, or a kind of ftonecoal, found in Shropfhire and other parts of England.

4. PETROLEUM BARBADENSE *Pharm. Edinb.* *Bitumen barbadenfe. Piffelæum indicum.* Barbadoes tar: of a reddifh black colour, and a thick confiftence, approaching to that of treacle or common tar. It is found in feveral of our American iflands, particularly, as is faid, in that from which it receives its name.

This bitumen, greatly efteemed by the Americans as a fudorific, in diforders of the breaft, and as an external difcutient and antiparalytic, is in fmell more difagreeable, and both in fmell and tafte lefs pungent, than the foregoing petrolea. It is likewife lefs inflammable, and leaves on being burnt a confiderable quantity of

---

(a) See *l'Hiftoire & les memoires de l'acad. roy. des fcienc. de Paris, pour les années* 1715 & 1726.

Ol. petrolei *Ph. Lond.*

Petroleum fulphurat. *Ph. Lond.*

ashes. In distillation, it yields an oil different, in regard to its colour, from those afforded by such of the other bitumens as have been examined; appearing, when placed betwixt the eye and the light, of an orange colour, in other positions blue; but losing this variability of aspect in long keeping, and then looking in all situations yellow. This oil, and a balsam prepared by boiling the petroleum itself with one fourth its weight of flowers of sulphur, are directed by the London college to be kept in the shops.

PETROSELINUM.

PETROSELINUM Pharm. Lond. & Edinb. Apium hortense seu petroselinum vulgo C. B. Apium Petroselinum Linn. Parsley: an umbelliferous plant, with deep green winged leaves, of which those that grow on the stalk are divided into fine oblong narrow segments: the seeds are small; somewhat crookedly planoconvex, of a dusky greenish colour, with four yellow ridges along the convex side; the root long, whitish, about the thickness of the finger. It is biennial, a native of moist grounds in the southern parts of Europe, and common in our culinary grounds.

The roots of parsley are sometimes used in apozems, and supposed to be aperient and diuretic, but liable to produce flatulencies. Their taste is sweetish, accompanied with a slight warmth or flavour, somewhat resembling that of a carrot. Rectified spirit takes up, by digestion, all their active matter, and on inspissating the tincture, leaves it entire in the extract; in which, the sweetness is very considerable,

the

the warmth very weak. Diftilled with water, they impregnate the firft runnings pretty ftrongly with their flavour: when large quantities are diftilled, there feparates a fmall portion, two or three drams from two hundred pounds, of effential oil, which partly fwims on the water, partly finks, and partly concretes about the nofe of the worm into a butyraceous matter.

The leaves of the plant have a greater warmth and lefs fweetnefs than the roots. In diftillation with water, they yield a greater quantity of effential oil, about ten drams from two hundred pounds, fmelling agreeably of the herb, and in tafte moderately pungent.

The feeds, faid to be carminative, refolvent, and diuretic, and commended in the German ephemerides for deftroying cutaneous infects in children, are in tafte warmer and more aromatic than any other part of the plant, and accompanied with a confiderable bitternefs. In diftillation, three pounds yielded above an ounce of effential oil, great part of which funk in the watery fluid. They give out little by infufion to watery menftrua, but readily impart all their virtue to rectified fpirit: the tincture lofes nothing confiderable in being gently infpiffated to the confiftence of an extract, which proves a moderately warm, pungent, bitterifh, not very grateful, aromatic.

PETROSELINUM MACEDONICUM.

APIUM MACEDONICUM C. B. Apium petræum & petrapium quibufdam. Bubon macedonicum Linn. MACEDONIAN PARSLEY: differing from the foregoing, in the upper and lower leaves being alike, the ftalks hairy and much branched, the feeds dark coloured and covered with rough hoarinefs; It is a native of ftony foils in Macedonia, and cultivated in fome of our gardens.

The Macedonian parfley is fimilar in quality to the common fort, but weaker and lefs grateful. The feeds are the only part made ufe of, and thefe only as ingredients in the mithridate and theriaca: hence the Edinburgh college, having now dropt thofe compofitions, has dropt alfo the Macedonian parfley.

PEUCEDANUM.

PEUCEDANUM *germanicum* C. B. *Pinaftellum, fœniculum porcinum, fœniculum filveftre, marathrum filveftre, marathrophyllum, & cauda porcina quibufdam.* Peucedanum *officinale* Linn. Hogs fennel, Horestrong, Sulphurwort: an umbelliferous plant, with large leaves divided and fubdivided tripartitely into fine oblong narrow fegments: the feed is fomewhat oval, flattifh, marked with three ftriæ, and furrounded with a leafy margin: the root long and thick, with a tuft of filaments on the top, blackifh on the outfide and pale coloured within. It is perennial, grows wild by the fea fhores and in moift fhady grounds, and flowers in July.

The roots of fulphurwort have a ftrong fetid fmell, fomewhat refembling that of fulphureous folutions; and an unctuous, fubacrid, bitterifh tafte. Wounded when frefh, in the fpring or autumn, particularly in the former feafon, in which they are moft vigorous, they yield a confiderable quantity of yellow juice, which foon dries into a folid gummy-refin, retaining the tafte and the ftrong fmell of the root. This gummy

PILOSELLA.

gummy-refin ftands recommended as an aperient, and antihyfteric.

PILOSELLA.

PILOSELLA, Myofotis, feu Auricula muris. Pharm. Parif. Pilofella major repens hirfuta C. B. Hieracium Pilofella Linn. Mouse-ear: a low creeping hairy plant; with oval leaves, in fhape like thofe of the daify, joined to the ftalks without pedicles, green above and white underneath: the flowers, which ftand folitary on upright naked ftalks, are compofed of a number of yellow flofcules, fet in fcaly cups, and followed by fmall black feeds, winged with down. It is perennial, grows wild in dry pafture grounds, and flowers in June and July.

Pilosella is one of the bitterifh lactefcent plants. Its leaves differ from thofe of dandelion, cichory, and the other herbs of that clafs, in being much lefs juicy, lefs bitter, accompanied with fome aftringency which feems to prevail above the bitter, and a flight fweetifhnefs very durable in the mouth: in the extracts made from them, both by water and fpirit, the aftringency is more manifeftly the prevailing principle, though even when thus concentrated it is not very ftrong. The roots are confiderably bitterer than the leaves, and lefs, if at all, aftringent.

PIMPINELLA.

PIMPINELLA SAXIFRAGA Linn. Burnet-saxifrage; a perennial umbelliferous plant; with naked umbels; the outermoft flowers compofed of unequal petals, the inner equal; the

feeds

feeds small, oblong, somewhat pointed, flat on one side, convex and striated on the other; the lower leaves roundish, indented, set in pairs along a middle rib with an odd one at the end; the upper leaves oblong and very narrow; the roots long, slender, and whitish.

1. PIMPINELLA ALBA *Germanorum: Pimpinella saxifraga major umbella candida* C. B. Greater or white burnet-saxifrage: with some of the leaves pretty deeply cut, the odd one into three sections. It is not very common in this country, and therefore our markets have been generally supplied with the following.

2. PIMPINELLA SAXIFRAGA: *Pimpinella saxifraga minor foliis sanguisorbæ Raii*; *Tragoselinum alterum majus Tourn.* Smaller burnet-saxifrage; with uncut leaves. It grows wild in dry pasture grounds.

3. PIMPINELLA SAXIFRAGA MINOR C. B. *Tragoselinum minus Tourn.* Small burnet-saxifrage; with the upper leaves divided into oblong narrow segments; taller than the others, but with smaller leaves. This is the most common sort in the fields about London.

ALL these plants appear to be possessed of the same qualities, and to differ little otherwise than in external appearance: and even in this, their difference is so inconsiderable and inconstant, that Linnæus has joined them into one species, under the name of *pimpinella foliis pinnatis, foliolis radicalibus subrotundis, summis linearibus:* he says he has seen the second sort produced from the seeds of the first sown in a richer soil. Instead of the first, which has been generally understood

PIMPINELLA.

as the officinal kind, our college allows either of the others to be taken indifferently.

The roots of pimpinella have a hot, pungent, not very durable taste; and emit, when fresh, an acrid halitus, of no particular smell, but affecting the eyes like that of horseradish or mustard-seed, though in a lower degree. In drying, they lose this subtile matter, and in long keeping the pungency of their taste is diminished. Their virtue is extracted, partially by water, and completely by rectified spirit. In distillation with water, a part of their pungency arises and impregnates the distilled fluid, and a part remains behind in the decoction: when large quantities are distilled, there separates from the water a small portion of a yellowish essential oil extremely acrid and fiery. On inspissating the spirituous tincture, little or nothing of the virtue of the pimpinella rises with the spirit: the remaining extract, small in quantity, is of great pungency and heat. The leaves and seeds of the plant have likewise a considerable acrimony; the leaves less than the seeds, and both less than the roots.

This pungent root is in great esteem among the Germans, as a warm stimulating resolvent, aperient, diaphoretic, &c. in weakness of the stomach from viscid phlegm, infarctions of the breast, tumours and obstructions of the glands, impurities of the blood, and in general wherever tenacious humours are to be attenuated, or the fluid secretions promoted. It is an useful ingredient in our officinal compound arum powder, supplying in good measure the pungency which the arum root loses in being reduced into that form. It is employed also as a masticatory for stimulating the salival glands; and in gargarisms for dissolving viscid mucus in the fauces.

PIN-

PINGUEDO.

PINGUEDO five adeps: Sevum ovillum & hircinum, Axungia porcina & viperina. Animal fats: sheeps suet, goats suet, hogs lard, and vipers fat.

The medical use of these substances is wholly external, as the basis of ointments and other unctuous applications. In their effects, they do not seem to differ materially from one another; all of them having one common emollient virtue, supplying and relaxing the part to which they are applied, and obstructing its perspiration. The principal difference to be considered in them is that of their consistence, by which they are adapted to different forms, or for receiving different admixtures; the solid *seva* serving to give the thick consistence of an unguent to oils and the more fluid resinous juices, while the softer *axungiæ* procure a like consistence to solid resins and powders. The fat of the viper is commonly preferred to the others in affections of the eyes; but its superiority, in these cases, to other soft fats, does not appear to have been sufficiently determined by experience. Nor indeed does it appear, that animal fats, and flavourless vegetable oils, of similar consistences, are materially different, respectively, from one another, in their effects when used in external applications. Even in regard to qualities, more remote than those, by which they can act when applied to the external parts of the body, the difference between the vegetable and animal fats is, perhaps, less than might be expected, and apparently less than that which is observed between the other corresponding substances of the

PINGUEDO.

the two kingdoms, as the gelatinous matters of the one and the gummy of the other: animal fats, in their refolution by fire, yield neither the peculiar ftench, nor the volatile alkaline falt, which fubftances completely animalized afford.

Lard and fuet are directed to be tried or pu- *Adipis fuillæ* rified, by chopping them into fmall pieces; *fevique ovilli* melting them by a gentle heat, with the addi- *curatio Ph.* tion of a little water, which fecures them from *Lond.* any danger of burning or turning black, this fluid not being fufceptible of a degree of heat fufficient for that effect; and then ftraining them from the membranes. Vipers fat, feparated from the heart, liver, and other bloody parts, is ordered to be melted without addition, and then ftrained through a linen cloth; the quantity of this fat, ufually purified at a time, being fo fmall, that the heat may be eafily regulated, fo as to prevent burning, without water.

Tried lard is formed into an elegant ointment, *Unguentum* commonly called pomatum, by beating it with *adipis fuillæ* rofewater, in the proportion of three ounces of *Ph. Lond.* the water to two pounds of the lard, till they are well mixed; then melting it over a very gentle fire, and after ftanding for a little while, that the watery part may fettle, pouring off the lard, and inceffantly ftirring and beating it about till it grows cold, fo as to reduce it into a light yielding mafs; and afterwards adding fo much effence of lemons as will be fufficient to give a grateful fmell. Some fcent it with oil of rhodium; and previoufly digeft the lard for ten days with common water, renewing the water every day, a procefs which does not appear to be of much ufe. Thefe ointments may be tinged of a fine red colour, for lip-falves, by a proper addition of alkanet root: the faculty of *Pomatum* Paris directs, for this purpofe, twenty-four parts *rubrum Ph.* of *Parif.*

of the white pomatum, eight of oxes marrow, and eight of white wax cut in small pieces, to be melted together by the heat of a water bath; one part of powdered alkanet root to be added; the mixture stirred at times till it appears tinged of a deep red colour, and then strained through a linen cloth.

Animal fats are not diffoluble by spirit of wine any more than by water: when scented with essential oils, the oil may be totally extracted by digestion in rectified spirit, so as to leave the fat inodorous. By the same menstruum, fats may be freed from their ill smell, and even those that have grown considerably rancid by keeping may be made sweet again as at first; the rancidity and smell seeming to consist in a part of the fat attenuated, or subtilized, into a state analogous to that of the oil into which fats are resolved by distillation, which oil is totally diffoluble in spirit.

PINGUICULA.

PINGUICULA: *Sanicula montana flore calcari donato* C. B. *Pinguicula five sanicula eboracensis* Gerard. *Viola palustris, liparis, cucullata, & dodecatheon plinii quibusdam. Pinguicula vulgaris* Linn. Butterwort or Yorkshire sanicle: a small plant, with a few, oblong, obtuse, uncut, pale, glossy, unctuous leaves, lying on the ground; among which rise naked pedicles, bearing, each, a purplish monopetalous flower divided into two lips (of which the upper is cut like a heart, the lower into three sections) with a slender cylindrical spur or tail at bottom: the flower is followed by a roundish capsule full of small seeds. It is perennial, grows wild in elevated

PIPER.

elevated marshy grounds, and flowers in the spring.

The remarkable unctuosity of this plant, and of some others of the same genus, seems to entitle them to a further examination than has yet been bestowed upon them *(a)*. It is said, that the unctuous and glutinous juice of the *pinguicula* is used in some places as a liniment for chaps *(b)*, and as a pomatum for the hair *(c)*: that new milk, poured upon the fresh leaves, on a strainer, and after quick colature, set by for a day or two, becomes thick, tenacious, very agreeable and salubrious, and throws off no whey unless long kept; and that a little of the milk, so thickened, serves for bringing fresh milk to the same state *(d)*: that a syrup made from the juice, and decoctions of the leaves in broth, are used among the common people in Wales as cathartics: and that the herb is hurtful to cattle that feed upon it *(e)*.

PIPER.

PEPPER: the small, round, aromatic fruit of a trailing plant growing in Sumatra, Java, and Malabar, *(Piper nigrum Linn.)* The pepper-corns adhere in clusters to the stalks, without pedicles: when ripe, they are firm, not juicy, of a red colour, which changes in drying to a black.

(a) Linnæus, *Flora lapponica,* p. 10.
(b) Simon Paulli, *Quadripartit. botanic.*
(c) Ray, *Historia plantarum,* i. 752.
(d) Gissler, *Suenska vetenskaps academiens handl.* 1749.
(e) Gerard, *Herbal emaculated,* p. 789.

1. Piper

1. PIPER NIGRUM *Pharm. Lond. & Edinb.* *Melanopiper.* Common or black pepper: the fruit gathered, probably, before perfect maturity, and dried in the fun.

2. PIPER ALBUM. *Leucopiper.* White pepper: the ripe fruit decorticated by maceration in water. Some of the grains, as brought to us, have pieces of a dark-coloured fkin ftill upon them.

Of thefe pungent hot fpices, the black fort is the hotteft and ftrongeft, and moft commonly made ufe of for medicinal as well as culinary purpofes. They both feem to heat the conftitution more than fome other fpices that are of equal pungency upon the palate; and from thofe fpices they differ in this, that their pungency does not refide in the volatile parts or effential oil, but in a fubftance of a more fixt kind, which does not rife in the heat of boiling water.

Pepper, infufed in water, impregnates the menftruum pretty ftrongly with its flavour, but weakly with its tafte: by boiling for fome time, a little more of its pungent matter is extracted, and its flavour diffipated. On collecting the fluid that exhales in the boiling, the water is found agreeably impregnated with the odour of the fpice, but fcarcely difcovers any tafte: the effential oil, which rifes to the furface of the water, thin, light, and limpid, in fmell ftrong and agreeable, is in tafte mild; a drop or two impreffing on the tongue only a moderate grateful warmth. On infpiffating the decoction, a part of the pungency of the pepper is found

in

in the mucilaginous extract, and a part is retained by the pepper itself.* *(a)*

Rectified spirit extracts completely the active matter of the pepper. The tincture is extremely hot and fiery, a few drops setting the mouth as it were in a flame: infpiffated, it leaves an extract still more fiery; the spirit carrying off in its exhalation a little of the flavour, but nothing of the heat or pungency of the spice. The quantity of extract is nearly the same from both kinds of pepper; the spirituous amounting to about one eighth, and the watery to near one half their weight: but those of the white, like the spice in substance, are weaker than those of the black sort.

PIPER LONGUM.

PIPER LONGUM *Pharm. Lond. & Edinb.* Macropiper *Pharm. Parif.* Piper longum orientale *C. B.* Long pepper: the fruit of an East Indian plant of the same genus with that which produces the black pepper, *(Piper longum Linn.)*; of a cylindrical figure, about an inch and a half in length, having numerous minute grains disposed round it in a kind of spiral direction.

This spice is hotter and more pungent than either of the preceding kinds, and its spirituous extract is proportionably more fiery. In pharmaceutic properties, it entirely agrees with them; its active matter being only partially dissoluble in watery menstrua, and its pungency not rising in the heat of boiling water. Decoc‑

---

* *(a)* The acrid matter of pepper is so strongly retained, that a quantity boiled succeffively in frefh parcels of water, had not loft all its tafte till the forty-third boiling. *Gaubii Adverfar.*

tions of it are very mucilaginous, rather more so than those of the black or white.

PIPER JAMAICENSE.

PIMENTO *Pharm. Lond.* *Pimenta sive piper jamaicense* Pharm. *Edinb.* *Amomum* Pharm. *Wirtemb.* JAMAICA PEPPER, PIMENTO, ALLSPICE: the dried aromatic berry of a large tree growing in the mountainous parts of Jamaica; reckoned a species of myrtle, and called by Sir Hans Sloane *myrtus arborea aromatica foliis laurinis*, by Linnæus *myrtus (pimenta) foliis alternis*.

PIMENTO is a moderately warm spice, of an agreeable flavour, somewhat resembling that of a mixture of cloves, cinnamon, and nutmegs. Distilled with water, it yields an elegant essential oil, so ponderous as to sink in the aqueous fluid, in taste moderately pungent, in smell and flavour approaching to oil of cloves, or rather a mixture of those of cloves and nutmegs: the remaining decoction, inspissated, leaves an extract somewhat ungrateful but not pungent, and the berry itself is now found to be almost wholly deprived of its taste as well as flavour; the warmth of this spice residing rather in the volatile than in the fixt parts. To rectified spirit it imparts, by maceration or digestion, the whole of its virtue, together with a brownish green tincture: in distillation, it gives over nothing confiderable to this menstruum, nearly all its active matter remaining concentrated in the inspissated extract; which is very warm and pungent, but not of a fiery heat like those obtained from the foregoing sorts of pepper.

This spice, at first brought over for dietetic uses, has been long employed in the shops as a

succedaneum

succedaneum to the more costly oriental aromatics; from them it was introduced into our hospitals, and is now received both in the London and Edinburgh pharmacopœias. A simple water is directed to be distilled from it, in the proportion of a gallon or ten pounds from half a pound: this is strongly impregnated with the flavour of the pimento, though it is less elegant than the spirituous water which the shops have been accustomed to prepare, by drawing off two or three gallons of proof spirit from the same quantity of the spice. The Edinburgh college directs only nine pounds from this quantity. The essential oil does not seem to be much known in practice; though it promises to be a very useful one, and might, doubtless, on many occasions, supply the place of many of the dearer oils. The quantity of oil afforded by the spice is very considerable: Cartheuser indeed says, that only about half a dram is to be got from sixteen ounces; a mistake, which probably has arisen from inadvertence in copying Neumann's proportion, of half a dram from an ounce, or one sixteenth. So large a proportion as this last cannot, however, be collected in its proper form, the oil that remains dissolved in the distilled water being here included.

Aq. pimento Ph. Lond.
Aq. piperis jamaicensis Ph. Ed.
Spir. pimento Ph. Lond.
Aq. piper. jamaic. spirit. Ph. Ed.
Ol. essent. pip. jamaic. Ph. Ed.

PIPER INDICUM.

PIPER INDICUM Pharm. Lond. & Edinb. Capsicum Pharm. Parif. Piper indicum, brazilianum, guineense, calecuticum, hispanicum, & lusitanicum, quibusdam. Capsicum siliquis longis propendentibus Tourn. Siliquastrum plinii J. Bauh. Capsicum annuum Linn. CAPSICUM or GUINEA PEPPER: long, roundish, taper, bright red pods, divided into two or three cells full of small whitish

whitish seeds: the fruit of an annual plant, with square stalks, oblong acuminated leaves, and white flowers growing in their bosoms divided into five segments in form of a star; a native of the East and West Indies, and raised in some of our gardens.

This fruit, when fresh, discovers to the organs of smell, a penetrating acrid halitus, which in drying is dissipated: its taste, whether fresh or dry, is extremely pungent and acrimonious, setting the mouth as it were on fire, and producing a painful burning vellication of long continuance, like that occasioned by arum root, but more of the warm aromatic kind. It gives out its pungency to rectified spirit, together with a pale yellowish red tincture: the spirit, gently distilled off, has no considerable impregnation from the capsicum: the remaining extract is insupportably fiery.

Capsicum is sometimes given, in minute quantities, as one of the highest stimulants, in cold sluggish phlegmatic temperaments, in some paralytic cases, in relaxations and insensibility of the stomach, and for promoting the efficacy of aloetic medicines and the deobstruent gums in uterine disorders. It is used principally at table: a species of it, called in the West Indies bird-pepper, is the basis of the powder brought from thence under the name of Cayenne pepper. It is observable that this fruit, perhaps the strongest of the aromatic stimulants, is used freely, as is said, by the natives even of the warm climates: possibly these pungent antiseptic kinds of substances may there be more salubrious than they are, in general, among us, as they seem qualified to resist or correct the putre-
dinous

dinous colliquation of the humours which immoderate heat produces.

*P I X.

PIX LIQUIDA Pharm. Lond. & Edinb.
TAR: a thick, black, refinous, very adhefive juice; melted out by fire from old pines and fir-trees. The trees, cut in pieces, are inclofed in a large oven, which being heated by a fire on the outfide, or the wood itfelf kindled and fmothered, the juice runs off by a canal at the bottom.

TAR differs from the turpentine or native refinous juice of the trees, in having received a difagreeable empyreumatic impreffion from the fire; and in containing, along with the pungent bitter terebinthinate matter, a portion of the acid which is extricated from the wood by the heat, and likewife of its gummy or mucilaginous matter. By the mediation of thefe principles, a part of the terebinthinate oil and refin becomes diffoluble in watery liquors, which extract little or nothing from the purer turpentines.

Water impregnated with the more foluble parts of tar has been recommended as a remedy for almoft all difeafes. The proportions that have been commonly followed are, two pounds of tar to a gallon of water; which are to be well ftirred together, then fuffered to fettle for two days, and the clear liquor poured off for ufe. It is obferved, that " tar water, when right, is
" not paler than French, nor deeper coloured
" than Spanifh white wine, and full as clear:
" if there be not a fpirit very fenfibly perceived
" in drinking, the tar-water is not good. It
" may be drank either cold or warm. As to
" the

"the quantity, in common chronical indifpo-
"fitions a pint a day may fuffice, taken on an
"empty ftomach, at two or four times: more
"may be taken by ftrong ftomachs. But thofe
"who labour under great and inveterate mala-
"dies, muft drink a greater quantity, at leaft
"a quart every twenty-four hours. In acute
"difeafes, it muft be drank in bed warm, and
"in great quantity (the fever ftill enabling the
"patient to drink) perhaps a pint every hour."
Though this medicine is undoubtedly very far
inferiour to the character that has been given
of it, it is apparently capable of anfwering important purpofes, as a deobftruent balfamic folution, moderately warm and ftimulating. It
fenfibly raifes the pulfe, and increafes either
perfpiration or the groffer evacuations. I have
been informed of fome late inftances of its good
effects in diforders of the leprous kind.

Some have imagined the acid to be the principle that gives virtue to tar-water; and hence
have endeavoured to introduce, inftead of the
infufion, an acid fpirit extracted from tar by
diftillation. But the effects of this, as of other
acids, are oppofite to thofe experienced from
tar-water; nor does the acid of tar differ from
that which is extricated by fire from all kinds of
recent wood. Tar-water, diftilled, yields a
liquor very confiderably impregnated with its
flavour, though more grateful than the infufion
itfelf both in fmell and tafte: there remains a
light, fpongy, blackifh fubftance, not acid but
bitter, partially diffoluble again in water.

Pil. piceæ
Nofocom. Ed.

This juice is fometimes given alfo in fubftance, mixed with fo much powdered liquorice,
or other like powdery matters, as is fufficient to
render it of a due confiftence for being formed
into

into pills. An ointment, made by melting it with an equal weight of mutton suet, and straining the mixture whilst hot†, or by melting together five parts of tar and two of yellow wax‡, is sometimes used as a digestive, and said to be particularly serviceable against scorbutic and other cutaneous eruptions.

Unguent. e pice † Ph. Lond.

‡ *Ph. Ed.*

On inspissating tar, or boiling it down to dryness without addition, it gives over an acid liquor in considerable quantity, and an ethereal oil of the same general nature with that of turpentine, but impregnated with the empyreumatic flavour of the tar. The solid residuum is the common pitch, *pix arida*. *Pix ficca, palimpiffa dioscoridis* C. B. This is less pungent, and less bitter than the liquid tar, and used only in some external applications, as a warm adhesive resinous substance. Neumann observes, that when melted with oils, resins, and fats, into ointments and plasters, the pitch is greatly disposed to separate and precipitate.

PLANTAGO.

PLANTAIN: a small perennial herb, common in fields and by road sides; with the leaves lying on the ground; and naked unbranched stalks, bearing on the top a spike of small imperfect four-leaved flowers, followed by little capsules, which, opening horizontally, shed numerous crooked seeds.

1. PLANTAGO *Pharm. Edinb. Plantago latifolia sinuata* C. B. *Plantago septinervia. Plantago major*, Linn. Common greater plantain: with oval leaves, having seven ribs, prominent on the lower side, running from end to end; and long slender spikes.

2. PLANTAGO MINOR *seu quinquenervia*: *Plantago major angustifolia* C. B. *Plantago lanceolata* J. B. *Plantago lanceolata Linn.* Narrow-leaved plantain or ribwort: with oblong, five-ribbed leaves; and short thick spikes.

The leaves and seeds of plantain, recommended as vulneraries, in phthisical complaints, spittings of blood, alvine fluxes, &c. appear to be of the milder kind of restringents or corroborants. The leaves, of no remarkable smell, are in taste slightly acerb: their expressed juice, depurated by settling and colature, or clarified with white of eggs, and inspissated to the consistence of honey, discovers a considerable saline austerity. The two sorts are not sensibly different in quality from one another, though the first has been generally directed for medicinal use in preference to the other. The leaves are, in some places, the usual application made by the common people to slight wounds.

* For the use of a species of plantain, with horehound, in the bite of the rattlesnake, see the art. *Marrubium*.

PLUMBUM.

PLUMBUM Pharm. Lond. LEAD: a pale, livid, soft, very flexible metal; above eleven times specifically heavier than water; fusible in a small heat, somewhat less than that in which expressed oils begin to boil. Continued in fusion it contracts a various-coloured pellicle on the surface, and if kept stirring, so as that fresh surfaces may be exposed to the air, it changes by degrees into a powdery dusky-coloured calx: this powder, calcined for some time in a stronger fire, in such a manner that the flame may reverberate

PLUMBUM.

verberate all over it, becomes firſt yellow, and afterwards of a deep red colour†: all theſe calces, if the fire be haſtily raiſed to a conſiderable degree, melt into the appearance of oil, and on cooling form a ſoft flaky pulverable ſubſtance called litharge‡, of a pale yellowiſh or reddiſh colour, according as the lead has been leſs or more calcined: if the calces be urged with a pretty ſtrong fire, they run into a yellowiſh glaſs, which, while in fuſion, powerfully diſſolves moſt kinds of earthy bodies, and corrodes the common crucibles till it has ſaturated itſelf with their earth.

† Minium *Ph. Lond.*

‡ Lithargyrus *Ph. Lond. & Ed.*

The ores of lead, in colour commonly reſembling lead itſelf, and of a cubical or parallelopipedal ſtructure, are plentiful in England and other parts of the world. The metal, extracted from the ore by fuſion, contains frequently a portion of ſilver, and ſometimes of gold: on keeping the compound melted in a due degree of heat, the lead calcines and turns to litharge, which is raked or blown off till the noble metals remain pure; all the other common metallic bodies being ſcorified and carried off by the lead. From the works, wherein ſilver is thus extracted from lead in the large way, the ſhops are ſupplied with litharge; which, when pale coloured, is called litharge of ſilver; when high coloured, litharge of gold. The latter is to be preferred, not as containing any of the metal by whoſe name it is diſtinguiſhed, but as being more thoroughly calcined than the pale ſort: the pale may be freed from the uncalcined lead it holds, by melting it; the uncalcined part falling to the bottom during the fuſion.

The nitrous acid, diluted with about an equal quantity of water, diſſolves lead pretty readily

into

into a gold-coloured liquor: by the vitriolic and marine acids it is very difficultly acted on; and when previously 'diffolved in the nitrous, it is by either of thefe precipitated. Vegetable acids, digefted on lead in fubftance, diffolve it exceeding fparingly: by certain managements they may be made to act more vigoroufly, and to fatiate themfelves with the metal.

Ceruffa *Ph. Lond. & Ed.*

Thin plates of lead, fufpended over vinegar in a proper veffel, and fet to digeft in a gentle heat, as that of horfe-dung, that the acid vapour may rife and circulate round the plates, are found, in about twenty days, covered with a white powdery or flaky matter: this being fcraped off, and the procefs repeated, the whole of the metal is thus corroded by degrees into ceruffe or white lead. This commodity, the preparation of which makes a confiderable trade, is frequently adulterated with a mixture of whiting: the entire flaky maffes, called flake lead, fhould be chofen, as not being liable to abufe. The adulteration may be difcovered by means of vinegar, which will effervefce with and diffolve the whiting or calcareous earth: the liquor being then poured off clear, or filtered, the addition of a little fpirit of falt will precipitate fuch part of the lead as the vinegar may have taken up; after which the calcareous earth will manifeft itfelf on adding a little vitriolic acid.

Acetum lithargyrites.

The calces of lead are much eafier of folution in vegetable acids than lead in its metallic form. On digefting four ounces of litharge about three days in a fand heat with a pint of ftrong vinegar, and now and then fhaking the veffel; the liquor, filtered, is found to have received a ftrong impregnation from the litharge, and to have diffolved about one tenth of it, whereas,

whereas, of the fame quantity of lead in fubftance, fcarcely one hundredth part would be diffolved. Lead even in its vitreous ftate, or in the glazing of the common earthen-ware veffels, is confiderably acted on by vegetable acids; which, by being boiled in thofe veffels receive from them the peculiar tafte, and pernicious qualities of faturnine folutions.—Lead may be difcovered in acid liquors by a reddifh, brown, or blackifh colour being produced in them on adding a few drops of a folution of orpiment or common fulphur made in lime-water, and by the colour not being deftroyed on the fuperaddition of a little fpirit of falt *(a)*: other metals, diffolved in vegetable acids, produce, as well as lead, a dark colour with the fulphureous folutions, but fpirit of falt rediffolves them, and totally difcharges the colour.

Of all the faturnine calces, the ceruffe, on account of the corrofion it has previoufly undergone from the fteam of vinegar, is the moft eafily diffoluble in frefh vinegar, and hence is made choice of where a faturated folution is required. The folution made in vinegar, in fpiffated to the confiftence of honey and fet in the cold, fhoots by proper management into cryftals, called, from their tafte, *fugar* of lead. All the folutions, and foluble preparations of this metal, have a remarkably fweet tafte, mixed with a confiderable aufterity.

Ceruffa acetât. *Ph. Lond.*

Sal plumbi *vulgo* facch. faturni *Ph. Ed.*

LEAD in its metallic form, or when calcined by fire, does not appear to have any medicinal operation: diffolved or rendered foluble

(a) The brownifh or reddifh colour produced by alkalies in cyder impregnated with lead, is totally difcharged by fpirit of falt. *M. S. of Dr. Lewis.*

by

by acids, it is one of the moſt powerful ſtyptics, but at the ſame time, for internal uſes, one of the moſt dangerous. A few grains of the ſugar have been ventured on for checking obſtinate hemorrhagies and other profuſe evacuations: a tincture drawn with rectified ſpirit, by maceration without heat, from ſugar of lead and green vitriol, in the proportion of three ounces of the ſugar and two of the vitriol, to a quart of ſpirit, has been given from fifteen to thirty drops, for reſtraining the colliquative ſweats attending phthiſes and hectic fevers. This practice has in ſome inſtances been ſuccefsful, but the hazard is very great: all the ſaturnine preparations that have any activity are in a peculiar manner injurious to the nervous ſyſtem, and ought never to be ventured on but in deſperate caſes as a laſt reſource. Obſtinate conſtipations, violent colics, pains and contractions of the limbs, tremors and reſolutions of the nerves, and ſlow waſting fevers, are the general conſequences of ſaturnines taken in any conſiderable quantities internally, and of the fumes to which the workmen are expoſed in the fuſion of the metal in the way of buſineſs *(a)*.

Tinct. ſaturnina *vulgo* antiphthiſica *Ph. Ed.*

Externally, this metal and its preparations are of ſufficient ſafety and of great utility. The plaſter, in general uſe for ſlight cutaneous injuries, and which makes the baſis of ſeveral other plaſters, is a ſolution of litharge in oil olive, in the proportion of five pounds of the litharge, ſubtilely powdered, to eight pints† or ten pounds‡ of the oil. The union is effected by

Empl. litharg. † *Ph. Lond.* commune ‡ *Ph. Ed.*

(a) Vide Hoffman, *Philoſophia corp. human. morboſi*, P. II. *cap.* viii. §. 20. & *ſeq.* Hundert mark, *Acta acad. cæſareæ nat. curioſ. vol.* vii. *Append. p.* 96.

boiling them together over a gentle fire, with the addition of about a quart of water to prevent their burning, and keeping them continually stirring, till they incorporate and acquire a due consistence: if all the water should be consumed before this happens, some more water, previously made hot, is added. A red plaster is prepared in the same manner with minium Emp. e minio. instead of litharge, but as it does not stick so well as the other, it is more rarely used: it is likewise more difficult of preparation, the compound being very apt, though a considerable quantity of water be used, to burn and grow black in the boiling.

The ceruffe and fugar, particularly the latter, are cooling, drying, and aftrictive: the fugar is used in collyria for inflammations and defluxions of the eyes, and in injections for restraining simple gonorrhœas; and both preparations in unguents and liniments, against cutaneous heats and excoriations, flight ferpiginous eruptions, and for anointing the lips of wounds or ulcers that itch much or tend to inflammation. Compositions for these purposes are made in the Ung. e ce- shops, by mixing one part of ceruffe with five ruffa, *vulgo* of the simple ointment made with oil and wax; album *Ph. Ed.* by grinding two ounces of litharge, and adding, alternately and by little and little, two ounces of vinegar and fix of oil†; or by boiling and † Ung. nustirring, over a gentle fire, four ounces of the tritum. common plaster, with one of vinegar, and two of oil where a thick unguent is required‡, or ‡ Ung. trifour of oil for a softer liniment‖: this last is a pharmacum lefs troublesome method of uniting the litharge pharmacum. with the oil and vinegar, than trituration; and the composition proves likewise more smooth and uniform, and less liable to grow hard in
keeping

keeping *(a). But the moſt elegant and effectual of all the ſaturnine unguents, are thoſe made with the ſugar; in the proportion of half an ounce† to a pint of oil and three ounces of white wax; or one part, to twenty parts of the ſimple oil and wax ointments‡*(b).

Ung. ceruſſ. acet. † *Ph. Lond.*
— ſaturni. ‡ *Ph. Ed.*

* Mr. Goulard, a ſurgeon of Montpellier, has been the means of greatly extending for ſome years paſt the external uſe of lead. The baſis of his preparations is what he calls the *extract of lead*, or a ſolution of litharge in ſtrong vinegar, boiled down to almoſt a ſyrupy conſiſtence. This, diluted in a large quantity of ſoft water, makes his *vegeto-mineral water*, which is employed as a lotion or fotus, or boiled with bread to make a cataplaſm. The extract is likewiſe combined with unguentous matters into a variety of forms. Theſe preparations have, in fact, been found of the greateſt utility in various caſes of inflammation, particularly of the eryſipelatous kind, and the conſequences of burns and ſcalds. Their moſt liberal application has not, in the opinion of moſt practitioners, been obſerved to produce any of thoſe affections of the nervous ſyſtem, which characterize the poiſonous effects of lead taken internally. At the ſame time, the abuſe of ſaturnine applications, on the ground of thoſe

*(a) The *ung. nutritum*, made without heat, though now expunged from our diſpenſatories, is much the beſt of the above preparations, and a very excellent application in many caſes. It ſhould not be long kept, but made freſh as wanted.

*(b) Theſe are by no means the efficacious preparations here repreſented. The oil and wax ſo cover the metallic ſalt, that its action is prevented; or, if it acts at all, it proves highly ſtimulating from the *undiſſolved* ſtate in which it is applied.

falſe

false and inconsistent ideas of their action which Mr. Goulard has supported, has not infrequently been attended with disagreeable consequences.

* The London college have now given a preparation similar to Goulard's *extract*, directing two pounds and four ounces of litharge to be boiled in a gallon of distilled vinegar till reduced to six pints, continually stirring the liquor, and straining it after subsidence. They have likewise given another, similar to Goulard's vegeto-mineral water, in which two drams of this preparation, with as much proof spirit, are mixed with a quart of distilled water. *Aq. litharg. acetat. Ph. Lond.*

Aq. litharg. acetat. comp. Ph. Lond.

POLIUM.

POLEY-MOUNTAIN: a small shrubby plant, with square stalks, oblong woolly leaves set in pairs; and labiated flowers wanting the upper lip and having the lower divided into five segments.

1. *Polium maritimum erectum monspeliacum* C. B. *Teucrium capitatum Linn.* Poley-mountain of Montpellier; with the leaves indented towards the end and joined to the stalk without pedicles, the flowers white and set in roundish spikes on the tops of the branches.

2. *Polium angustifolium creticum* C. B. *Teucrium frutescens, stoechados arabicae folio, & facie Tourn. Teucrium creticum Linn.* Poley-mountain of Candy; with the leaves not indented and set on short pedicles, the flowers standing in loose clusters, each on a separate foot-stalk.

SEVERAL other species, or varieties of *polium*, erect and procumbent, with white, yellow, and purplish

purplish flowers, have been received in the shops. The second above described has been commonly understood as the true officinal sort, and procured dry from the island Candy, of which it is a native: the first, which better bears the winters of our own climate, appears to be of the same quality; and hence the college allow either sort to be taken indifferently.

The leaves and tops of poley-mountain have a moderately strong aromatic smell, and a disagreeable bitter taste: distilled with water, they yield a small quantity of yellowish essential oil, of a pungent taste, in smell less agreeable than the herb itself; the remaining decoction, inspissated, leaves a strongly bitter extract. They stand recommended as corroborants, aperients, and antispasmodics; but are at present scarcely otherwise made use of than as an ingredient in mithridate and theriaca.

POLYGALA.

MILKWORT: a small perennial plant; with the leaves alternate, uncut, and those on the upper parts of the stalks larger than on the lower; the flowers irregular, tubulous, tripetalous, labiated, set in loose spikes on the tops; the cup composed of five leaves, the two larger of which continue after the flower has fallen, and embrace, like wings, a flat bicellular seed-vessel.

1. SENEKA *Lond. & Pharm. Edinb. Polygala (Senega) floribus imberbibus spicatis, caule erecto herbaceo simplicissimo, foliis lato-lanceolatis* Linn. Seneka or Senegaw milkwort, rattlesnake-rooted milkwort: with oblong, somewhat oval, pointed leaves;

POLYGALA.

leaves; upright unbranched ftalks; white flowers; and a varioufly bent and divaricated jointed root, about the thicknefs of the little finger, with a membranous margin running its whole length on each fide, externally of a yellowifh or pale brownifh colour, internally white. It is a native of Virginia, Penfylvania, and Maryland, and cultivated in fome of our gardens.

The root of this plant is faid to be the fpecific of the Senegaw Indians againft the poifon of the bite of the rattlefnake; and to be effectual, when ufed early, even in the middle of the fummer heats, when the poifon is in its higheft vigour, and when all their other antidotes fail. The powder or a decoction of the root is taken internally; and either the powder, or cataplafms made with it, applied to the wound.

Dr. Tennent, obferving that this poifon produces fymptoms refembling thofe of pleurifies and peripneumonies (a difficulty of breathing, cough, fpitting of coagulated blood, and a ftrong quick pulfe) conjectured that it might be ferviceable in thofe diftempers alfo: and from the trials made by the gentlemen of the French academy, as well as thofe mentioned by him, its virtues appear to be great. It made the fizy blood fluid, procured a plentiful fpitting, increafed perfpiration and urine, and fometimes purged or vomited. The ufual dofe was thirty or thirty-five grains of the powder; or three fpoonfuls of a decoction prepared by boiling three ounces of the root in a quart of water till near half the liquor was confumed.

The feneka root has been tried likewife in hydropic cafes, and found in fome inftances to procure a copious evacuation by ftool, urine, and perfpiration, after the common purgatives and diuretics had failed. Monf. Bouvart observes, that though dropfies were thus removed by the feneka, the cure did not feem complete, a fwelling and hardnefs of the fpleen remaining, which fometimes occafioned a frefh extravafation: that the medicine fometimes acts by liquefying the blood and juices, without producing a due difcharge; and that in thefe cafes it does harm unlefs affifted by proper additions, but that fo long as it proves cathartic, nothing is to be feared from it. It is faid to have been found ferviceable alfo in the rheumatifm and gout.

This root, of no remarkable fmell, has a peculiar kind of fubtile pungent penetrating tafte. Its virtue is extracted both by water and fpirit, though the powder in fubftance is fuppofed to be more effectual than either the decoction or tincture. The watery decoction, on firft tafting, feems not unpleafant, but the peculiar pungency of the root quickly difcovers itfelf, fpreading through the fauces, or exciting a copious difcharge of faliva, and frequently, as Linnæus obferves, a fhort cough: thofe to whom I have directed this medicine, have generally found a little Madeira moft effectual for removing its tafte from the mouth, and making it to fit eafy on the ftomach.

Decoct. fe- The Edinburgh college direct a decoction
nekæ *Ph. Ed.* made with one ounce of the root boiled in two pounds of water to fixteen ounces. A tincture of the root in rectified fpirit is of a more fiery pungency, extremely durable in the mouth and throat, and apt to promote vomiting or reaching.

2. POLYGALA: *Polygala vulgaris* C. B. &
Linn. Flos ambarvalis. Common milkwort: with the ftalks procumbent; the lower leaves roundifh, the upper oblong, narrow, and pointed;

POLYPODIUM.

ed; the flowers blue, purplish or red, sometimes white, with a kind of fringed appendix on the lower lip; the roots slender and hard. It grows wild in dry pasture grounds.

The roots of this species are somewhat similar in taste to those of the preceding, but far weaker: they have been found likewise to produce the same effects in pleurisies, in a lower degree. The leaves of the plant are very bitter: Gesner, who from this quality gives it the name of *amarella*, relates, that an infusion of a handful of them in wine is a safe and gentle purgative.

POLYPODIUM.

POLYPODIUM *vulgare* C. B. & Linn. Polypody: a plant with long leaves issuing from the root, divided on both sides, down to the rib, into a number of oblong segments, broadest at the base: it has no stalk, or manifest flower: the seeds are a fine dust, lying on the backs of the leaves, in roundish specks, which are disposed in rows parallel to the rib: the roots are long and slender, of a reddish brown colour on the outside, greenish within, full of small tubercles, which are resembled to the feet of an insect, whence the name of the plant. It grows in the clefts of old walls, rocks, and decayed trees: that produced on the oak has been generally accounted the best, though not sensibly different from the others. It is found green at all seasons of the year.

The leaves of polypody have a weak ungrateful smell, and a nauseous sweet taste, leaving a kind of roughness and slight acrimony in the mouth. They give out their smell and taste, together with a yellow colour, both to water and rectified fpirit: the fpirituous tincture is fweeter than the watery, but in infpiffation its fweetnefs is in great part deftroyed or covered by the other matter; the fpirituous extract, as Cartheufer obferves, being to the tafte only fubaftringent and fubacrid, with very little fweetnefs, while the watery extract feems to retain the full fweetnefs of the polypody. The root is fuppofed to be aperient, refolvent, and expectorant: it was formerly ranked among the purgatives, but operates fo weakly, a decoction of an ounce or two fcarcely moving the belly, that it has long been expunged from that clafs: the prefent practice pays very little regard to it in any intention.

POPULUS.

POPULUS NIGRA C. B. & Linn. Populus nigra five aigeiros J. B. Black poplar: a large tree; with dark green, fomewhat rhomboidal acuminated leaves; producing imperfect flowers, in catkins: in fome of the individuals, called male, the flowers are barren; in others, called female, they are followed by membranous pods, containing a number of feeds winged with down. It is indigenous in watery places, and quick of growth.

The young buds or rudiments of the leaves, which appear in the beginning of the fpring, were formerly employed in an officinal ointment, which received its name from them. At prefent, they are almoft entirely difregarded; though they fhould feem, from their fenfible qualities, to be applicable to purpofes of fome importance. They abound with a yellow, unctuous,

PRUNELLA.

tuous, odorous, balfamic juice, which they readily impart, by maceration or digeftion, to rectified fpirit. The tincture, infpiffated, yields a fragrant refin, fuperiour to many of thofe brought from abroad, and approach to the nature of ftorax.

* A fpecies of poplar growing in Siberia and in North America, called by Linnæus *Populus balfamifera*, is faid to be much more abundant in balfamic juice than the former, infomuch that the buds give it out on mere expreffion (a).

PRUNELLA.

PRUNELLA *five Brunella*. *Prunella major folio non diffecto* C. B. *Confolida minor. Symphitum minus. Prunella vulgaris* Linn. SELF-HEAL: a fmall plant; with fquare ftalks; oval uncut leaves fet in pairs on pedicles; and fhort thick fpikes of purplifh labiated flowers. It is perennial, grows wild in pafture grounds, and flowers in June and July.

THIS herb is recommended as a mild reftringent and vulnerary, in fpittings of blood, and other hemorrhagies and fluxes: and in gargarifms againft aphthæ and inflammations of the fauces. Its virtues do not appear to be very great: to the tafte it difcovers a very flight aufterity or bitterifhnefs; which is more fenfible in the flowery tops than in the leaves; though the latter are generally directed for medicinal ufe.

PRUNUS.

PRUNUS: a tree with pentapetalous white flowers; each of which is fucceeded by a round-

* (a) Bergius, *Mat. Med.* 804.

ish or oval fruit, standing on a long pedicle, composed of a fleshy pulp including a flat stone pointed at both ends.

1. PRUNUS HORTENSIS. *Prunus domestica Linn.* Garden plum tree: without prickles; bearing a sweet fruit. Three sorts of this fruit are ranked among the articles of the materia medica: they are all met with in our gardens, but the shops are supplied with them, moderately dried, from abroad. 1. PRUNA BRIGNOLENSIA. *Pruna ex flavo rufescentia mixti saporis gratissima* C. B. The Brignole plum or prunelloe, brought from Brignole in Provence, of a reddish yellow colour, and a very grateful sweet subacid taste. 2. PRUNA GALLICA *Ph. Lond. & Edinb. Pruna parva dulcia atro-cærulea* C. B. The common or French prunes, called by our gardeners the little black damask plum. 3. PRUNA DAMASCENA: *Pruna magna dulcia atro-cærulea* C. B. Damsons, the larger damask violet plum of Tours: this is seldom kept in the shops, and has been generally supplied by the common prune.

All these fruits possess the same general qualities with the other summer fruits. The prunelloes, in which the sweetness has a greater mixture of acidity than in the other sorts, are used as mild refrigerants in fevers and other hot indispositions, and are sometimes kept in the mouth for alleviating thirst in hydropic cases. The French prunes and damsons are the most emollient, lubricating and laxative: they are taken by themselves for gently loosening the belly in costive habits and where there is a tendency to inflammation: decoctions of them afford an useful basis for laxative or purgative mixtures,

PRUNUS.

mixtures, and the pulp in fubftance for electuaries.

2. PRUNUS SILVESTRIS *Pharm. Lond. & Edinb. & C. B. Acacia Germanorum. Prunus fpinofa Linn.* Black thorn or floe: a prickly bufh, common in hedges, producing auftere fruit, fomewhat fmaller than an ordinary cherry.

THE fruit of the floe bufh is fo harfh and auftere, as not to be eatable till thoroughly mellowed by frofts. The juice expreffed from it while unripe, or before it has been thus mellowed, infpiffated by a gentle heat to drynefs, is called German acacia, and has been ufually fold in the fhops for the Egyptian juice of that name; from which it differs in being harder, heavier, darker coloured, of a fharper or tarter tafte, and more remarkable in this, that it gives out its aftringency in good meafure to rectified fpirit as well as to water, whereas that of the Egyptian acacia is not at all diffoluble in fpirit. A conferve of this fruit is likewife prepared in the fhops, by mixing the pulp with thrice its weight of double-refined fugar, the floes being previoufly fteeped in water, over the fire, with care that they do not burft, till they are fufficiently foftened to admit of the pulp being preffed out through a fieve. In fome places, the unripe floes are dried in an oven, and then fermented with wines or malt liquors, for a reftringent diet drink in alvine and uterine laxities.

Conferv. pruni fylveftrif. *Ph. Lond.* pruni fylveftrium *Ph. Ed.*

The bark, both of the branches and of the roots, is faid to have been given with fuccefs in intermitting fevers, and by fome ftands recommended as equal to the Peruvian bark. It is apparently a ftrong ftyptic; and its ftyptic matter is of that kind which is not eafily extracted by watery menftrua.

The

The flowers, in smell very agreeable, and in taste bitterish, appear to have a laxative virtue, like those of the peach tree or the damask rose. They impregnate water, by distillation, strongly with their fragrance; and give out their active matter, by infusion, both to water and spirit. The watery infusion, sweetened with sugar, or made into a syrup, is said to be a very useful purgative for children.

PSYLLIUM.

PSYLLIUM Pharm. Parif. Pulicaris herba Lugdun. Pfyllium majus erectum C. B. Plantago Pfyllium Linn. Fleawort: an herb of the plantain kind, differing from the common plantains in being annual, and having its stalks branched, with leaves upon them, which are long, slender, and somewhat hairy. It grows wild in the warmer parts of Europe, and is sometimes raised in our gardens. The seeds have been usually brought from the south of France: they are small, smooth, slippery, of a shining brown colour, of an oblong flattish figure supposed to resemble that of a flea, whence the name of the plant.

The seeds of fleawort have a nauseous mucilaginous taste, and no remarkable smell: a dram renders near a pint of water slimy and yellowish: the decoction, inspissated, leaves a strong dark brown mucilage, which impresses on the palate an unpleasant, weak, but penetrating acrimony. This mucilage has been employed chiefly in emollient glysters, in gargarisms for hoarseness and asperity of the fauces, and in external applications for chaps of the lips and inflammations of the eyes. Prosper Alpinus relates,

PTARMICA.

lates, that among the Egyptians, the mucilage or an infusion of the seeds is given internally, in ardent fevers; and that it generally either loosens the belly or promotes sweat. The particular virtue of these seeds, or whatever virtue they may have distinct from that of mucilaginous substances in general, appears to reside in the acrid matter, which may be separated from the mucilaginous by rectified spirit: the seeds, digested in rectified spirit, give out their acrimony and ill taste, and yield afterwards to water an almost insipid mucilage.

PTARMICA.

PTARMICA, *Pseudopyrethrum, Pyrethrum silvestre, Draco silvestris, Tarchon silvestris, Sternutamentoria: Dracunculus pratensis serrato folio* C. B. *Achillæa Ptarmica Linn.* SNEEZEWORT or BASTARD PELLITORY: a plant with long narrow leaves finely serrated about the edges, and radiated discous white flowers set in form of umbels on the tops of the branches. It is perennial, grows wild on heaths and in moist shady grounds, and is found in flower from June to the end of summer.

THE roots of this plant have a hot biting taste, approaching to that of pyrethrum, with which they nearly agree also in their pharmaceutic properties, and to which they have been sometimes substituted in the shops. They are by some recommended internally as a warm stimulant and attenuant; but their principal use is as a masticatory and sternutatory.

PULEGIUM.

PULEGIUM.

PENNYROYAL: a plant of the mint kind; differing from the mints strictly so called, in the flowers being disposed, not in spikes on the tops, but in thick clusters, at distances, round the joints of the stalks; and the upper segment of the flower not being nipped at the extremity.

1. PULEGIUM *Pharm. Lond. & Edinb. Pulegium latifolium C. B. Mentha palustris sive pulegium Pharm. Parif. Pulegium regium Ger. emac. Mentha pulegium Linn.* Common pennyroyal: with somewhat oval obtuse leaves, and trailing stalks, striking root at the joints. It grows wild on moist commons and in watery places, and flowers in June.

2. PULEGIUM ERECTUM: *Pulegium erectum officinarum Dale: Pulegium mas Ger. emac.* Upright pennyroyal; with the stamina standing out from the flowers; said to be a native of Spain, common in our gardens, and usually substituted in our markets to the foregoing species.

3. PULEGIUM CERVINUM: *Pulegium angustifolium C. B. Mentha aquatica satureiae folio Tourn. Mentha cervina Linn.* Harts pennyroyal, with small oblong narrow leaves; said to grow wild about Montpellier.

ALL the pennyroyals are warm pungent herbs, somewhat similar to mint, but more acrid and less agreeable both in smell and taste, less proper in common nauseae and weakness of the stomach,

mach, more efficacious as warm carminatives and deobstruents in hysteric cases and disorders of the breast: the last species is the strongest, though least ungrateful, of the three. Their active principle is an essential oil; of a more volatile nature than that of mint, coming over hastily with water at the beginning of the distillation, and rising also in great part with highly-rectified spirit; in taste very pungent, and of a strong smell; when newly drawn, of a yellowish colour with a cast of green; by age turning brownish. The oil, and a simple† and spirituous‡ water strongly impregnated with it, by drawing off a gallon of water or proof spirit from a pound and a half of the dry leaves, are kept in the shops.

Ol. pulegii essent. *Ph. Lond.*

Aq. pulegii *Ph. Lond.* & *Ed.* † Spirit. pulegii *Ph. Lond.* ‡

PULMONARIA.

PULMONARIA MACULOSA. Symphytum maculosum five pulmonaria latifolia C. B. *Pulmonaria officinalis Linn.* SPOTTED LUNGWORT, JERUSALEM COWSLIPS, JERUSALEM SAGE: a hairy scabrous plant, with the leaves of a dark brownish green colour on the upper side and spotted for the most part with white, underneath of a paler green, the lower oval and set on broad pedicles, those on the stalks narrower, long-pointed, set alternately, without pedicles: the flowers are monopetalous, cut into five sections, of a purple or blue colour, and sometimes white, followed each by four seeds inclosed in the cup. It is perennial, grows wild in several parts of Europe, and flowers in our gardens in April and May.

The leaves of pulmonaria, recommended in hemoptoës, tickling coughs, asperities of the fauces, &c. appear to be of little medicinal virtue.

tue. The dried leaves have hardly any smell; and their taste is just perceptibly mucilaginous, sweetish, and roughish. They seem to be nearly of the same nature with the *adianthum* and *trichomanes*.

*PULSATILLA.

PULSATILLA NIGRICANS Stærck, Pharm. Edinb. *Pulsatilla flore minore nigricante* C. B. *Anemone pratensis* Linn. A species of anemone, much resembling the *pulsatilla vulgaris*, or pasque flower, but its flower is less, and of a darker hue. It is a native of the south of Germany, and other neighbouring countries.

All the anemonies have a considerable degree of acrimony; but this seems to possess the largest share. The whole plant when chewed impresses the tongue with a sharp, burning, durable taste. The root is milder than the other parts. On distilling the plant with water, the liquor which comes over is strongly impregnated with its virtues; and the remaining extract is also considerably active.

Dr. Stœrck of Vienna, to whom the introduction of so many of the more powerful vegetables is owing, has likewise recommended this to the medical practitioner. From numerous trials, he celebrates its efficacy in various chronic diseases of the eye; in venereal nodes and nocturnal pains; in foul ulcers with caries; in serpigo; and suppressed menses. He relates instances of its curing blindness of many years continuance, by dissipating and dissolving films and obscurities of the cornea. In these cases, its good effects were first indicated by considerable pain excited in the eye. The sensible operation of the medicine was nausea and vomiting,

PYRETHRUM.

vomiting, particularly when the diftilled water was ufed; an increafed flow of urine; and sometimes gripes and loofenefs; with increafed pain at firft in the affected part. From all thefe circumftances, the pulfatilla feems to be endued with very active and penetrating powers, yet fuch as may be employed with perfect fafety if proper caution be ufed. The dofe of the diftilled water to adults is about half an ounce, twice or thrice a day; of the extract, reduced to powder with the addition of fugar, five or fix grains. Bergius mentions having given the extract copioufly, efpecially in difeafes of the eyes, but without any effect *(a)*.

The Edinburgh college had adopted the diftilled water of pulfatilla, but has now changed it for the extract.
<small>Extract. folior. pulfa- tillæ nigricantis *Ph. Ed.*</small>

PYRETHRUM.

PYRETHRUM Pharm. *Lond. & Edinb.* Pyrethrum flore bellidis C. B: Chamæmelum fpeciofo flore, radice longa fervida *Shaw afr.* Dentaria, herba falivaris, & pos alexandrinus quibufdam. Anthemis Pyrethrum Linn. PELLITORY of SPAIN: a trailing perennial plant; with finely divided leaves fomewhat like thofe of camomile or fennel; and naked thick ftalks, bearing each a large flower, which confifts of a yellow difk furrounded with petala of a pure white colour on the upper fide, and a fine purple underneath: the root, which finks deep in the ground like a carrot, is of a brownifh colour on the outfide and whitifh within. It is a native of the warmer climates, but bears the cold of our own, and often produces flowers in fucceffion from January to May: the roots alfo, as Parkinfon

(a) Mat. Med. 491.

obferves,

obferves, grow larger with us, than thofe which the fhops are fupplied with from abroad.

PELLITORY root has a very hot pungent tafte, without any fenfible fmell. Its pungency refides in a refinous matter, of the more fixt kind; being extracted completely by rectified fpirit, and only in fmall part by water; and not being carried off, in evaporation or diftillation, by either menftruum. The fpirituous extract is extremely fiery, but in fmall quantity, fcarcely amounting to one twentieth of the weight of the root. The watery infufion is naufeous, but fcarcely difcovers any acrimony till concentrated by infpiffation; when reduced to the confiftence of an extract, it proves confiderably pungent: the quantity of this extract is commonly five or fix times as large as that of the fpirituous. The root remaining, after the action of water, yields ftill with rectified fpirit a very fiery extract; whereas that, which has been digefted in fpirit, yields with water only an infipid mucilaginous fubftance.

The principal ufe of pyrethrum, in the prefent practice, is as a mafticatory, for ftimulating the falival glands, &c. and evacuating vifcid humours from the head and parts adjacent: by this means it frequently relieves tooth-achs, fome kinds of head-achs, lethargic complaints, and paralyfes of the tongue. It has fometimes likewife been given internally, from a few grains to a fcruple, as a hot ftimulant and attenuant, in paralytic and rheumatic diforders.

PYRITES.

MARCASSITA Pharm. Parif. PYRITES or MARCASITE: a hard foffil; ftriking fire with fteel,

PYRITES.

steel, copiously, and in large sparks; becoming vitriolic, either by simple exposure to the air, or by calcination and subsequent exposure.

This mineral varies extremely in its appearances. It is found of a bright brass yellow, of a greenish, of a grey or whitish colour, and of different intermediate or mixt shades: in masses, rarely of any great size, globular, oblong and flattish, cubical, octoedral, dodecaedral; sometimes covered with a coat or crust, but oftener bare; internally sometimes striated, and sometimes of an even and simple structure *(a)*. It is met with in different places of this kingdom, and in most parts of the world; on the surface of the earth, on the sea shores, in clay pits, embedded in earthy and stony bodies of various kinds.

The pyritæ consist, in general, of sulphur, iron, and unmetallic earth: in some, a little copper is joined to the iron; and in some, copper is the prevailing metal. In some, particularly the yellow kind, the quantity of sulphur is large: in others, particularly the white, both the sulphur and metal are in small proportion.

If artificial mixtures of sulphur with iron or copper be gently calcined, the inflammable principle of the sulphur exhales, and its acid remains united with the metal, forming therewith a saline vitriolic compound: a mixture of iron filings and sulphur, moistened with water, suffers a like change without external heat, and if the quantity is large, bursts spontaneously into fire. A resolution of the same kind happens in the natural pyritæ on exposing them to the air and rain; provided, where they are very sul-

(a) See Henckel's *Pyritologia oder kiefs-historie.*

phureous,

phureous, a part of the sulphur be previously diffipated by calcination. On this expofure, they all become powdery and acquire a vitriolic tafte, the ferrugineous much more eafily than thofe which have any admixture of copper: fome fhoot out efflorefcences of vitriol upon the furface: from others, the faline matter, wafhed off by rain, is found to confift chiefly of the fulphureous or vitriolic acid. If the pyritæ, even fuch as have the leaft fulphureous and metallic impregnation, as thofe from which the Englifh vitriol is made, be laid in large heaps, they grow hot, and take fire, and emit, during the burning, ftrong diffufive fulphureous vapours. *(a)*.

The pyritæ, in fubftance, are never ufed medicinally; but in their products they are very important. It is from thefe, that common fulphur is extracted, in Sweden and Saxony; that the native vitriols are produced in caverns of the earth or on its furface; that the greateft quantities of artificial vitriol are prepared; and that the chalybeate mineral waters are fuppofed to receive their impregnation: fee the refpective articles.

* *QUASSIA.*

QUASSIA Pharm. Lond. & Edinb. Lignum Quaffiæ Amænit. Acad. vol. vi. *Bois de Coiffi Fermin Surinam.* QUASSY ROOT: the woody root of a tree growing in Surinam, called by Linnæus, *Quaffia amara,* of the clafs and order decandria monogynia in his fyftem. This root is as thick as a man's arm. Its wood is whitifh, hard, folid, and tough, becoming yellowifh on

(a) Dr. Slare, *Philofophical tranfactions,* numb. 213.

expofure

exposure to the air. It is covered by a thin, grey, fissured, and brittle bark.

Quassi root has no sensible odour. Its taste is that of a pure bitter, more intense and durable than that of almost any other known substance. Its watery infusions and decoctions, and its spirituous tinctures, are all almost equally bitter, of a pale yellowish hue, which is not blackened by the addition of martial vitriol. The watery extract is from a sixth to a ninth of the weight of the wood; the spirituous about a twenty-fourth. The bark of the root is reckoned in Surinam more powerful than the wood. The flowers also are a strong bitter.

The medical use of the quassi has been a considerable time known in Surinam. The flowers were long ago employed by the natives as an excellent stomachic. The root was a secret remedy used by a negro, named Quassi, in the fatal fevers of that country, from whom it was purchased by Dan. Rolander, a Swede, who returned from thence in 1756. Some specimens of the wood and of the fructification were, in 1761, presented by M. Dahlberg to the celebrated Linnæus; who drew up a botanical description of the plant, with an account of its virtues, and published it in the sixth vol. of the *Amæni. Acad.* A confirmation of its medical powers appeared in a letter from Mr. Farley, a practitioner in Antigua, printed in the *Phil. Transact.* vol. lviii. He found it remarkably efficacious in suppressing vomitings, stopping a tendency to putrefaction, and removing fevers. It seemed capable of producing all the good effects of Peruvian bark, without heating. Some further experiments on the quassi are contained

contained in a late medical thesis by Dr. Ebeling. He confirms the general account of its virtues, with this additional circumstance, that though its general antiseptic powers were inferiour to those of Peruvian bark, yet it preserved bile a longer time from putrefaction. In this circumstance it agrees with another pure bitter, the columbo root.

From these relations, the quassi appears to be a valuable addition to our tonic remedies, and has therefore obtained a place in the last Edinburgh and other pharmacopœias. It may be used either in infusion, or extract: the latter, made into pills, on account of the intense bitterness of the drug, is preferable for delicate stomachs.

QUERCUS.

QUERCUS Lond. & Pharm. Edinb. Quercus cum longis pediculis C. B. *Quercus Robur Linn.* OAK: a large tree, with oblong leaves, widening from the bottom to the extremity, and sinuated or bluntly indented about the edges: the fruit is an acorn, or kernel with a coriaceous covering, inclosed at bottom in a scaly cup. It is a common forest tree in most parts of Europe.

THE bark of the oak is a strong astringent, accompanied with a moderate bitterness, but no remarkable smell or particular flavour: with solution of chalybeate vitriol it strikes an inky blackness. It is said to have been employed with success, not only for restraining hæmorrhagies, and other immoderate evacuations, but likewise in intermitting fevers and in gleeting gangrenous wounds and ulcers; in which
cases,

cafes, an extract made from it is faid by fome to be equal to that of the Peruvian bark. A decoction of it ufed as a fomentation is faid to have cured a procidentia recti *(a)*. It gives out its virtue both to water and rectified fpirit.

QUERCUS MARINA.

QUERCUS MARINA five fucus veficulofus: Fucus maritimus five quercus maritima veficulas habens C. B. Fucus veficulofus Linn. SEA WRACK or SEA OAK: a 'foft, very flippery, marine plant; common upon rocks that are left dry at the ebb tide; with the leaves fomewhat refembling in fhape thofe of the oak tree; the ftalks running along the middle of the leaves, and terminated by warty bladders containing either air or a flippery fluid. The veficles begin in March to fill with a thin juice; and about the end of July they burft, and difcharge a matter as thick as honey.

DR. RUSSEL relates, that he found this plant an ufeful affiftant to fea water in the cure of diforders of the glands: that he gave it in powder to the quantity of a dram, and that in large dofes it naufeated the ftomach: that by burning in the open air it was reduced into a black faline powder, which feemed, as an internal medicine, greatly to excel the officinal burnt fponge; which was ufed with benefit, as a dentifrice, for correcting laxities of the gums; and which fhewed a notable degree of detergent virtue by its effect in cleaning the teeth: that the juice of the veficles, after ftanding to putrefy, yielded, on evaporation, an acrid pungent falt, amount-

Æthiops vegetabilis *Dr. Ruffel.*

(a) Med. Eff. Edinb. ii. 257.

ing to about a fcruple from two fpoonfuls: that the putrefied juice, applied to the fkin, finks in immediately, excites a flight fenfe of pungency, and deterges like a folution of foap: that one of the beft applications for difcuffing hardnefs, particularly in the decline of glandular fwellings, is a mixture of two pounds of the juicy veficles, gathered in July, with a quart of fea water, kept in a glafs veffel for ten or fifteen days, till the liquor comes near to the confiftence of very thin honey: the parts affected are to be rubbed with the ftrained liquor twice or thrice a day, and afterwards wafhed clean with fea water.

* RADIX LOPEZIANA.

RADIX INDICA LOPEZIANA *Pharm. Edinb.* Radix Indica a Joanne Lopez denominata *Gaubii Adverfar.* Cap. vi. *Rais di Juan Lopez Lufitanis.* The root of an unknown tree, growing, as fome affert, at Goa, as others fuppofe, in Malacca, from whence it is fometimes brought to Batavia. It is met with in pieces of different thicknefs, fome, at leaft, of two inches diameter. The woody part is whitifh, and very light; fofter, more fpongy, and whiter next the bark, including a denfer fomewhat reddifh medullary part. The bark is rough, wrinkled, brown, foft, and as it were woolly, pretty thick, covered with a thin paler cuticle.

Neither the woody nor cortical part has any remarkable fmell or tafte, nor any appearance of refinous matter. On boiling in water, no odour is emitted; and the ftrained liquor, which is of a yellow hue, is almoft infipid, only impreffing the tongue with a very light obfcure bitterifhnefs;

RADIX LOPEZIANA.

bitterifhnefs; and without vifcidity. The extract obtained by evaporating the decoction is equally void of fenfible activity. Rectified fpirit is tinged by the root of a brown colour, but acquires no particular tafte. After drawing off the fpirit from the tincture, a matter remains refembling balfam, which bubbles and inflames in the fire, and has a bitterifh tafte, like that of opium.

Though the preceding examination of this root is not favourable to the opinion of its medical powers, yet it is regarded in the Eaft Indies as a medicine of extraordinary efficacy in diarrhœas; and the learned Gaubius, in his *Adverfaria*, has publifhed an account of fome experiments made with it, which in fome degree confirm its reputation. From his own trials, and thofe of his friends, it appeared moft remarkably effectual in ftopping colliquative diarrhœas which had refifted the ufual remedies. Thofe attending the laft ftage of confumptions were particularly relieved by its ufe. It feemed to act not by any aftringent power, but by a faculty of reftraining and appeafing fpafmodic and inordinate motions in the inteftines. Gaubius compares its action to that of fimarouba, but thinks it more efficacious than this medicine.

The mode of exhibiting it in India, is to levigate the root with water on a porphyry till reduced to a fine pulp. In Europe the powder of it has been given with any proper vehicle, in dofes from fifteen to thirty grains, repeated three or four times a day: one practitioner found a tincture of it in common fpirits equally effectual with the root in fubftance. Of this, a teafpoonful was given thrice a day in red wine. The colleges of Edinburgh and Brunfwick have received

received this root into their catalogues; but it is scarcely yet to be met with in the shops.

RANUNCULUS.

CROWFOOT: a plant with pentapetalous flowers set in five-leaved cups; followed each by a round cluster of naked feeds. It is perennial.

1. RANUNCULUS. *Ranunculus pratensis radice verticilli modo rotunda* C. B. *Ranunculus bulbosus Linn.* Bulbous crowfoot, butter flower, gold cup: with a round tuberous root about the size of an olive; the leaves divided commonly into three segments, and these further subdivided; the stalks erect: the flowers of a bright glossy yellow, and their cups turned downwards. It is common in pasture grounds, and flowers in May.

2. FLAMMULA: *Ranunculus longifolius palustris minor* C. B. *Ranunculus Flammula Linn.* Smaller water crowfoot or spearwort: with fibrous roots, long narrow leaves acuminated at both ends, and leaning or procumbent stalks. It grows in watery places or moist meadows, and flowers in June.

The roots and leaves of these plants are of no considerable smell, but in taste highly acrid and fiery. Taken internally, they appear to be deleterious, even when so far freed from the caustic matter, by boiling in water, as to discover no ill quality to the palate. The effluvia likewise even of the less acrid species or varieties cultivated in gardens, when freely inspired, have occasioned headachs, anxieties, vomitings, and spasmodic

RAPHANUS.

fpafmodic symptoms. The leaves and roots, applied externally, inflame and exulcerate or veficate the part, and are liable to affect alfo the adjacent parts to a confiderable extent*(a)*: they have fometimes, particularly among empyrics and the common people, fupplied the place of the far fafer and not lefs effectual veficatory, cantharides, for procuring an ulcer and difcharge of ferum, in fciaticas and fome fixt pains of the head. Their pungency is diminifhed by drying, and by long keeping feems to be diffipated or deftroyed.

RAPHANUS.

RAPHANUS RUSTICANUS Pharm. Lond. & Edinb. & C. B. *Cochlearia folio cubitali Tourn. Cochlearia Armoracia Linn.* HORSERADISH: a plant refembling fcurvygrafs in the flowers and feeds, but differing in the leaves being very large and long, and indented about the edges. It is fometimes found wild about the fides of ditches and rivulets, but for medicinal and culinary ufes is cultivated in gardens. It is perennial, flowers in June, rarely perfects its feeds, and is propagated from tranfverfe cuttings of the roots.

HORSERADISH root affects the organs both of tafte and fmell with a quick penetrating pungency: it neverthelefs contains in certain veffels a fweet juice, which fometimes exudes in little drops upon the furface. Its pungent matter is of a very volatile kind; being totally diffipated in drying, and carried off in evaporation or

(a) Willis, *Pharmaceutice rationalis*, P. II. *fect.* iii. *cap.* 3.

diftillation

diftillation both by water and rectified fpirit: as the pungency exhales, the fweet matter of the root becomes more fenfible, though this alfo is in great meafure diffipated or deftroyed. It impregnates both water and fpirit, by infufion or by diftillation, very richly with its active matter: in diftillation with water it yields a fmall quantity of effential oil exceedingly penetrating and pungent. This root appears therefore to agree with fcurvygrafs and creffes, and to differ from muftard feed to which it is by fome refembled, in the volatility of its pungent matter, and its folubility in fpirit.

Horferadifh is a moderately ftimulating, aperient, and antifeptic medicine: it fenfibly promotes perfpiration, urine, and the expectoration of vifcid phlegm, and excites appetite when the ftomach is weakened or relaxed, without being fo liable to produce immoderate heat, or inflammatory fymptoms, as the ftimulants of the aromatic kind. It is principally ufed in paralytic and rheumatic complaints, in fcurvies and fcorbutic impurities of the humours, in cachectic diforders, and in dropfies, particularly in thofe which follow intermitting fevers. Taken in confiderable quantities, it provokes vomiting.

RAPUM.

RAPA *fativa rotunda* C. B. *Braffica Rapa Linn.* TURNEP; a plant with a round root; jagged leaves, rude to the touch; tetrapetalous flowers, commonly yellow; and fmall round fmooth reddifh or blackifh feeds lodged in long pods. The garden turnep is fuppofed to be a variety produced by culture from the fmaller fort which grows wild in fandy grounds in fome parts of England. It is biennial.

TURNEPS

RHABARBARUM.

Turneps are accounted a falubrious food; demulcent, detergent, fomewhat laxative and diuretic; but liable in weak ftomachs to produce flatulencies, and prove difficult of digeftion: the liquor preffed out from them after boiling, is fometimes taken medicinally, in coughs and diforders of the breaft. The feeds have been accounted alexipharmac or diaphoretic: they have no fmell, but difcover to the tafte a mild acrimony, feemingly of the fame nature with that of muftard feed, though far weaker.

RHABARBARUM.

RHUBARB: a plant with large dock-like leaves, among which arifes a fingle thick ftalk bearing loofe clufters of naked monopetalous bell-fhaped flowers divided into fix fegments: each flower contains nine ftamina (whereof the docks ftrictly fo called have but fix), and is followed by a triangular feed furrounded about the edges with a leafy margin.

1. Rhabarbarum *Pharm. Lond. Rheum Pharm. Edinb. Lapathum orientale folio latiffimo undulato & mucronato Mill. dict. Rheum undulatum foliis fubvillofis, petiolis aequalibus Linn.* Rhubarb: with the leaves fomewhat heart-fhaped, acuminated, and flightly hairy, and the pedicles plano-convex. It is a native of China and Siberia, and has lately been raifed in fome of our gardens, where it is found to grow with vigour in the open ground *(a)*.

Two

(a) The plant above defcribed is that which is generally reckoned the true rhubarb plant, having been produced from the feeds, fent from Ruffia, as thofe of the true rhubarb to Juffieu at Paris, Rand at Chelfea, and Linnæus at Upfal.

Two sorts of rhubarb roots are met with in the shops. The first is imported from Turkey and Russia, in roundish pieces, freed from the bark, with a hole through the middle of each, externally of a yellow colour, internally variegated with lively reddish streaks. The other, which is less esteemed, comes immediately from the East Indies, in longish pieces, harder, heavier, and more compact than the foregoing. The first sort, unless kept very dry, is apt to grow mouldy and worm-eaten: the second is less subject to these inconveniencies. Some of the more industrious artists are said to fill up the worm holes with certain mixtures, and to colour the outside of the damaged pieces with powder of the finer sorts of rhubarb, and sometimes with cheaper materials. The marks of the goodness of rhubarb are, the liveliness of its colour when cut; its being firm and solid, but not flinty or hard; its being easily pulverable, and appearing when powdered of a fine bright yellow colour; its imparting to the spittle, on being chewed, a deep saffron tinge, and not proving slimy or mucilaginous in the mouth. Its taste is subacrid, bitterish, and somewhat styptic; the smell, lightly aromatic.

Rhubarb is a mild cathartic, and commonly looked upon as one of the safest and most innocent of the substances of this class. Besides

---

Upsal. Dr. Hope received lately rhubarb seeds from the same country, which being sown in the open ground at Edinburgh, produced a different species, *Rheum palmatum Linnæi*, with the leaves deeply cut into pointed segments. He observes that the root of this plant, though taken up too young, and at an improper season, viz. in July, agreed perfectly with the best foreign rhubarb, in colour, smell, taste, and purgative quality. See *Philosoph. Transact.* vol. lv. for the year 1765.——Perhaps the roots of both species may be of the same quality, and taken promiscuously.

its

RHABARBARUM.

its purgative virtue, it has a mild aftringent one, difcoverable by the tafte, and by its ftriking an inky blacknefs with chalybeate folutions: hence it is found to ftrengthen the tone of the ftomach and inteftines, to leave the belly coftive, and to be one of the moft ufeful purgatives in diarrhœas, dyfenteries, and all diforders proceeding from a debility and laxity of the fibres: it is frequently indeed given with a view rather to this ftomachic and corroborating virtue, than to its producing any confiderable evacuation. It tinges the urine of a high yellow colour.

Rhubarb in fubftance purges more effectually than any preparation of it: the dofe is from a fcruple to a dram. By roafting it with a gentle heat, till it becomes eafily friable, its cathartic power is diminifhed, and its aftringency fuppofed to be increafed. *Rhabarb. torrefactum.*

In its habitude to menftrua, it differs remarkably from moft of the other cathartic drugs, its purgative virtue being extracted far more perfectly by water than by rectified fpirit: the root remaining after the action of water is almoft, if not wholly, inactive; whereas, after repeated digeftion in fpirit, it proves ftill very confiderably purgative: the colour of both tinctures is a fine deep yellow, that of the fpirituous paleft; when the rhubarb has given out to fpirit all that this menftruum can extract, it ftill imparts a deep colour, as well as a purgative impregnation, to water. The watery infufion, in being infpiffated by a gentle heat, has its virtue fo much diminifhed, that a dram of the extract is faid to have fcarcely any greater effect than a fcruple of the root in fubftance: the fpirituous tincture loofes lefs; half a dram of this extract proving moderately purgative, though fcarcely more fo than an equal quantity

of

of the powder. The fpirituous extract diffolves almoft wholly in water; and hence the tincture does not, like the fpirituous infufions of moft other vegetables, turn milky on being mixed with aqueous liquors: of the watery extract, fcarce above one fourth is diffolved by rectified fpirit, and the part that does not diffolve proves more purgative than that which does.

Infuf. rhei
Ph. Ed.

* A watery infufion is directed in the Edinburgh pharmacopœia, made by infufing half an ounce of rhubarb for a night in eight ounces of boiling water, and adding to the ftrained liquor one ounce of fpirituous cinnamon water. The

Tinct. rhabarb. comp.
Ph. Lond.

London has a more compound and warmer infufion, in which an ounce of rhubarb, with a dram of ginger, a dram of faffron, and two drams of liquorice root, are digefted for a fortnight in half a pint of water and fix ounces of proof fpirit.

Tinctures of this root are drawn in the fhops with proof fpirit and with mountain wine. The London college directs an ounce of rhubarb with two drams of cardamom feeds, and one of

† Tinct. rhabarb. vinum.
Rhabarb.
Ph. Lond.

faffron *(a)*, for a pint of each tincture†: that of Edinburgh, orders for the vinous tincture, two ounces of rhubarb and one dram of canella alba to be infufed in fifteen ounces of mountain

‡ Vinum rhei
Ph. Ed.

wine, and two of proof fpirit‡; for the fimple fpirituous tincture, three ounces of rhubarb, and half an ounce of leffer cardamom feeds, to

‖ Tinct. rhei.

two pounds and a half of proof fpirit‖; in which, fometimes are diffolved, four ounces of

§ Tinct. rhei dulcis.

fugar candy§; and a compound tincture, com-

(a) Saffron does not appear to be a vety proper ingredient in thefe preparations, as it renders the tafte rather more unpleafant; nor indeed does rhubarb feem, for general ufe, to want any aromatic addition.

pofed

posed of two ounces of rhubarb, half an ounce of gentian, one dram of snakeroot, and two pounds and a half of proof spirit ††. These preparations are used chiefly as mildly laxative corroborants, in weakness of the stomach, indigestion, diarrhœas, colicky and other like complaints. The last tincture is, in many cases, an useful assistant to the Peruvian bark in the cure of intermittents.

†† Tinct. rhei amara *Ph. Ed.*

The Turkey rhubarb is, among us, universally preferred to the East India sort, though this last appears to be for some purposes at least equal to the other. It is manifestly more astringent, but has somewhat less of an aromatic flavour. Tinctures made from both with equal quantities of rectified spirit, have nearly the same taste: on drawing off the menstrua, the extract left by the tincture of the East India rhubarb proves in taste considerably stronger than the other. Both sorts appear to be the produce of the same climate, and the roots of the same species of plant, taken up probably at a different season, or cured in a different manner.

2. RHAPONTICUM *Pharm. Parif.* Rhabarbarum dioscoridis & antiquorum. *Tourn.* Rhaponticum folio lapathi majoris glabro *C. B.* Rheum Rhaponticum *Linn.* Rhapontic: with smooth roundish leaves, and somewhat channelled pedicles. It grows wild on the mountain Rhodope in Thrace, from whence it was brought into Europe by Alpinus about the year 1610: it bears the hardest winters of this climate.

The root of this plant, which appears to have been the true rhubarb of the ancients, is by some confounded with the modern rhubarb, though considerably different from that root in appearance as well as in quality. The rhapontic

tic is of a dusky colour on the surface, and a loose spongy texture; more astringent than rhubarb, and less purgative: in this last intention two or three drams are required for a dose.

*RHODODENDRON.

RHODODENDRON CHRYSANTHE-MUM, Pharm. Edinb. Rhododendron Chrysanthum Linn. syst. veg. ed. xiv. This plant, which is a new species of the *rhododendron* of Linnæus, discovered by Professor Pallas, is a shrub growing near the tops of the high mountains named Sajanes, in the neighbourhood of the river Jenisea in Siberia.

It is called by the natives of the place *chei*, or tea, from their commonly drinking a weak infusion of it, as we do the Chinese plant of that name. A stronger preparation of it is, however, used by them as a powerful medicine in arthritic and rheumatic disorders. For this purpose, they take about two drams of the dried shrub, stalk and leaves together, and infuse it in nine or ten ounces of boiling water for a night, in the heat of an oven. This is drunk next morning for a dose; which occasions heat, a degree of intoxication, with a singular uneasy kind of sensation, and a sort of *vermiculation* in the affected parts. The patient is not permitted to quench the thirst this medicine occasions, as liquids, especially cold water, would produce vomiting, and diminish the effect of the remedy. In a few hours, all disagreeable symptoms go off, commonly with two or three stools; and the patient finds his disease greatly relieved. A repetition of the dose twice or thrice generally completes the cure. This is the substance of the account given in a letter from
Dr.

RICINUS.

Dr. Guthrie, of Petersburgh, to Dr. Duncan, *Med. Comment.* vol. v. p. 434.

The rhododendron has been since tried by Dr. Home in the infirmary at Edinburgh; and the result of his trials, as published in his *Clinical Cases and Experiments*, is, that it is a very powerful sedative, remarkably diminishing the frequency of the pulse; but that it was not peculiarly efficacious in removing the acute rheumatism.

RICINUS.

RICINUS & ricinoides: large plants, with small flowers in clusters, and the fruit growing at a little distance from, or succeeding only a few of, the flowers: the fruit consists of three capsules, containing each a single seed, flatted on one side, generally about the size of a small bean, composed of a thin skin or shell including an oily kernel.

1. RICINUS *Pharm. Lond. & Edinb.* PALMACHRISTI *Pharm. Paris. Cataputia major, cherva major, kiki, & granum regium quibusdam: Ricinus vulgaris* C. B. *Ricinus communis Linn.* Palma-christi, Mexico seed: with the fruit triangular, the seed furnished with a little knob at one end, externally variegated with blackish and whitish streaks, resembling both in shape and colour the insect *ricinus* or tick.

2. RICINOIDES, *seu pineus purgans, vel pinhones indici Pharm. Paris. Carcas, nux barbadensis, & faba purgatrix quibusdam: Ricinus americanus major femine nigro* C. B. *Jatropha Curcas Linn.* Barbadoes nut: with an oval walnut-like fruit, and oblong black seeds.

3. AVELLANA

3. Avellana purgatrix C. B. *Nuces purgantes* Ger. *Jatropha multifida Linn.* Purging nut: with oval fruit, and roundish, somewhat triangular, pale brownish seeds.

4. Tiglium, *grana tiglia*. *Pinus indica nucleo purgante, & lignum moluccense foliis malvæ, fructu avellanæ minore, cortice molliore & nigricante, pavana incolis* C. B. *Croton Tiglium Linn.* Grana tilia: with roundish fruit, and dark greyish seeds in shape nearly like those of the first species.

The first of these plants is said to be found wild in some of the southern parts of Europe: it is biennial. The others are middling sized trees, natives of America and the East Indies, from whence the seeds are sometimes brought to us.

The two first of these seeds are sweetish, nauseous, and acrid: the third has scarcely any acrimony, and tastes nearly like almonds: the fourth is intensely hot and acrimonious. They are all strong evacuants, operating, in doses of a few grains, both upwards and downwards; the sweet species not excepted. The grana tilia are the most violent, too much so to be taken with any tolerable safety; and indeed they all appear too drastic to be ventured on in substance.

They yield upon expression a considerable quantity of oil, impregnated more or less with the taste and the purgative quality of the seeds: of the oil of the grana tilia, Geoffroy limits the dose to one grain, which is probably an error of the press for one dram: that of the Barbadoes nut is said to be taken in America in larger quantities,

quantities, and to purge without much inconvenience.

The oil of the palma-christi, vulgarly called in America castor oil, has been often given from two to four spoonfuls, and found to act as a sufficient mild laxative: it is said to be particularly useful in the dry-belly-ach, and in other disorders where irritating purgatives cannot be borne, and where the common laxatives, on account of the large dose in which they require to be given, are apt to be rejected by the stomach. From such trials as I have made of this medicine, it did not seem to have any peculiar good qualities, or to produce any other effects than may be equally obtained by combining the more common purgatives, as tincture or infusion of sena, with common oil.—It is said that some, or all, of the above oils act as purgatives, when applied externally to the umbilical region.

The wood and leaves of the plants are likewise strong cathartics: Hermann relates, that the wood of the tilia, called *panava* or *pavana*, operates violently, when fresh, in the dose of a scruple or half a dram: that when dried and long kept, it is given to the quantity of a whole dram as a purgative, and in smaller doses as a sudorific. Among us, all these substances are entire strangers to practice (except that the oil of the first species has of late been sometimes made use of;) and, so far as can be judged from the accounts given of them, they have little claim to be received.

* Since the above was written, the *Oleum Ricini* has increased in reputation, and is now a common remedy in calculous, nephritic, bilious, and various other cases. It is, however, liable to the inconvenience of much uncertainty

in its operation, owing, probably, to the different modes of preparing the oil, or its different degrees of genuineness. One spoonful, or half an ounce, is the dose usually begun with.

Dr. Wright, in a paper containing an account of the medicinal plants growing in Jamaica, printed in the *London Medical Journal for* 1787, *part* iii. gives the following information concerning this oil.

" Castor oil is obtained either by expression or decoction. The first method is practised in England; the latter in Jamaica. It is best prepared in the following manner. A large iron pot or boiler is first prepared, and half filled with water. The nuts are then beaten in parcels in deep wooden mortars, and, after a quantity is beaten, it is thrown into the iron vessel. The fire is then lighted, and the liquor is gently boiled for two hours, and kept constantly stirred. About this time the oil begins to separate, and swims on the top, mixed with a white froth, and is skimmed off till no more rises. The skimmings are heated in a small iron pot, and strained through a cloth. When cold, it is put up in jars or bottles for use. Castor oil thus made is clear and well flavoured, and, if put into proper bottles, will keep sweet for years. The expressed castor oil soon turns rancid, because the mucilaginous and acrid parts of the nut are squeezed out with the oil.

As a medicine, castor oil purges without stimulus, and is so mild as to be given to infants soon after birth to purge off the meconium. All oils are noxious to insects, but the castor oil kills and expels them. It is generally given as a purge after using the cabbage-bath some days. In constipation and belly-ach this oil

oil is used with remarkable succefs. It fits well on the ftomach, allays the fpafm, and brings about a plentiful evacuation by ftool, efpecially if at the fame time fomentations, or the warm bath, are ufed."

ROSA.

ROSE: a prickly bufh; with oval ferrated leaves, fet in pairs along a middle rib, which is terminated by an odd one; producing large elegant flowers, whofe cup is divided into five long fegments, with a knob at the bottom, which becomes an umbilicated foft fruit full of hairy feeds.

1. ROSA DAMASCENA *Pharm. Lond.* *Rofa pallida Pharm. Edinb.* *Rofa purpurea* C. B. *Rofa centifolia Linn.* The damafk rofe: with double flowers, of the fine pale red called from them rofe-colour.

The pleafant fmell of damafk rofes is of a lefs perifhable kind than that of many other odoriferous flowers, not being much diminifhed in drying, nor foon diffipated in keeping. They impart their odorous matter to watery liquors both by infufion and diftillation: fix pounds of the frefh rofes impregnate, by diftillation, a gallon or more of water ftrongly with their fine flavour. On diftilling large quantities, there feparates from the watery fluid a fmall portion of a fragrant butyraceous oil, which liquefies by heat and appears yellow, but concretes in the cold into a white mafs: an hundred pounds of the flowers, according to the experiments of Tachenius and Hoffman, afford fcarcely half an ounce of oil. The oil and water, ufed chiefly as perfumes and flavour-

Aq. rofæ *Ph. Lond.*
— rofar. pallid. *Ph. Ed.*

ing materials, are recommended by Hoffman as excellent cordials, for raising the strength and spirits, and allaying pain. They appear to be of a very mild nature, and not liable to irritate or heat the constitution; even the essential oil discovering to the taste but little pungency.

Syrup. rosæ *Ph. Lond.* — rosar. pallid. *Ph. Ed.*

These flowers contain likewise a bitterish substance; which is extracted by water along with the odoriferous principle; which, after this last has been separated by distillation or evaporation, is found entire in the remaining decoction; and which appears to be of a gently purgative nature. The decoction, or a strong infusion of the flowers, made into a syrup with a proper quantity of sugar, proves an useful laxative for children, in doses of a spoonful: of the extract obtained by inspissating the decoction, from a scruple to a dram is said to be sufficient for adults. The college of London directs the syrup to be made, by pressing out the liquor remaining after the distillation of six pounds of damask roses, and boiling it down to three pints; then, after it has settled for a night, adding five pounds of fine sugar, and boiling the mixture to the weight of seven pounds and a half: a spoonful of this syrup appears to be equivalent to about three drams of the fresh flowers. The solutive matter of the flowers is combined also in the same manner, for the purposes of glysters, with brown sugar and honey:

Mel. solutiv.

towards the end of the boiling down of the strained decoction, an ounce of cummin seeds, bruised a little and tied in a linen cloth, is added; and the liquor afterwards boiled with four pounds of brown sugar and two of honey.

Rectified spirit extracts both the odoriferous and the purgative matter of the damask rose; equally

equally with water, or rather more completely. The spirit, distilled off from the filtered tincture, proves lightly impregnated with the fragrance of the flowers, and the inspissated extract retains likewise a part of their flavour along with the bitterish matter. This extract, in quantity smaller, and in taste stronger, may be presumed to be more purgative, than that made with water.

2. ROSA RUBRA *Pharm. Lond. & Edinb.* *Rosa rubra multiplex* C. B. *Rosa gallica* Linn. The red rose: with double flowers of a deep red colour.

The red rose has very little of the fine flavour admired in the pale sort: to the taste, it is bitterish and subastringent. The astringency is greatest before the flowers have opened, and, in this state, they are chosen for medicinal use as a mild corroborant: the full-blown flowers are probably as laxative as those of the foregoing species; for Poterius relates, that he found a dram of powdered red roses occasion three or four stools, and this not in a few instances, but constantly, in an extensive practice, for several years. The astringency of the buds is improved by hasty exsiccation in a gentle heat: by slow drying, both the astringency and the colour are impaired.

The fresh buds, clipt from the white heels, and beaten with thrice their weight of fine sugar, form an agreeable and useful conserve; which is given in doses of a dram or two, dissolved in warm milk, in weaknesses of the stomach, coughs, and phthisical complaints. Instances are mentioned in the German ephemerides, and in Riverius's praxis, of very dangerous phthisical disorders being cured by

Conserva rosarum *Ph. Lond. & Ed.*

the

the continued use of this medicine: in one of these cases, twenty pounds of the conserve were taken in the space of a month, and in another upwards of thirty pounds. Mixtures of the roses with a larger proportion of sugar are made in the shops into lozenges: one part of the buds clipt from the heels and hastily dried, and twelve parts of fine sugar, are separately reduced into powder, then mixed, and moistened with so much water as will render them of a due consistence for being formed: or the conserve is mixed with as much fresh sugar as is sufficient to bring it to a like consistence, that is, about thrice its own weight.

These flowers give out their virtue both to water and rectified spirit, and tinge the former of a fine red colour, but the latter of a very pale one: the extract obtained by inspissating the watery infusion, is moderately austere, bitterish, and subsaline; the spirituous extract is considerably stronger both in astringency and bitterness. In the shops, seven ounces of the dried rosebuds are infused in five pounds of boiling water; and the infusion made into a syrup with six pounds of fine sugar †, or boiled to a syrupy consistence with seven ‡ pounds of clarified honey: the syrup is valued chiefly for its gratefulness and fine red colour: the mixture with honey is used as a mild cooling detergent, particularly in gargarisms for inflammations and ulcerations of the mouth and tonsils. The infusions acidulated with a little vitriolic acid, and sweetened with sugar, make a grateful, cooling, restringent julep, which is sometimes directed in hectic cases and hemorrhagies, and along with boluses or electuaries of Peruvian bark, and sometimes is used as a gargarism: the college of London orders two pints and a half of

Saccharum rosaceum.

† Syrup. e rosis siccis *Ph. Ed.*
Mel. rosæ ‡ *Ph. Lond.*

Infus. rosæ *Ph. Lond.*

of boiling water, mixed with three drams of dilute vitriolic acid to 'be' poured on half an ounce of the frefh buds, and an ounce and an half of fine fugar to be diffolved in the ftrained infufion: that of Edinburgh orders two pounds and a half of water, and half a dram of the acid, to half an ounce of the dry buds and an ounce of fugar. *Infufum vulgo Tinct. rofarum Ph. Ed.*

3. Cynosbatus *Pharm. Lond. Rofa filveftris vulgaris flore odorato incarnato C. B. Rofa filveftris inodora feu canina Park. Cynorrhodon. Rofa canina Linn.* Dog-rofe, wild briar, hipp-tree: with fingle pentapetalous flowers, of a whitifh colour mixed with various fhades of red. It is one of the largeft plants of the rofe kind; grows wild in hedges; and flowers, as the garden forts, in June.

The flowers of this fpecies, of an agreeable but weak fmell, and in tafte bitterifh and roughifh, are faid to have a greater degree of laxative virtue than thofe of the damafk rofe, together with a mild corroborating or reftringent quality. The fruit, the only part of the dog-rofe made ufe of in medicine among us, is agreeably dulco-acid, and ftands recommended as a cooling reftringent, in bilious fluxes, fharpnefs of urine, and hot indifpofitions of the ftomach: the frefh pulp is made in the fhops into a conferve, by mixing three ounces of it with five of fine fugar. *Conf. cynofbati Ph. Lond.* The pulp fhould be feparated with great care from the rough prickly matter, which inclofes the feeds; a fmall quantity of which, retained in the conferve, is apt to occafion an uneafinefs at the ftomach, pruritus about the anus, and fometimes vomiting.

ROSMARINUS.

ROSMARINUS *Pharm. Lond. & Edinb.* *Rosmarinus hortensis angustiore folio C. B. Libanotis coronaria quorundam. Rosmarinus officinalis Linn.* Rosemary: a large shrubby plant, clothed with long narrow stiff leaves, set in pairs, of a dark green colour above and hoary underneath; producing pale bluish labiated flowers, which stand in clusters round the stalk in the bosoms of the leaves. It is a native of the southern parts of Europe, common in our gardens, and seems to grow larger and more woody in this than in most other countries. It flowers in April and May, and sometimes again about the end of August.

Rosemary is a warm pungent aromatic; particularly useful in phlegmatic habits and debilities of the nervous system; of the same general nature with lavender, but with more of a camphorated kind of pungency, and of a stronger, and to most people less grateful, smell. The tender tops are the strongest both in smell and taste, and next to these the cups of the flowers; which last, though somewhat weaker than the leaves or tops, are nevertheless the most pleasant, and hence are generally preferred: it is chiefly, if not wholly, in the cup, that the active matter of the flower resides; for the bluish petalum, carefully separated, has very little smell or taste. The fragrance of these flowers is greatly diminished, or in great measure destroyed, by bruising or beating; and hence the officinal conserve, formerly made by beating them with thrice their weight of sugar, had very little of the flavour of the rosemary.

The

The leaves and tops of rosemary give out their virtues completely to rectified spirit, but only partially to water: the spirituous tinctures are of a yellowish-green colour, the aqueous of a dark greenish brown. Distilled with water, they yield a thin, light, pale coloured essential Ol. rorismar. oil, inclining a little to yellowish or greenish, effent. *Ph. Lond. & Ed.* of great fragrancy; though not quite so agreeable as the rosemary itself: from one hundred pounds of the herb in flower were obtained eight ounces of oil: the decoction, thus divested of the aromatic part of the plant, yields on being inspissated an unpleasant weakly bitterish extract. Rectified spirit likewise, distilled from rosemary leaves, becomes considerably impregnated with their fragrance, leaving however in the extract the greatest share both of their flavour and pungency. The active matter of the flowers is somewhat more volatile than that of the leaves, greatest part of it arising with spirit. The Hungary water, used as a perfume, and sometimes medicinally in nervous complaints, and which is said to have received its name from its being first made public by an empress of that nation who was cured by its continued use of a paralytic disorder, is a strong spirit distilled from fresh rosemary flowers: the college of Edinburgh directs a gallon of rectified spirit to be drawn over in the heat of a water bath from two pounds of the flowers as soon as they are gathered †: that of London takes the tops, and a spirit not quite so strong; putting a gallon of proof spirit to a pound and a half of the fresh tops, and drawing off in the heat of a water bath five pints ‡. The hungary water brought ‡ *Ph. Lond.* from France is more fragrant than such as is generally prepared among us.

Sp. rorismar. † *Ph. Ed.*

RUBIA.

RUBIA.

RUBIA Pharm. Lond. & Edinb. Rubia tinctorum fativa C. B. *Radix rubra, & erythrodanum quibufdam. Rubia tinctorum Linn.* MADDER: a rough procumbent plant, with fquare jointed ftalks, and five or fix oblong pointed leaves fet in form of a ftar, at every joint: on the tops come forth greenifh yellow monopetalous flowers, deeply divided into four, five, or fix fegments, followed by two black berries: the root is long, flender, juicy, of a red colour both externally and internally; with a whitifh woody pith in the middle. It is perennial, and cultivated in different parts of Europe (in fome of which it is faid to be indigenous) for the ufe of the dyers: the roots have been brought to us chiefly from Zealand; but thofe which have for fome years paft been raifed in England, appear fuperiour to the foreign, both as a colouring and a medicinal drug.

THE roots of madder have a bitterifh, fomewhat auftere tafte, and a flight fmell, not of the agreeable kind. They impart to water a dark red tincture, to rectified fpirit and diftilled oils a bright red: both the watery and fpirituous tinctures tafte ftrongly of the madder. The root taken internally tinges the urine and milk red; and in the Philofophical Tranfactions, and the Memoirs of the French Academy of Sciences, there are accounts of its producing a like effect upon the bones of animals with whofe food it had been mixed; all the bones, particularly the more folid ones, were changed both externally and internally to a deep red, though neither the flefhy nor the cartilaginous parts suffered any alteration. The bones, so tinged, gave out nothing of their colour either to water or spirit of wine.

This root appears therefore to be possessed of great subtility of parts, which may possibly render its medical virtues more considerable than they are now in general supposed to be. It has been chiefly recommended as a resolvent and aperient, in obstructions of the viscera, particularly of the urinary organs, in coagulations of blood from falls or bruises, in jaundices, and in beginning dropsies. * It has lately come into reputation as an emmenagogue, and is said to be a very efficacious medicine of this class *(a)*. From a scruple to half a dram of the powder, or two ounces of the decoction, may be given three or four times a day in this intention.

RUSCUS.

RUSCUS C. B. *Ruscus myrtifolius aculeatus* Tourn. *Bruscus oxymyrsine, myrtacantha, myacantha, & scopa regia quibusdam. Ruscus aculeatus* Linn. BUTCHERS-BROOM or KNEEHOLLY: a low woody plant, with oblong stiff prickly leaves joined immediately to the stalks: from the middle ribs of the leaves, on the upper side, issue small yellowish flowers succeeded by red berries: the root is pretty thick, knotty, furnished with long fibres matted together, of a pale brownish colour on the outside and white within. It grows wild in woods and on heaths, is perennial and evergreen, flowers in May, and ripens its berries in August.

(a) Home's *Clinical Cases and Experiments.*

THE

The root of butchers-broom has a sweetish taste, mixed with a slight bitterishness. It stands recommended as an aperient and diuretic, in urinary obstructions, nephritic cases, dropsies, &c. Riverius tells us of an hydropic person who was completely cured by using a decoction of butchers-broom for his only drink, and taking two purges of sena. The virtues of the root are extracted both by water and spirit, and on inspissating the liquors, seem to remain entire behind: neither of the extracts is very strong in taste; the watery the least so.

RUTA.

RUTA Pharm. Lond. & Edinb. Ruta hortensis latifolia C. B. Ruta graveolens Linn.
Rue: a small shrubby plant, with thick bluish green leaves divided into numerous roundish segments: on the tops of the branches come forth yellowish tetrapetalous (sometimes pentapetalous) flowers, followed each by a capsule, which is divided into four partitions full of small blackish rough seeds. It is cultivated in gardens, flowers in June, and holds its leaves all the winter. The markets are frequently supplied with a narrow-leaved sort, which is cultivated in preference to the other, on account of its appearing variegated during the winter with white streaks.

This herb has a strong unpleasant smell, and a penetrating pungent bitterish taste: much handled, it is apt to inflame and exulcerate the skin. It is recommended as a powerful stimulant, aperient, antiseptic, and as possessing some degree of an antispasmodic power; in crudities and indigestion, for preserving against contagious

ous diseases and the ill effects of corrupted air, in uterine obstructions and hysteric complaints, and externally in discutient and antiseptic fomentations. Among the common people, the leaves are sometimes taken with treacle, on an empty stomach, as an anthelmintic. A conserve, made by beating the fresh leaves with thrice their weight of fine sugar, is the most commodious form for the exhibition of the herb in substance.

The virtues of rue are extracted both by water and rectified spirit, most perfectly by the latter: the watery infusions are of a greenish yellow or brownish; the spirituous, made from the fresh leaves, of a deep green, from the dry of a dark yellowish brown colour: the leaves themselves, in drying, change their bluish green colour to a yellow. On inspissating the spirituous tincture, very little of its flavour rises with the menstruum; nearly all the active parts of the rue remaining concentrated in the extract, which impresses on the palate a very warm, subtile, durable pungency, and is in smell rather less unpleasant than the herb in substance. In distillation with water, an essential oil separates; in colour yellowish or brownish, in taste moderately acrid, and of a very penetrating smell rather more unpleasant than that of the herb: a very considerable part of the virtue of the rue remains behind; the decoction, inspissated, yielding, a moderately warm, pungent, bitterish extract. The active matter of this plant appears therefore to be chiefly of the more fixt kind: the essential oil itself is not very volatile, or at least is so strongly locked up by the other principles, as not to be readily elevated in distillation. The seeds and their capsules appear to contain more oil than

Extractum rutæ *Ph. Lond. & Ed.*

than the leaves: from twelve pounds of the leaves, gathered before the plant had flowered, only about three drams were obtained; whereas the same quantity of the herb with the seeds almost ripe yielded above an ounce.

SABINA.

SABINA *Pharm. Lond. & Edinb. Sabina folio tamarisci diofcoridis C. B. Savina quibusdam. Juniperus Sabina Linn.* SAVIN: an ever-green shrub or small tree, clothed with very short narrow leaves so stiff as to be prickly; producing small imperfect flowers, and sometimes, when grown old, bluish black berries like those of juniper, of which the modern botany reckons it a species. It is a native of some of the southern parts of Europe, and raised with us in gardens.

Ol. essent. sabinæ *Ph. Ed.*

Extractum sabinæ *Ph. Lond.*

THE leaves and tops of savin have a moderately strong smell, of the disagreeable kind; and a hot, bitterish, acrid taste. They give out great part of their active matter to watery liquors, and the whole to rectified spirit; tinging the former of a brownish, and the latter of a dull dark green colour. Distilled with water, they yield a large quantity of essential oil: Hoffman says, that from thirty-two ounces he obtained full five ounces of oil, and observes that there is no other known vegetable substance, except some of the resinous juices, as turpentine, that affords so much. The oil smells strongly, and tastes moderately of the savin: decoctions of the leaves, freed from this volatile principle by inspissation to the consistence of an extract, retain a considerable share of their pungency and warmth along with their bitterishness, and have likewise some degree of smell,

smell, but not resembling that of the plant itself. On inspissating the spirituous tincture, there remains an extract consisting of two distinct substances; one yellow, unctuous or oily, bitterish and very pungent; the other black, resinous, tenacious, less pungent, and subastringent.

Savin is a warm stimulant and aperient; supposed particularly serviceable in uterine obstructions, proceeding from a laxity or weakness of the vessels, or a cold sluggish indisposition of the juices. The distilled oil is accounted one of the most potent emmenagogues: it is likewise a strong diuretic, and, as Boerhaave observes, impregnates the urine with its smell. * The powdered leaves have been recommended as a very effectual escharotic for consuming warty venereal excrescences [a].

SACCHARUM.

SUGAR: a sweet substance, of a saline nature; prepared from the juice of an elegant large cane or reed, *arundo saccharifera* C. B. which grows spontaneously in the East Indies and some of the warmer parts of the West, and is cultivated in large plantations in several of the American islands. The expressed juice of the cane is clarified with the addition of lime water, and boiled down to a somewhat thick consistence: being then removed from the fire, the saccharine part concretes into brown coloured masses, *saccharum non purificatum Pharm. Lond.* leaving an unctuous liquid matter called melasses or treacle, from which a little more solid sugar, but of a coarser kind, is obtainable by a repetition of the boiling and clarification. The

[a] *Med. Ess. Edinb.* III. 395.

brown

brown 'fugar is purified in conical moulds, by spreading, on the upper broad surface, some moist clay; whose watery moisture, slowly percolating through the mass, carries with it a considerable part of the remains of the treacly matter. The clayed sugar, imported from America, is by our refiners dissolved in water, the solution clarified with whites of eggs, and after due inspissation, poured as before into conical moulds, where, as soon as the sugar has concreted, and the fluid part is drained off by an aperture at the bottom, the surface of the loaf is again covered with moist clay. The sugar, thus once refined, *saccharum album* becomes, by a repetition of the process, the double-refined sugar of the shops, *saccharum purissimum Pharm. Lond. & Edinb.* Solutions of the brown or white sugars, boiled down till they begin to grow thick, and then removed into a very hot room, shoot, upon sticks placed across the vessels for that purpose, into brown or white crystals or candy, *saccharum crystallinum*.

Sugar dissolves by the assistance of heat, in rectified spirit; but greatest part of it separates again in the cold, and concretes into a crystalline form: on this foundation, saccharine concretions are obtained from saturated spirituous tinctures of several of the sweet plants of our own growth; the saccharine part separating when the tincture is set in the cold, while the resinous or other matter extracted from the plant, remains dissolved in the spirit. Solutions of sugar mingle uniformly with those of other saline substances, whether acid, alkaline, or neutral; and make no visible alteration in the infusions of the coloured flowers of vegetables, or other

liquors,

SACCHARUM.

liquors, in which acids or alkalies produce a change of colour or a precipitation. This sweet saline substance appears on all trials completely neutral*(a); and unites with most kinds of humid bodies, without altering their native qualities: it serves as an intermedium for uniting together some bodies naturally repugnant, as distilled oils and water. On the same principle it impedes the coagulation of milk, and the separation of its butyraceous part.

Sugar, in consequence of this property, is supposed to unite the unctuous part of the food with the animal juices. Hence some have concluded, that it increases corpulence or fatness; others, that it has a contrary effect, by preventing the separation of the oily matter, which forms fat, from the blood; and others, that it renders the juices thicker and more sluggish, retards the circulation, obstructs the natural secretions, and thus occasions or aggravates scorbutic, cachectic, hypochondriacal and other disorders. General experience, however, has not shewn, that sugar produces any of these effects in any remarkable degree: its moderate use appears to be innocent; and perhaps, of all that have yet been discovered, it is the most universally innocent and inoffensive, as well as the most simple, sweet.

Sugar preserves both animal and vegetable substances from putrefaction, and appears to possess this power in a higher degree than the common alimentary salt: I have seen animal flesh preserved by it untainted for upwards of three years. From this property it has been

*(a) An acid of a peculiar kind has been separated from it in small proportion and by a laborious process, in which the nitrous acid is employed as the separating medium.

sometimes applied externally as a balsamic and antiseptic.

The impure brown sugars, by virtue of their oily or treacly matter, prove emollient and gently laxative. The crystals or candy are most difficult of solution, and hence are properest where this soft lubricating sweet is wanted to dissolve slowly in the mouth, as in tickling coughs and hoarseness. The uses of sugar in medicinal compositions, whether for their preservation, for procuring the intended form and consistence, or for reconciling to the stomach and palate substances of themselves disgustful, are too obvious to require being enlarged on.

SAGAPENUM.

SAGAPENUM Pharm. Lond. & Edinb. Serapinum quibusdam. SAGAPENUM: the concrete gummy-resinous juice of an oriental plant, of which we have no certain account, but which appears, from the seeds and pieces of stalks sometimes found among the juice as brought to us, to be of the ferulaceous or umbelliferous kind. The sagapenum comes immediately from Alexandria, either in distinct tears, or run together into large masses; outwardly of a yellow colour, internally somewhat paler and clear like horn; growing soft on being handled, so as to stick to the fingers. It is sometimes supplied in the shops by the larger and darker coloured masses of bdellium broken in pieces; which greatly resemble it in appearance, but may be distinguished by their much weaker smell.

SAGAPENUM has a strong disagreeable smell, somewhat of the leek kind, or like that of a
mixture

mixture of galbanum with a little afafetida; and a moderately hot biting tafte. It is one of the ftrongeft of the deobftruent gums, and frequently prefcribed, either by itfelf, or in conjunction with ammoniacum or galbanum, in hyfteric cafes, uterine obftructions, afthmas, and other diforders. It may be commodioufly taken in the form of pills, from two or three grains to a fcruple or half a dram: in dofes of a dram, it loofens the belly.

On boiling this gummy-refin in water, about three-fourths of it are refolved into a turbid yellowifh white liquor, which fmells and taftes weakly of the fagapenum. Rectified fpirit fcarcely takes up above one half, and receives very little colour: the folution fmells weakly, and taftes pretty ftrongly. Both the watery and fpirituous folutions lofe much, in evaporation, of their tafte as well as their fmell; the watery lofes moft, the extract being very confiderably weaker than the fagapenum in fubftance. It is probable that the more active parts are carried off by the watery vapour, but that in the fpirituous extract they are only invifcated by the groffer refinous matter: for the water, collected by diftillation, is notably impregnated with the flavour of the fagapenum, and difcovers likewife a fmall portion of effential oil; whereas the diftilled fpirit is almoft flavourlefs.

SALES ALKALINI.

ALKALIES, or ALKALINE SALTS: fubftances of a very pungent tafte; diffoluble in cold water; changing the colours of the blue flowers of vegetables to a green; deftroying the acidity of four liquors, and forming with the acid a neutral compound; precipitating earthy bodies diffolved

MATERIA MEDICA.

diſſolved in acids *(a)*; producing no precipitation or turbidneſs in ſolutions of the lixivial ſalts of vegetables. Theſe lixivial ſalts are themſelves alkalies: and to mingle uniformly with theſe bodies of their own kind, in a liquid ſtate, is the moſt commodious and ſure mark I can recollect, for diſtinguiſhing alkalies, univerſally, from certain ſolutions of earthy bodies in acids; ſome of which have, in a greater or leſs degree, all the common characters of alkalicity; but on being examined by this criterion, readily betray their compoſition, by rendering the limpid lixivial liquor milky, and depoſiting their earth; the acid, which before held the earth diſſolved, being abſorbed from it by the lixivial ſalt.

1. SAL ALKALINUS FIXUS. Fixt alkaline or lixivial ſalt: obtained from the aſhes of vegetables, by macerating or boiling them in water, and afterwards evaporating the lye till the ſalt remains dry. It is fixt and fuſible in the fire * *(b);* deliquiates in a moiſt air, diſſolves in equal

---

(a) To this character of alkalicity there is one exception or limitation. Volatile alkaline ſpirits made completely cauſtic by quicklime, on being mixed with a ſolution of calcareous earth in the nitrous or marine acids, occaſion no precipitation or cloudineſs. If the mixture be expoſed for ſome time to the air, the alkaline ſpirit gradually loſes its cauſticity; and then precipitates the earth: on blowing into it air from the lungs, through a glaſs pipe, the precipitation began immediately.

* *(b)* If in fuſion a coal falls in, the alkali is reſolved into denſe white fumes, which act prodigiouſly on the brain and nervous ſyſtem, rendering the head weak and benumbed, as in convaleſcence from ſome great diſeaſe, occaſioning impatience and inquietude in every member. *Beaumé.*

its

SALES ALKALINI.

its weight or lefs of water, and does not affume a cryftalline form *(a)*.

Fixt alkaline falts have an acrid fiery tafte, and leave in the mouth a kind of urinous flavour. Saturated folutions of them in water corrode the folid parts of animals, diffolve fats and oils into faponaceous compounds, and liquefy almoft all the animal humours, except perhaps only milk, which, when heated, they coagulate. Diluted largely with water, and drank warm in bed, they generally excite fweat: if that evacuation is not favoured by external warmth, they operate chiefly by urine, of which, in many cafes, as in maniacal and hydropic ones, they frequently procure a copious and falutary difcharge: they likewife loofen the belly, and in coftive habits, where the direct purgatives or laxatives give only temporary relief, they render the benefit more lafting. They feem in general to act by ftimulating and deterging the folids, and refolving the vifcidities of the humours; and by thefe means opening obftructions, or promoting fecretion, in all the organs through which they pafs. The dofe is from two or three grains to fifteen or twenty; in fome cafes it has been extended to a dram. That they may be given, and continued for fome time, with fafety, in very confiderable dofes, appears from the experience of thofe, who have taken the ftrong folution of them called foap-lyes for the relief of calculous complaints.

(a) Though thefe falts, as commonly prepared, are never found to fhoot into cryftals, they do cryftallize in part when folutions of them have been expofed for a length of time to the open air. The cryftals are far milder in tafte, and effervefce more ftrongly with acids, than the alkali in its common ftate.

MATERIA MEDICA.

In putrid diforders, and a colliquated ftate of the humours, thefe falts have been generally, and I think juftly, condemned: for though they have lately been difcovered to refift putrefaction both in the fluids and folids of dead animals, yet in living ones they apparently increafe the colliquation, with which all putrid difeafes are accompanied.

Fixt alkaline falts are obtainable, in greater or lefs quantity, from almoft all vegetables; excepting perhaps only a few of the volatile acrid kind, as muftard feed. The falts of different plants, in the ftate wherein they are firft extracted from the afhes, are found to differ in degree of ftrength, and in fome other refpects, from one another; many of them containing a portion, and fome a very confiderable one, of neutral falts of the vitriolic or marine kind *(a)*. Purified by calcination, fo as that all remains of the oil of the vegetable may be burnt out; and by deliquiation

(a) The readieft way of difcovering neutral falt in the lixivial falts of vegetables is, by fhaking a ftrong folution of them in a vial with about an equal quantity of rectified fpirit of wine. If the falt is purely alkaline, the two liquors, on ftanding for a moment, will feparate from one another; the fpirit rifing to the top, and the alkaline folution collecting itfelf at the bottom, both of them tranfparent as at firft. If neutral falts are mixed with the alkali, though in very fmall proportion, the fpirit produces inftantly an opake milkinefs in the lye; and on ftanding for a few minutes, a faline matter feparates and falls to the bottom, in greater or lefs quantity, according as the alkali has a greater or lefs admixture of the neutral falt.

The exact quantity of pure alkali in any kind of lixivial falt or potafh may be determined by means of acids. Some alkaline falt known to be pure, as good falt of tartar, is to be melted in an iron ladle, that all remains of watery moifture may be expelled: a certain quantity of this falt, as a dram, weighed out while warm, is to be diffolved in a little

deliquiation in the air, by which only the alkali diffolves; they are all, except thofe of fome marine plants (fee *Natron*), fo much alike, as not to be diftinguifhable, by any known method of trial, from one another.

The falts of the leaves and other herbaceous parts of plants are more difficultly brought to a ftate of perfect purity than thofe of the more woody and compact; a portion of oily matter being tenacioufly retained, minute indeed, yet fufficient to give a brownifh tinge. A falt of this kind is generally prepared, or expected Sal abfinth. to be prepared from wormwood, fometimes from broom, and fometimes from bean ftalks, all which are fufficiently well adapted to this ufe, their afhes yielding as large a proportion as moft of the common herbaceous matters, and their falt feeming to be almoft merely alkaline, or free from any confiderable quantity of the other kinds of faline matter, of which the afhes of fome vegetables contain more than they do of alkaline falt. About London, the fhops are ufually fupplied from the country with the afhes of wormwood ready burnt; but that more of the oil may be confumed than the fimple burning

little water, and faturated with any convenient acid, as diluted fpirit of falt: the point of faturation is readily and accurately obtained by means of ftained paper, as directed in page 19. For the greater facility in trials of this kind, a quantity of the fpirit of falt may be fo diluted, that fixteen drams of it for inftance may exactly faturate the one dram of pure alkali. If then a folution of one dram of any given falt be faturated with the fame acid liquor, fo many drams or parts of a dram of the acid as are required for the faturation, fo many fixteenths or parts of fixteenths of pure alkali does the given falt contain. This appears to be the moft fimple and commodious, as well as the moft accurate way, that has yet been contrived, for determining the alkalicity, or degree of purity, of all kinds of lixivial falts.

of the herb has diffipated, they are further calcined with a red heat, and occafionally ftirred, for fome hours: the white afhes are then boiled in water, and the filtered lye evaporated to drynefs.

Some have endeavoured to retain in the falt as much as poffible of the oil, by burning the plant with a clofe fmothering heat, continued no longer than till it is reduced fully to afhes: that is, till the alkaline falt is generated, for thefe falts do not appear to exift naturally in vegetables. The alkalies thus prepared are of a dark brown colour, and fuppofed to be much milder and lefs acrimonious, and more of a faponaceous nature, than thofe which have been farther divefted of oil. But as we now have, in the foda or natron, an alkali as mild as can be wifhed for, this inelegant, precarious and unfrugal method of fuppreffing the acrimony of the common alkalies, becomes unneceffary.

Among all the known vegetables, or vegetable productions, there are none from which a pure alkaline falt is obtainable fo eafily, and in fo large a quantity, as from tartar. If red or white tartar be burnt with a moderately ftrong fire, either in a proper veffel, or wrapped up in wetted brown paper, to prevent the fmaller pieces from dropping down through the interftices of the coals on being firft injected into the furnace, it foon turns to white afhes, which yield on the firft elixation a ftrong fiery falt[†], of a fnowy whitenefs, amounting to about one fourth the weight of the tartar. The ftrength of the falt is fomewhat further increafed, by keeping it melted for fome hours in an intenfe fire; in which operation, if the crucible cracks or is left uncovered, fo as that the flame may have any accefs to the falt, or if a minute portion

[†] Sal tartari *Ph. Ed.*

tion of any inflammable matter is introduced, it assumes, in part at least, a greenish or blue colour, which is commonly looked upon as a mark of its strength.—A pure and strong alkaline solution is obtained, by exposing to the air, in a moist place, either the salt, or the white ashes‡ of tartar: the alkali imbibes in a few days so much of the aereal moisture, as to run wholly into a liquor, leaving, how highly soever the salt has been purified before, a considerable quantity of earthy matter. If the liquor be inspissated to dryness, and the dry salt again deliquiated in the air or dissolved in pure water, an earthy matter is still left: and even if the filtered solution be kept for a length of time in a close stopt glass vessel, an earthy substance gradually separates and falls to the bottom.

‡ Lixivium tartari.

Alkaline salts are prepared for common uses, in the way of trade, chiefly from wood; of which, in the forests of Germany, Russia, and Sweden, large piles are burnt on purpose. To save the trouble of boiling down the lye, the finer part of the ashes unelixated is in some places tempered with it into the consistence of mortar, which is afterwards stratified with some of the more inflammable kinds of wood, and burnt a second time: in others, the lye is soaked up in dry straw, and this drained and burnt. The impure saline masses, obtained by these or similar methods, are called *Potashes*; the strongest of which has been generally reckoned that brought from Russia *(Cineres russici)* in dark-coloured hard masses, of a very pungent taste, yet containing so much earthy matter as not readily to liquefy or grow moist in the air. This potash is said to be prepared in the first of the ways above-mentioned: but it appears from some late experiments, that another ingredient

gredient is made use of in the process; the masses, as brought to us, being found to contain more quicklime than alkaline salt *(a)*, and on this depends the great strength and corrosiveness of the Russian potash. For a purer salt, the lye is boiled down in large iron vessels; and the dark-coloured dry salt, which concretes into a hard crust on the sides and bottom of the vessel, is beaten off with a mallet and chisel, and calcined in an oven, with a gradual fire, till it becomes white; in which state it is called, from its pearly appearance, *pearl-ash*. For some years past, we have been supplied chiefly from our American colonies, with compact alkaline masses, much more pure that the above *pot-ashes*, though less so than the *pearl-ash*; prepared by boiling down the lye to dryness, and then increasing the fire till

(a) See Dr. Home's *Experiments on bleaching.*—It has been suspected that the matter in Russia potash, which seemed from Dr. Home's experiments to be quicklime, is no other than the earth of the vegetable ashes themselves, which earth, by strong calcination, such as this kind of potash is said to undergo, assumes some of the most striking characters of true quicklime. Since the establishment of the American manufacture, the Russia sort has in this country fallen so much into disuse, that it is very difficult to procure a specimen that can be depended on as genuine. What has been sent to me as true Russia potash (and which indeed has greatly the appearance of what used to be sold under that name) on being elixated with water, leaves a large quantity of earthy matter, greatest part of which dissolves readily in aquafortis. This solution has exactly the same taste with a solution of chalk made in the same acid: on dropping into it a little vitriolic acid, the liquor grows instantly milky, and a copious precipitation ensues. This precipitability by the vitriolic acid is one of the properties of calcareous earths, which the earth of vegetables has not been found to acquire by any degree of calcination; and therefore we may conclude that in the making of this potash real quicklime is mixed, in very large proportion.

the

SALES ALKALINI.

the falt becomes red-hot, and melts, so as to be conveniently laded out with iron ladles: the troublesome operation of getting off the indurated falt from the boiler is thus avoided; and the strong melting heat, though of short continuance, supplies in great measure the tedious calcination of the falt; for though the inflammable matter, on which the colour depends, is in fusion not consumed, it is burnt to an indissoluble coaly state, so that lyes made from these melted potashes with water are nearly as colourless as those of the whitest pearl-ashes. * The college of Edinburgh, which has discarded the oily alkalies of wormwood, broom, &c. now directs a pure incinerated alkali to be made from pearl-ashes, first burned with a red heat in a crucible, then dissolved in water, cleared by subsidence, and evaporated to dryness in an iron pot. This falt will dissolve in equal its weight in water, and the solution is analogous to the former *oleum tartari per deliquium*. The London college, in their last pharmacopœia, have given the specific apellation of *kali* to the fixed vegetable alkali, in all its varieties; and they order it to be got pure from pearl-ashes, or any other vegetable ashes, by lixiviation in water, colature, evaporation to a pellicle, separation of the neutral falts which will then crystallize, and lastly, evaporation to dryness. It is also allowed to be similarly prepared from burned tartar. For a liquid preparation, this pure falt is set apart in a moist place till it spontaneously deliquefces.

Quicklime remarkably increases the activity of all these falts; enabling them, in a liquid or dilute form, to dissolve oils, fats, &c. far more powerfully than either the lime or alkali by themselves; and in a solid or more concentrated
one,

Sal alkalinus fixus vegetabilis purificatus Ph. Ed.

Kali præpar. Ph. Lond.

Aqua kali Ph. Lond.

Aqua kali
Ph. Lond.

Lixivium caufticum
Ph. Ed.

one, to act as cauftics. For thefe purpofes, the London college directs four pints of water to be poured to fix pounds of quicklime, which are to ftand together for an hour: then as much more water as will make the whole four gallons, with four pounds of alkaline falt, are to be added, and the whole boiled for a quarter of an hour, when it is to be fuffered to cool, and ftrained off. The liquor is known by the name of foap lye, and ought to be of fuch a ftrength, that an exact wine pint may weigh juft fixteen ounces troy. If it excites any fermentation with acids, more quicklime is to be added. The common lyes of our foft-foap makers are confiderably ftronger than this: Dr. Pemberton obferves, that their lyes will be reduced to the ftrength here propofed, by diluting them with fomewhat lefs than an equal meafure of water. * In the Edinburgh pharmacopœia this preparation is thus directed. Eight ounces of frefh quicklime are put into an iron or earthen veffel with twenty-eight ounces of warm water. When the ebullition is over, fix ounces of pure vegetable fixt alkali are added, and after perfect mixture, the veffel is covered and fuffered to cool. The matter is then poured into a glafs funnel, lined with a linen rag, and is fet to drain into a glafs bottle as long as any liquor will run. Some more water is then to be poured to the matter in the funnel, which will drain through it; and this is to be repeated till three pounds of liquor are procured, which is to be fhaken together, and kept in a well ftopped phial. *(a)*

The

(a) Pure alkaline falt requires commonly about twice its weight of quicklime to render it completely cauftic. Complete caufticity is known by the lye making no effervefcence

SALES ALKALINI.

The dry salt obtained by evaporating these Kali purum lyes is a strong and sudden caustic: for the *Ph. Lond.* greater convenience of using, it is urged in a crucible with a strong fire, till it flows like oil, then poured upon a flat plate made hot, and Causticum whilst the matter continues soft, cut into pieces commune of a proper size and figure, which are kept in acerrimum *Ph. Ed.* a glass vessel closely stopt. It deliquiates much sooner in the air, and dissolves more readily in watery liquors, than the milder alkalies, and in this consists its principal inconvenience; being apt to liquefy so much upon the part to which it is applied, as to spread beyond the limits in which it is intended to operate. This inconvenience is avoided, by boiling down the soap lye only to one third† or fourth‡ part, and † Causticum then, while the liquor continues boiling, sprink- commune ling in, by little and little, so much powdered ‡ Calx cum quicklime as will absorb it so as to form a kind kali puro *Ph.* of paste†: or, more accurately, in the pro- *Lond.* portion of five pounds four ounces, to sixteen pints of the original lye‡: the addition of the lime in substance renders the preparation less apt to liquefy, and hence more easily confinable within the intended limits, but at the same time proportionably more flow in its operation.

2. SAL ALKALINUS VOLATILIS. Volatile alkaline salt: obtainable, by distillation with a strong fire, from all animal matters, from soot, and in small quantity from most vegetables: producible also in animal substances, very plentifully in urine, by putrefaction, and in this

---

vescence with acids. A redundance of lime is known, by the lye growing milky on dropping into it a little common alkaline lye, or on blowing into it with the breath through a glass pipe.

case

cafe feparable by diftillation with a gentle heat. When the falt is once formed, whether by ignition or putrefaction, it gradually exhales in moderately warm air; and rifes fooner in diftillation than highly-rectified vinous fpirits, condenfing about the fides of the recipient into cryftalline concretions. It requires for its folution three or four times its weight of water.

Thefe falts are in fmell as well as tafte very penetrating and pungent: they are the only concrete falts that in their pure ftate emit fenfible effluvia. They diffolve oils, refins, fats, &c. more languidly than the fixt alkalies, on account perhaps of their not being fufceptible of any confiderable heat, by which their menftrual power might be promoted. In the bodies of animals, they operate more powerfully than the fixt, both as refolvents and ftimulants; are more difpofed to direct their force to the cutaneous pores, and lefs to the groffer emunctories; and act more remarkably upon the nervous fyftem. They are particularly ufeful in lethargic and apoplectic cafes; in hyfterical and hypochondriacal diforders, and the languors, head-achs, inflations of the ftomach, flatulent colics, and other fymptoms attending thofe diftempers, efpecially in aged perfons and thofe of a phlegmatic habit: in languors and faintings, their ftimulating fmell gives oftentimes immediate relief. In fome kinds of fevers, particularly thofe of the low kind, accompanied with a cough, hoarfenefs, redundance of phlegm, and lentor of the blood, they are of great utility; liquefying the thick juices, raifing the pulfe, and exciting a falutary diaphorefis. In putrid fevers, fcurvies, and wherever the mafs of blood is thin and acrimonious, they are hurtful: for though they powerfully refift the putrefaction

SALES ALKALINI.

putrefaction, of animal fubftances, that are detached from the vital œconomy, yet, in living animals, one of their primary effects is a colliquation of the humours, which in its advanced ftate is very nearly allied to the advanced ftate of putrefactive colliquation: their immoderate ufe has brought on high fcorbutic fymptoms, refembling thofe of the true putrid fcurvy *(a)*. Thefe falts are moft commodioufly taken in a liquid form, largely diluted; or in that of a bolus, which fhould be made up only as wanted, the falt foon flying off. The dofe is from two or three grains to ten, twelve, or more.

The volatile alkalies obtained from different fubftances appear, like the fixt, to be, in their ftate of perfect purity, one and the fame thing. But as firft diftilled from the fubject, they are largely impregnated with its oil rendered fetid or empyreumatic by the procefs in which the falt was generated; and as thefe oils differ from one another in degree of fubtility and fetidnefs, the falts partake of the fame differences, till repeated diftillations or other proceffes have either feparated the adhering oils, or fubtilized and purified them to the fame degree. By repeated diftillations, all animal oils become limpid as water, lofe their fetor, acquire a penetrating fragrant fmell, and a gratefully pungent tafte: thus rectified, they are faid, by Dippelius, Hoffman, and others, to act, in dofes of half a drop, as diaphoretics, anodynes, and antifpafmodics. The volatile alkalies ufed in medicine are expected to be either pure from oil, or to have their oil in this fubtilized ftate. Oleum animale *vulgo.*

The oily volatile alkalies have been chiefly prepared from hartfhorn, by diftillation in large

---

(a) See Huxham's *Differtation on the malignant fore throat.*

iron

iron pots, with a fire increased by degrees to a strong red heat. At first there arises an aqueous liquor, then the volatile salt, along with a yellowish and at length a dark reddish oil: if the aqueous liquor is not removed before the salt begins to come over, a part of the salt dissolves in it, and thus forms what is called 'spirit.

The oil, excepting so much of it as is incorporated with the alkali, may be separated from the spirit by filtration through wetted paper, which transmits the spirit and retains the oil. The salt and spirit are then distilled again together, with a very gentle heat, in a glass retort; and the distillation carefully repeated several times, till the salt becomes exceedingly white, and the spirit limpid as water, and of a grateful smell. The salt becomes the sooner pure, if it be separated from the spirit, and sublimed first from an equal weight of pure chalk and afterwards from a little rectified spirit of wine. If the whole of the volatile salt is required in a solid form, it may be recovered from the spirit by sublimation in a tall narrow cucurbit, the salt rising into the head, while the watery fluid remains behind. In all the distillations of the spirit, greatest part of the salt comes over before the phlegm; and the process should be continued no longer than till so much of the phlegm has followed as is nearly sufficient to dissolve it; that a part of the salt remaining undissolved may be a criterion, to the purchaser, of the saturation or strength of the spirit.—A spirit, salt, and oil, are sometimes distilled in the same manner from wood soot, but here more labour is required to render the salt and spirit pure.

Liquor volatilis, sal, & oleum cornu cervi Ph. Lond.

Though

SALES ALKALINI.

Though the whitenefs and limpidity which the falts and fpirits of hartfhorn, foot, and other like fubftances, acquire by the above methods of purification, feem to fhew that they are divefted of oil; they are neverthelefs found to participate ftill of that principle in no fmall degree. In long keeping they contract a yellow colour, and at length become again naufeous and fœtid; the oil feeming to be more and more extricated, or to lofe by degrees of the fubtility and gratefulnefs which it received from the rectification. The oftener the diftillation is repeated, the more permanent is the fubtilization of the oil.

The moft effectual purification of thefe falts is obtained, by combining them with mineral acids, and afterwards feparating the acid. It is not needful to make fuch a combination on purpofe: for fuch a one is produced more compendioufly, in the way of trade, and called in the fhops fal ammoniac: fee the following article.

If fal ammoniac be mingled with any fixt alkaline falt, either in the form of powder or folution, its acid will be abforbed by the fixt alkali; and the volatile alkali, thus fet at liberty again, will immediately difcover itfelf by its pungent odour, and may be collected perfectly pure by diftillation. Eighteen ounces of fixt alkali, and one pound of fal ammoniac, may be diftilled with four pints of water in a gentle heat, till two pints are drawn off†: or fixteen ounces each of the two falts may be diftilled with two pounds of water, to drynefs‡. The volatile alkaline falt may be extricated likewife by means of chalk, but with this difference, that the chalk does not begin to act upon the fal ammoniac, or abforb its acid, till the mixture is confiderably heated: one part of the fal ammoniac

Aqua ammoniæ *Ph. Lond.*
Spirit. fal. ammoniac.
‡ *Ph. Ed.*

<small>Ammonia præp. Ph. Lond.
Alcali volatile ex sale ammon. vulgo sal ammon. volat. Ph. Ed.</small>

ammoniac may be mixed with two of chalk, and the mixture set to sublime in a retort with a strong fire.

Quicklime, which heightens the pungency of fixt alkalies even to causticity, has a like effect upon the volatile: it renders the fixt more easily liquefiable, and the volatile permanently liquid, preventing their concretion into a solid form: the volatile alkali, like the fixt, in having its activity thus increased by quicklime, loses its power of effervescing with acids; from whence it may be presumed, that the lime acts, on one alkali as on the other, by absorbing their air (see page 255). This pungent volatile spirit may be prepared, by flaking two pounds of quicklime in two pints of water, letting it stand an hour, and then adding a pound of powdered sal ammoniac with six pints of warm water: immediately adapt a recipient, (for the pungent vapours begin to arise on the first contact) and with a gentle heat draw off one pint. *Or two pounds of quicklime may be flaked with one pound of water, and covered till it falls to a powder, which is then to be put in a retort with sixteen ounces of sal ammoniac dissolved in four times its weight of water. The vessels are to be very carefully luted, and with a very gentle heat twenty ounces are to be drawn off. This spirit is held too acrimonious for internal use, and has therefore been chiefly employed in smelling-bottles. It is an excellent menstruum for certain vegetable substances, as Peruvian bark, which the milder spirits extract little from; and when saturated with bodies of this kind, its pungency is so far sheathed, that it may be taken inwardly with as great safety as tinctures made in the other spirits. In long keeping, unless the bottle is quite full

<small>Aqua ammoniæ puræ Ph. Lond.</small>

<small>Alcali volat. caust. Ph. Ed.</small>

SALES ALKALINI.

full and very closely secured, it gradually imbibes air, as appears from the effervescence which it raises with acids; and loses proportionably of its pungency.

Some have mixed a quantity of this caustic spirit with the officinal spirits both of sal ammoniac and of hartshorn; which thus become more pungent, so as to bear an addition of a considerable proportion of water without danger of discovery from the taste or smell. This abuse may be detected, by adding to the suspected spirit a quantity of spirit of wine; which, if the volatile spirit is genuine, will precipitate a part of its volatile salt, but has no such effect either on the caustic spirit itself, or on such as is sophisticated with it.

Some have substituted to the spirit a solution of sal ammoniac and fixt alkaline salt: this liquor eludes the above method of trial, as it deposites a saline matter on the addition of spirit of wine, in the same manner as the genuine volatile spirit: it may be distinguished by the matter, thus deposited, being not volatile but fixt; or by a fixt salt being left upon evaporating a little of the liquor; or more compendiously, by adding a drop or two of solution of silver made in aquafortis, which immediately produces a milkiness in the counterfeit, but makes no apparent change in the genuine spirit.

The addition of spirit of wine to volatile alkaline spirit affords means of judging, in some degree, of their strength or saturation as well as of their purity. If the volatile spirit be fully saturated with salt, a quantity of highly-rectified spirit of wine, poured on it slowly down the sides of the glass, in a cool place, produces immediately an opake dense coagulum on the surface where the liquors touch: on shaking them together, the whole becomes a confiftent mafs, which foon refolves by warmth into a fluid and a folid part. This is fuppofed by fome to be a volatile foap, compofed of the alkaline falt of the one fpirit and the oily principle of the other; though in effect it is no more than the alkaline falt itfelf diflodged, by the vinous fpirit from the watery fluid in which it was diffolved: the quantity of falt, thus feparated, will be in proportion to the ftrength of the volatile fpirit.

<small>Offa alba Helmontii.</small>

Though volatile alkalies, not cauftic, appear from the above experiments to be averfe to any union with vinous fpirits; a folution of them even in rectified fpirit is neverthelefs obtainable, by adding it, along with water, in the procefs by which they are extricated from the fal ammoniac. For this purpofe, three pints of proof fpirit are put to four ounces of fal ammoniac, and fix (or lefs) of any fixt alkaline falt, and one half of the liquor drawn off with a gentle heat†. Or fixteen ounces of quicklime and eight of fal ammoniac are powdered together, and put into a retort with thirty-two ounces of rectified fpirit, and the diftillation continued till all the fpirit is come over‡. This preparation has lately come into efteem both as a medicine and a menftruum.

<small>† Spirit. ammoniæ Ph. Lond.</small>

<small>‡ Spiritus fal. ammon. vinofus Ph. Ed.</small>

Mixtures of volatile and vinous fpirits, flavoured with different aromatic oils, and other like materials, have long been in general ufe under the name of *fal-volatile*. The college of London orders a quart of the above dulcified fpirit, two drams of effential oil of nutmegs, and the fame quantity of effence of lemons, to be mixed together: that of Edinburgh directs a dram and a half of oil of rofemary, and a dram of effence of lemons, to be diffolved in eight ounces of the dulcified fpirit. Volatile fpirits are impregnated

<small>Spir. ammoniæ comp. Ph. Lond.</small>

<small>Spirit. volat. aromat. *vulgo* Spir. volat. oleos. & Spir. falinus aromat. Ph. Ed.</small>

ted alfo in the fhops with afafetida, in the pro- Spir. ammon. portion of four ounces to five pints†, or one fœtid.
ounce to fixteen ounces‡.
† *Ph. Lond.*
‡ — volat.

Thefe kinds of compofitions may be made fœt. *Ph. Ed.* extemporaneoufly, by dropping any proper effential oils into the dulcified fpirit of fal ammoniac, (as now directed above by the London and Edinburgh colleges) which will readily diffolve them without the affiftance of diftillation. By this method, a fal-volatile may be occafionally prepared, of any particular flavour, or adapted to particular purpofes: thus, in hyfterical diforders, where the uterine purgations are deficient, a preparation of this kind, made with the oils of rue or favin, proves an ufeful remedy: in weaknefs of the ftomach, oil of mint may be ufed; and in flatulent cafes, thofe of anifeeds or fweet fennel feeds: thefe laft remarkably cover the pungency of the volatile fpirit, and render it fupportable to the palate. The dofe of thefe compounds is from a few drops to fixty or more.

The cauftic fpirit made with quicklime appears in fome cafes preferable, for thefe kinds of compofitions, to the other volatile fpirits; as being perfectly mifcible with rectified fpirit of wine, in any proportions, without any feparation of its volatile alkaline part; and as being a more powerful menftruum for fome difficultly foluble oils. The very penetrating pungent volatile fpirit, which has lately come into vogue under the name of *eau de luce*, is faid to be made with this cauftic fpirit, and oil of amber that has been rectified or rediftilled till it becomes limpid and lofes its fmell: one fcruple of the oil, fo rectified, and ten grains of foap, Spir. ammoare directed to be diffolved in an ounce of recti- niæ fuccinat, fied fpirit of wine, and four ounces of the cauftic *Ph. Lond.*

fpirit

spirit added gradually to this folution: the mixture generally appears milky, and if required limpid may be rendered so by diftillation: fome tinge it of a fine blue colour, when defigned only for fmelling to, by adding a drop or two of folution of copper*(a)*.

SAL AMMONIACUS.

SAL AMMONIACUS Pharm. Lond. & *Edinb.* SAL AMMONIAC: a neutral falt; volatile in a moderate heat, but not in that of boiling water; formed by the coalition of volatile alkaline falt with marine acid. On mixing it with a fixt alkaline falt or calcareous earth, and expofing the mixture to the fire, its ingredients are difunited: the volatile alkali exhales, and may be collected in proper veffels (fee the foregoing article): and the acid remains combined with the fixt alkali or earth, forming therewith the fame compounds as if the pure marine acid had been poured upon them. Hence, in the preceding operations, where the volatile alkali of the fal ammoniac is feparated by the intervention of fixt alkalies, the refiduum, diffolved and cryftallized, is found to be the fame with *regenerated fea falt*; and when chalk or lime is ufed for the intermedium, the refiduum affords *calcareous marine falt*: fee *Sal communis*.

The other mineral acids form likewife ammoniacal falts with volatile alkali; and it is faid, that one made with the vitriolic acid is often fubftituted to the true officinal one with the marine. — The moft obvious character of ammoniacal falts in general is, their yielding a pungent urinous odour on being ground with a

(a) Malouin, *Chimie medicinale,* i. 202. ii. 431.

little quicklime. The *marine* fal ammoniac may be diftinguifhed from thofe made with the other acids, by its emitting white fumes on dropping upon it fome oil of vitriol; and by a folution of it in purified aquafortis being able to diffolve gold leaf or a mark made with gold on a touchftone. The *nitrous* fal ammoniac is diftinguifhed, by its deflagrating or flaming when thrown into a red-hot veffel, diffolving in fpirit of wine, and yielding red fumes with oil of vitriol. The *vitriolic* is diftinguifhed by a folution of it rendering folution of chalk in aquafortis milky; and by its not being acted upon by oil of vitriol.

Sal ammoniac has been hitherto prepared chiefly in Egypt: it is faid, that the earth abounds there with marine falt; that grafs and other vegetables are fenfibly impregnated with this falt; that the dung of graminivorous quadrupeds is ufed as fuel, and the foot carefully collected; and that from this foot, fal ammoniac is extracted, by fublimation without addition. The falt is brought to us, fometimes in conical loaves, moft commonly in large round cakes, convex on one fide and concave on the other, appearing when broken of a needled texture, or compofed of ftriæ running tranfverfely and parallel to one another: the internal part is generally pure, and of an almoft tranfparent whitenefs; the outfide for the moft part foul and of a yellowifh grey or blackifh hue. It is purified, either by fublimation, with a gradual fire, in an earthen cucurbit having a blindhead adapted to it; or, perhaps more perfectly, by folution in water, filtration, and cryftallization. It diffolves, in temperately warm weather, in about thrice its weight of water,

water, and by the affiftance of heat in a much fmaller quantity; and cryftallizes into long fhining fpicula, or thin fibrous plates like feathers. In fublimation, efpecially if the fire is haftily raifed, it remarkably volatilizes many kinds of bodies, perhaps all thofe that are foluble by the marine acid.

This falt has a very fharp penetrating tafte. It is a powerful attenuant and deobftruent, feeming to liquefy the animal juices almoft like alkaline falts: Boerhaave obferves that its liberal and continued ufe renders the blood fo thin as to burft through the veffels, particularly thofe of the lungs and the urinary organs. In dofes of half a dram or a dram, diffolved in water, if the patient is kept warm, it generally proves fudorific: by moderate exercife, or walking in the open air, its action is determined to the kidneys: in larger dofes it loofens the belly. It has by fome been held a fecret for the cure of intermittents; and is undoubtedly, in many cafes, as an aperient, an excellent affiftant to the Peruvian bark, where that aftringent drug by itfelf would produce dangerous obftructions, or aggravate thofe already formed. This falt is employed likewife externally as an antifeptic, and in lotions and fomentations for œdematous tumours; as alfo in gargarifms, for inflammations of the tonfils, and for attenuating and diffolving thick mucus in the mouth and fauces. Saturated folutions of it are faid to confume warts.

SAL CATHARTICUS.

PURGING SALT: a falt of a bitter tafte; foluble in twice its weight or lefs of water, and fhooting into long prifmatic cryftals like thofe
of

SAL CATHARTICUS.

of nitre; liquefying and bubbling up in a moderate heat, emitting a large quantity of aqueous vapours, and changing to a white spongy mass, bitterer than the salt at first. It is of two kinds; one a combination of the vitriolic acid with the earth called *magnesia*; the other, a combination of the same acid with the fixt alkali called *natron*. The most obvious criterion of their acid being the vitriolic is, their precipitating chalk dissolved in aquafortis or in other acids.

1. SAL AMARUS *Pharm. Lond.* Sal catharticus amarus *Pharm. Edinb.* Nitrum calcareum *Listero & hydrologis quibusdam.* Purging bitter salt: composed of the vitriolic acid and magnesia; distinguishable from that whose basis is an alkali, by solutions of it being turned milky, and depositing their earth, on the addition of any alkaline salt.

This is the salt with which the purging mineral waters are principally impregnated, and on which their purgative quality depends. It was first extracted from the Epsom waters, and has been commonly distinguished, both in this and other countries, by the name of Epsom salt: but those waters yielding the salt very sparingly, and their quantity being insufficient for its great demand, it was sought for elsewhere, and found, in plenty, in the bitter liquor remaining after the crystallization of common salt from sea water; from which it is now generally prepared.

This salt is a gentle purgative, operating in general with ease and safety, yet with sufficient efficacy, and quickly finishing its operation: its passing off hastily, and not extending its action so far as most other purgatives, seems to be its principal imperfection. For a full dose, eight or ten drams may be diffolved in a proper quantity of common water, or four or five drams in a pint or quart of the purging waters; to which may be added a little tincture of cardamom feeds, or fome other grateful aromatic, to render the liquor more acceptable to the ftomach. Thefe liquors, in fmaller dofes, pafs further into the habit, promote the fecretions in general, and prove excellent aperients in fundry chronical diforders.

2. SAL CATHARTICUS GLAUBERI, *vulgo fal mirabile.* Glauber's cathartic falt: compofed of the vitriolic acid and the mineral alkali *natron,* and hence fuffering no change from an admixture of frefh alkali.

This falt was difcovered, by the chemift whofe name it bears, in extracting the acid fpirit of fea falt by means of the vitriolic acid. When oil of vitriol is poured on fea falt, the marine acid, thereby difengaged from its own alkaline bafis, begins immediately to exhale, and by applying heat may be totally expelled; the vitriolic acid remaining combined with the natron or marine alkali. This combination is ftill procured chiefly in the fame manner: to the fea falt is added fix tenths† or half‡ its weight of oil of vitriol diluted with water, and the marine acid being diftilled off, the refiduum is diffolved and cryftallized. The fmalleft of thefe proportions of oil of vitriol appears to be fufficient for expelling the acid and faturating the alkali of the fea falt; but the larger is more eligible, as the Glauber's falt does not well cryftallize unlefs the acid prevails in the folution.

This falt is nearly of the fame medicinal qualities with the foregoing, which frequently supplies

Natron vitriolat*a* † *Ph. Lon*.
Soda vitriolata *vulgo* fal cath. glauberi ‡ *Ph. Ed.*

SAL COMMUNIS.

fupplies its place in the fhops. The Glauber's falt, fomewhat the leaft unpleafant to the tafte, is fuppofed to be the mildeft of the two, and to operate the moft kindly.

SAL COMMUNIS.

COMMON or CULINARY SALT; called, from its moft obvious fource, fea falt; though found alfo, in immenfe quantities, in the bowels of the earth. It is a perfectly neutral falt, compofed of a peculiar acid denominated from it the *marine acid,* and of the mineral alkali *natron.* It diffolves in lefs than thrice its weight of boiling water, and does not, like the other neutral falts, concrete again in the cold, fo long as the evaporation of the fluid is prevented; cold water diffolving nearly as much of this falt as boiling water. By gentle continued evaporation it fhoots into cubical cryftals, feveral of which unite together into the form of hollow truncated pyramids. The cryftals, expofed to the fire, burft and crackle,† foon after melt, and appear thin and limpid as water: if the falt be melted along with other fufible falts or with vitreous matters, it does not perfectly unite with them, but flows in part diftinct upon the furface. After fuffering a confiderable heat, it liquefies in the air.

† Sal decrepitatum.

1. SAL GEMMÆ. Sal gem, rock falt, foffil common falt. This is met with in feveral parts of the world, but in greateft plenty in certain deep mines, of prodigious extent, near Cracow in Poland: fome is likewife found in England, particularly in Chefhire. It is for the moft part very hard; fometimes pure, tranfparent and colourlefs; more commonly mixed with earthy

earthy or stony matters, of an opake whiteness, or of a red, green, blue, or other colours. These last sorts are purified, for the common uses of salt, by solution and crystallization.

2. SAL MURIATICUS *Pharm. Lond. Sal marinus hispanus Pharm. Edinb.* The salt extracted from sea water and saline springs. Sea waters yield from one fiftieth to about one thirtieth their weight of pure salt: from several springs much larger quantities are obtained: those in our own country at Nantwich, Northwich, and Droitwich, afford from one sixth to one third their weight. Sea water contains, besides the common salt, a portion of purging bitter salt, and of another saline substance which remains dissolved after the crystallization of the latter, of a very pungent taste, scarce reducible into a crystalline form, composed of marine acid and calcareous earth: from both these salts the spring waters are usually free. There are two general methods of extracting the common salt from these natural solutions of it: the one, a hasty evaporation, continued till the salt concretes and falls in grains to the bottom of the pan, from whence it is afterwards raked out, and set to drain from the bittern: the other, a slow and gradual evaporation, effected in the warmer climates by the sun's heat, by which the salt is formed, not into small grains, but into large crystals, called bay-salt. The salts obtained by these two processes differ in some respects from one another: that got by hasty evaporation, especially if a boiling heat, or one approaching to it, be continued during the time of the salts concreting, is apt to liquefy in a moist air; an inconvenience which the crystallized sort is not subject to: the crystals are

found

SAL COMMUNIS.

found likewife to be ftronger than the other, and to anfwer better for preferving provifions. Both forts prove impure and brown-coloured if the folutions are evaporated directly, but of perfect whitenefs if previoufly clarified by boiling with a little ox blood, or other like fubftances, which concreting by the heat, invifcate the unctuous matter, and carry it to the furface in form of fcum. Both forts generally retain a portion of the bitter falt; whofe bafis being an earth, folutions of them depofite this earth on the addition of any alkali.

COMMON SALT differs from other faline fubftances in occafioning drought, and tending, not to cool, but rather to heat the body. It prevents putrefaction lefs than moft others, and in fmall quantities, fuch as are taken with food, promotes it: by this quality it probably promotes alfo the refolution of the aliment in the ftomach, at the fame time that it proves a mild ftimulus to that vifcus itfelf. Salted animal foods are generally, perhaps juftly, accounted one of the principal caufes of the fcurvy at fea; not that the falt is of itfelf prejudicial, but on account of its being incapable of preferving the animal fubjects, for a length of time, in a perfectly uncorrupted ftate. Pure fea falt, and fea water, are rather falubrious than hurtful, both in the true fcurvy, and in impurities of the blood and humours in general. In confiderable dofes, they act as purgatives: Hoffman obferves, that an ounce of the falt, diffolved in a proper quantity of water, occafions commonly fix ftools or more, without uneafinefs; that this falt checks the operation of emetics, and carries them off by ftool; that in glyfters it is more effectual, though ufed only in the quantity of a

dram,

dram, than any of the purgatives; and that where other glyſters fail of opening the belly, a ſolution of common ſalt takes place.

* A remarkable inſtance of the efficacy of common ſalt given in very large doſes in a worm caſe, is related in the *Medical Tranſactions of the London College*, vol. i. A perſon reduced to the utmoſt extremity with pain in the ſtomach, obſtinate conſtipation, and contracted limbs, was adviſed, after many remedies had been uſed in vain, to drink ſalt and water. He drank within an hour two pounds of ſalt mixed in two quarts of water. It occaſioned violent vomiting, which brought up a quantity of ſmall worms, and its operation ended with purging and a profuſe ſweat. Great rawneſs and ſoreneſs of the gullet and ſtomach, with unquenchable thirſt, and dyſury, remained. Theſe ſymptoms went off by free dilution; and he ventured the third day after to repeat the doſe of ſalt, which had effects ſimilar to the former. A perfect cure was the conſequence of this ſingular practice.

The common ſorts of ſea ſalt, contrary to other neutral ones, part with a little of their acid in the boiling down of ſolutions of them to dryneſs *(a)*. To this cauſe are attributed the weakneſs of the ſalt prepared by that proceſs,

---

(a) This acid appears to proceed, not from the pure marine ſalt, but from the calcareous muriatic ſalt, or a combination of marine acid with earth, which all the common ſorts of ſea ſalt are found to partake of. On dropping into a ſolution of common ſea ſalt a little alkaline lye, the earth precipitates; and the acid, being thus ſaturated with alkali, is no longer diſpoſed to evaporate on boiling down the liquor.

SAL COMMUNIS.

and its difpofition to deliquiate in the air; both which imperfections are faid to be corrected by a fmall addition of frefh acid when the falt begins to concrete. Hence alfo diftilled fea water is manifeftly impregnated with acid, fo as to be unfit for drinking or for the common purpofes of life; unlefs a little chalk, vegetable afhes, or other like fubftances, be added in the diftillation, to abforb and keep down the acid extricated by the heat *(a)*: by this means the diftilled fluid proves perfectly fweet.

The acid of fea falt is completely difengaged from its alkaline bafis by the more powerful acid of vitriol; and may now be collected, in a concentrated ftate, by diftillation; but as, in this concentrated ftate, its fumes very difficultly condenfe, a little water is commonly added to promote that effect. On ten pounds of dry fea falt, the college of London directs fix pounds of oil of vitriol diluted with five pints of water, that of Edinburgh one pound of oil of vitriol diluted with equal its quantity of warm water, to be poured by little and little, under a chimney, that the operator may not be incommoded by the noxious fumes: the retort is placed in fand, and the diftillation performed with a fire gradually increafed till nothing more will arife. The fpirit may be freed from its fuperfluous water, by a fecond diftillation in a glafs cucurbit; the phlegmatic part rifing in a proper degree of heat, while the ftronger acid

Acidum muriaticum *Ph. Lond.*
Acidum muriaticum *vulgo* fpir. fal. marin. *Ph. Ed.*

(a) This affertion is contradicted by the fuccefs of fome late attempts to fupply fhips with frefh water by the diftillation of fea water. In thefe, good fweet water was obtained merely by fitting an apparatus to the fhips' boilers in which falt water was ufed for common purpofes. See, particularly, *Phipp's Voyage towards the North Pole.*

remains

remains behind. The distilled spirit proves nearly the same, whether the larger or smaller of the above proportions of oil of vitriol are used, the difference affecting chiefly the residuum: see the foregoing article. Its specific weight to that of water is stated by the London college at 1170 to 1000.

The marine acid is distinguished from the others, by its rising in white fumes; by its peculiar pungent smell; by its enabling the nitrous acid to dissolve gold, preventing its dissolving silver, and precipitating silver previously dissolved, but producing no precipitation in solutions of calcareous earths. It is sometimes given, from ten to sixty or seventy drops, properly diluted, as an antiphlogistic, diuretic, and for promoting appetite, and applied externally to chilblains, which are said by Linnæus to be radically cured by it in a short time, without fear of a return; but its principal use is in combination with other bodies.

Combined with volatile alkalies, it produces the officinal sal ammoniac. With the mineral fixt alkali, it regenerates common salt. With vegetable fixt alkalies, it forms a neutral salt of a sharper taste, and somewhat more difficult of fusion and solution, than common salt: this combination is prepared in the shops, by dropping into the marine spirit a lixivium of the fixt alkali till all effervescence ceases, and then evaporating the mixture to dryness: the same salt may be obtained from the matter which remains after the distillation of spirit of sal ammoniac with fixt alkali.

Sal marin. regeneratus vulgo.

With calcareous earths, it forms a very pungent saline compound, which difficultly assumes a crystalline form, deliquiates in the air, dissolves

Sal ammon. fixum vulgo. Sal muriatic. calcareus.

SAL COMMUNIS.

folves not only in water but in rectified fpirit of wine, and changes the colour of blue flowers of vegetables to a green. This falt is contained, in confiderable quantity, in fea water, and remains fluid after the cryftallization of its other faline matters: it is found alfo in fundry common waters, to which, like the calcareous nitre, it communicates, according to its quantity, a greater or lefs degree of hardnefs and indifpofition to putrefy: it is far more antifeptic than the perfect marine falt. It is faid to be diuretic and lithontriptic: the medicine commonly fold as a lithontriptic under the name of *liquid-fhell*, appears to be no other than a combination of this kind, confifting of calcined fhells diffolved in marine acid. Thefe combinations have been chiefly prepared, by mixing the calcareous earth with fal ammoniac, and urging the mixture with a gradual fire, till the volatile alkali of the fal ammoniac is either diffipated in the air or collected by diftillation, and only its acid left incorporated with the earth: fo much of the earth, as is fatiated with the acid, may be feparated from the reft by elixation with water.

This acid diffolves, among metallic bodies, zinc and iron pretty readily; copper and tin languidly; bifmuth and arfenic very difficultly and fparingly; lead, mercury, regulus of antimony, and filver, not at all, unlefs highly concentrated and applied in the form of fume: it diffolves, by digeftion, all metallic bodies when reduced to a ftate of calx, gold not excepted. Though it difficultly unites with metals, it adheres to moft of them more ftrongly than any other acid, and in part volatilizes them: it renders them likewife more fufible in the fire than the other acids do, and more difpofed to folution in fpirit of wine.

Of itself it is nevertheless the most averse of all acids to a perfect union with vinous spirits. If poured gradually into thrice its quantity of rectified spirit of wine, and the mixture, after digestion for some days, submitted to distillation in a sand heat; the spirit that comes over, appears to be little other than the acid simply diluted with the vinous spirit; whereas, when the nitrous or vitriolic acids are treated in the same manner, a new compound is formed by the intimate coalition of the acid spirit with the inflammable *(a)*. The dulcified marine acid has by some been held in great esteem against weakness of the stomach, indigestion, and other like complaints brought on by irregularities.

S A L I X.

SALIX Pharm. Edinb. Salix vulgaris alba arborescens C. B. Salix Ger. Salix fragilis Linn.
COMMON WHITE WILLOW: a pretty large tree, frequent in moist woods and hedges; producing loose spikes or catkins, either of imperfect barren flowers, or of seeds inclosed in down: it is the largest of the willows; and differs from the others, in the oblong pointed serrated leaves

---

(a) It is said that the marine acid may be combined with vinous spirits as intimately as the others, and an ethereal fluid produced from the mixture, by applying the acid spirit to the vinous while both are resolved by fire into vapour; or more commodiously, by using the acid in a high degree of concentration, such as is obtained by distillation from a mixture of mercury sublimate with tin, commonly called the *smoking spirit of Libavius*, and proceeding with this spirit and spirit of wine, in the same manner as with the other acids. It is supposed that this acid, in distillation from metallic substances, takes up a portion of the inflammable principle of the metal, which promotes its union with the vinous spirit.

being

being hoary on both sides, though moft fo' on the lower, and in the branches not being tough.

THE bark of this tree has lately been found an ufeful medicine in agues, of which many perfons have been cured by taking a dram of the powdered bark every four hours during the intermiffions, though in fome cafes it was neceffary to join to it a little Peruvian bark (fee the *Philofophical Tranfactions* for the year 1763). To the tafte this bark difcovers a pretty ftrong bitternefs and aftringency: with folution of chalybeate vitriol, it ftrikes an inky blacknefs. Thefe obfervations ferve therefore to confirm what has been remarked under the head of the Peruvian bark, that in the diftempers where that valuable medicine takes place, other bitters and aftringents are likewife ufeful, though in an inferiour degree. The aftringency of the willow bark is extracted both by water and fpirit; and the black matter, produced by adding vitriol to the watery infufion, is not difpofed to precipitate.

SALVIA.

SAGE: a low fhrubby plant, with fquare ftalks; obtufe wrinkled dry leaves fet in pairs, and large bluifh labiated flowers, in loofe fpikes on the tops of the branches: the upper lip of the flower is nipt at the extremity, the lower divided into three fegments. It is a native of the fouthern parts of Europe, common in our gardens, and flowers in June.

1. SALVIA *Pharm. Lond. & Edinb. Salvia major* C. B. *Salvia officinalis Linn.* Common fage: with the leaves nearly oval, but acuminated, fometimes green and fometimes red: both

both the green and red forts rife from the feeds of one and the fame plant.

2. Salvia hortensis minor. *Salvia minor aurita & non aurita* C. B. Small fage, or fage of virtue: with narrower leaves, generally whitifh, never red: moft of them have at the bottom a piece ftanding out at each fide in the form of ears. This is a variety of the former.

The leaves and tops of fage are moderately aromatic and corroborant, and ufed in debilities and relaxations both of the nervous and vafcular fyftem. Their fmell is pretty ftrong and not difagreeable; their tafte fomewhat warm, bitterifh, and fubaftringent: with folution of chalybeate vitriol, they ftrike a deep black colour. The fecond fort is both in fmell and tafte the ftrongeft, the firft moft agreeable. Of both kinds, the flowers are weaker and more grateful than the leaves; and the cup of the flower ftronger, and obvioufly more refinous, than any other part.

The leaves of fage give out their virtue both to water and rectified fpirit, moft perfectly to the latter; to the former they impart a brownifh, to the latter a dark green tincture. The watery infufion is often ufed as tea, and often acidulated with a little lemon juice for a diluent in febrile diftempers: the fpirituous tincture is in tafte ftronger than the watery, but the fmell of the fage is by this menftruum covered or fuppreffed. The leaves and flowery tops, diftilled with water, yield a fmall quantity of effential oil, fmelling ftrongly and agreeably of the herb, in tafte very warm and pungent, when newly diftilled of a fine greenifh colour, by age turning yellow or brown: the remaining decoction, divefted

SAMBUCUS.

vested of this aromatic and most active principle of the sage, yields an extract weakly bitterish, subastringent, and subsaline. The spirituous extract, in smell weak and somewhat different in kind from that of the herb itself, discovers to the taste a considerable aromatic warmth and pungency, resembling that of camphor, but milder.

SAMBUCUS.

ELDER: a plant, with finely serrated sharp-pointed leaves, set in pairs on a middle rib, with an odd one at the end; producing, on the tops of the branches, umbel-like clusters of small white flowers, followed each by a juicy berry, containing generally three seeds.

1. 'SAMBUCUS *Pharm. Lond. & Edinb. Sambucus fructu in umbella nigro* C. B. *Acte. Sambucus nigra Linn.* Elder tree: with nearly oval leaves, of which five or seven stand on one rib. It is a small tree or shrub, covered with an ash-coloured chapt bark, under which lies a thinner green one, and under this a white: it grows wild in hedges, flowers in May, and ripens its black berries in September.

The bark of this tree is recommended as a strong hydragogue in hydropic cases. Sydenham directs three handfuls of the inner bark to be boiled in a quart of milk and water till only a pint remains, of which one half is to be taken in the morning, and the other at night, and this repeated every day: he observes, that this medicine operates both upwards and downwards; and that if it does not vomit or purge at all, or but gently, it does no service. Boerhaave says, that the expressed juice of the middle bark, given from a dram to half an ounce (some go as far an ounce), is the best of hydragogues where the viscera are sound; and that it so powerfully dissolves the humours, and procures so plentiful watery discharges from all the emunctories, that the patient is ready to faint from the large and sudden inanition. The decoction and juice are recommended also, in smaller doses, as useful aperients and deobstruents in different chronical disorders. This bark has scarcely any smell, and very little taste: on first chewing, it impresses a kind of sweetishness, which is followed by a very slight but very durable acrimony, in which its medical activity seems to reside, and which it imparts both to watery and spirituous menstrua.

The leaves, of a faint unpleasant smell, and a strong, very nauseous, bitter kind of taste, are said to be purgative and emetic like the bark. They are celebrated externally against burns and inflammations, and for these purposes an ointment has been prepared for them in the shops: four ounces of the leaves, and the same quantity of the inner bark, fresh, were thoroughly bruised, and boiled in a quart of linseed oil till the watery moisture was consumed and the oil tinged of a green colour: the oil was then pressed out, and brought to the consistence of an ointment by melting in it six ounces of white wax.

The flowers of elder have an agreeable flavour, which they give over in distillation with water, and impart by infusion both to water and rectified spirit: on distilling with water a large quantity of the flowers, a small portion of a butyraceous essential oil separates. Infusions made from them while fresh are gently laxative and aperient: when dry, they are said to promote

SAMBUCUS.

mote chiefly the cuticular excretion, and to be particularly ferviceable in eryfipelatous and eruptive diforders. From thefe alfo an unguent is prepared, probably of equal efficacy with the other, and preferred by fome as being more elegant, by melting three pounds of mutton fuet with a pint of oil-olive, and boiling in this mixture four pounds of the full blown flowers till they are almoft crifp. *Unguentum fambuci Ph. Lond.*

The berries, in tafte fweetifh and not unpleafant, yield on expreffion a fine purplifh juice, which infpiffated to the confiftence of honey, either by itfelf or with the addition of half a pound of fine fugar to two pounds and an half ‡, proves an ufeful aperient and refolvent in recent colds and fundry chronical diforders, gently loofening the belly, and promoting urine and perfpiration. *Succ. bacc. fambuci fpiffat. Ph. Lond. Rob baccar. fambuci ‡ Ph. Ed.*

2. EBULUS. *Sambucus humilis five ebulus* C. B. *Chamæacte. Sambucus Ebulus Linn.* Dwarf-elder or Danewort: an herbaceous plant, dying to the ground in winter; with longer leaves than thofe of the elder tree, and nine leaves on one rib. It grows wild in fome parts of England, flowers in July, and produces ripe black berries in the beginning of September.

It is faid that this fpecies has the fame virtues with the preceding, but differs fomewhat in degree: that the bark (that of the root has been chiefly ufed) and the berries, are refpectively more efficacious, and the leaves lefs fo: that the rob or infpiffated juice of the berries, in dofes of half an ounce to an ounce, acts as a ftrong hydragogue, and in fmaller dofes as a powerful refolvent and deobftruent.

SANGUIS

SANGUIS DRACONIS.

SANGUIS DRACONIS *Pharm. Lond. & Edinb. Cinnabaris Græcorum.* Dragons-blood, so called: a resin, obtained from certain large palm-like trees *(Palmijuncus Dracò Rumph. amb. Calamus Rotang Linn.)* growing in the East Indies; brought over in oval drops wrapt up in flag-leaves, or in large and generally more impure masses composed of smaller tears; of a deep dark red colour, which changes, in pulverization, to a crimson. Sundry artificial compositions, coloured with the true dragons-blood, or with brazil wood or other materials, have been sometimes sold in the room of this commodity: some of these dissolve like gums in water, and others crackle in the fire without proving inflammable; whereas the genuine dragon's-blood readily melts and catches flame, and is scarcely acted on by watery liquors. It dissolves almost totally, by the assistance of heat, in rectified spirit, and tinges a large quantity of the menstruum of a deep red colour: it is likewise soluble in expressed oils, and imparts to them a red tincture, less beautiful than that which anchusa communicates.

This resin, in substance, has no perceptible smell or taste: when dissolved, whether in vinous spirits or in oils, it discovers some degree of pungency and warmth. It is usually looked upon as a gentle incrassant, desiccative, and restringent; and sometimes prescribed in these intentions against alvine and uterine fluxes, and ulcerations both internal and external.

SANICULA.

SANICULA officinarum C. B. *Sanicula europæa Linn.* SANICLE: an umbelliferous plant, with shining dark green roundish leaves, cut into five segments, serrated about the edges; and rough seeds, which stick to the clothes. It is perennial and evergreen, grows wild in woods on hilly grounds, and flowers in May.

This herb, recommended both externally and internally as a vulnerary or mild restringent, and supposed to have received its name from the sanative virtues ascribed to it, discovers to the taste a kind of bitterishness and roughness, followed by an impression of acrimony which affects chiefly the throat: in the fresh leaves, the taste is very weak; in the dry leaves, considerable; in the extracts made from them, by water and spirit, moderately strong.

SANTALUM.

SAUNDERS. Three different woods are brought under this name from the East Indies, in large billets: they are said to be the produce, chiefly, of the island Timor in the Indian ocean.

1. SANTALUM CITRINUM *Pharm. Edinb. Santalum album Linn.**(*a*). Yellow saunders: of a pale yellowish or brownish colour, and a close even grain. This wood has a pleasant smell, and a bitterish aromatic taste accompanied with an agreeable kind of pungency.

* (*a*) The Santalum album, below, is said not to be a wood of a different species, but the *alburnum* of the trunk of the same tree, the medullary part of which is the citrinum.

Distilled

Diftilled with water, it yields a fragrant effential oil, which thickens in the cold into the confiftence of a balfam, approaching in fmell to ambergris, or a mixture of ambergris and rofes: the remaining decoction, infpiffated to the confiftence of an extract, is bitterifh and flightly pungent. Rectified fpirit extracts, by digeftion, confiderably more than water: the colour of the tincture is a rich yellow. The fpirit, diftilled off, is lightly impregnated with the fine flavour of the wood: the remaining brownifh extract has a weak fmell, and a moderate balfamic pungency. This wood therefore, though at prefent among us difregarded, promifes to have a good claim to the corroborant virtues afcribed to it by Hoffman and others.

2. SANTALUM ALBUM. White faunders: of a clofe texture and ftraight fibres like the preceding, but of a paler whitifh colour. This fpecies, far weaker than the yellow, both in fmell and tafte, promifes very little medicinal virtue: it has long been entirely neglected, and is now rarely to be met with in the fhops.

3. SANTALUM RUBRUM *Pharm. Lond. & Edinb. Pterocarpus Santolinus Linn. Suppl.* Red faunders: of a dull red almoft blackifh colour on the outfide, and a deep brighter red within: its fibres are now and then curled, as in knots. This alfo, recommended as an aftringent and corroborant, appears to be of very little virtue, as it has no manifeft fmell, and little or no tafte: even of extracts made from it, with water or with fpirit, the tafte is inconfiderable. Its principal ufe is as a colouring drug. To watery liquors it communicates only a yellowifh tinge, but to rectified fpirit a fine deep red: a fmall quantity

SANTONICUM.

quantity of an extract made with this menstruum tinges a large one of fresh spirit of the same elegant colour; though it does not, like most other resinous bodies, dissolve in expressed oils, or communicate its colour to them: of distilled oils, there are some, as that of lavender, which receive a red tincture both from the wood itself and from the resinous extract, but the greater number does not.

Geoffroy and others take notice that the brazil woods are sometimes substituted to red saunders, and the college of Brussels doubts whether all that is sold among them for saunders is not really a wood of that kind. According to the account which they have given of their red saunders, it is plainly the brasil wood of the dyers; the distinguishing character of which is, that it imparts its colour to common water. Of the same kind also is the wood examined by Cartheuser under the name of red saunders, the watery infusion and extract of which were both of a dark red.

SANTONICUM.

SANTONICUM SEMEN Pharm. Lond. & Edinb. Semen cinæ, semen sanctum, semen contra, sementina. WORMSEED: a small light oval seed; composed as it were of a number of thin membranous coats; of a yellowish-greenish colour with a cast of brown; easily friable, by rubbing between the fingers, into a fine chaffy kind of substance. The seeds have commonly mixed with them a considerable quantity of this chaffy matter, and small bits of stalks and leaves. They are brought from the Levant, and supposed to be the produce of a species of *artemisia*, resembling in its general appearance our fine-
leaved

leaved mugwort, called by Linnæus *Artemisia (Santonicum) foliis caulinis linearibus pinnato-multifidis, ramis indivisis, spicis secundis reflexis*; the *Artemisia austriaca* of Jacquin.

These seeds have a moderately strong, not agreeable smell, somewhat of the wormwood kind; and a very bitter subacrid taste. They have been chiefly recommended as anthelmintics; and commonly taken, in this intention, either along with melasses, or candied with sugar. They might be used also for other purposes; as they appear (at least the specimens which I examined) to be a not inelegant strong bitter. They give out their virtue both to water and spirit, together with a brownish hue, which in the watery tincture has an admixture of reddish, in the spirituous of yellow: the spirituous is less ungrateful in taste, and discovers less also of the ill smell of the santonicum than the watery infusion. In evaporation, water carries off greatest part of the disagreeable flavour of the seeds, the inspissated extract being little other than simply bitter. An extract made by rectified spirit retains a considerable share of the flavour: this extract appears to be the most eligible preparation of the santonicum for the purposes of an anthelmintic; and the watery extract, or a tincture drawn from it, for the more general intentions of bitter medicines.

S A P O.

SOAP: a composition of oils or fats with alkaline salts, incorporated so as to dissolve together in water into a milky semitransparent liquid.

1. Sapo

1. SAPO DURUS. Hard foap. The fineft hard foap is prepared with frefh-drawn oil of almonds, by digefting it with thrice its meafure of the foap-lyes, formerly defcribed (fee *Sales alkalini*) in fuch a heat that they may juft fimmer. In a few hours they unite into a turbid fluid, which, on being boiled a little, becomes more tranfparent, and ropy, fo that if a little be fuffered to cool, it will concrete like jelly. Some fea falt is now thrown in, till the boiling liquor lofes its ropinefs; and the coction continued till, on receiving fome drops upon a tile, the foap is found to coagulate, and the water to freely feparate from it. The fire being then removed, the foap rifes gradually to the furface; from whence it is taken off before it grows cold, and put into a wooden mould, or frame, with a cloth bottom: being afterwards feparated from the mould, it is fet by till it has acquired a due confiftence. After the fame manner a hard foap is made with oil-olive, which fhould be of the fineft kind, that the foap may prove as little ungrateful as poffible either to the palate or ftomach. By the fame or fimilar proceffes this commodity is prepared for common ufes in the way of trade. The fineft of the common foaps is that called Spanifh or Caftile foap, which is made with oil-olive, and the alkaline falt called foda or barilla: our foap-boilers find that this alkali gives a better confiftence, or greater hardnefs to the foap, than the other potafhes or common vegetable alkalies. [Sapo amygdalinus. Sapo Ph. Lond. mat. med. Sapo albus hifpanus Ph. Ed. mat. med.]

Hard foap, triturated with vegetable refins and thick balfams, incorporates with them into a compound, foluble, like the foap itfelf, in watery liquors: hence it proves an ufeful ingredient in refinous pills, which of themfelves are apt to pafs entire through the inteftines, but by the

the admixture of soap become diffoluble in the stomach. It renders unctuous and thick mucous animal matters diffoluble in like manner in aqueous fluids, and hence may be presumed to act as a menstruum for these kinds of substances in the body, that is to attenuate viscid juices and resolve obstructions: such, in effect, are the virtues which it appears to exert in cachectic, hydropic, and icteric cases, in which last, particularly, its aperient and resolvent powers have been often experienced. Solutions of it have been found likewise to dissolve certain animal concretions of the harder kind, as the filaments which are sometimes seen floating in the urine of rheumatic and arthritic persons, the matter secreted in gouty joints, and the more compact urinary calculus: on these substances (at least on the latter) though soap of itself acts more languidly than limewater, yet when joined to that menstruum it remarkably increases its activity, the dissolving power of a composition of the two being, according to Dr. Whytt's experiments, considerably greater than that of the soap and limewater unmixed: of the good effects of these medicines in calculous cases there are several instances; but what their effects may be in gouty and rheumatic ones, is not yet well known.

The usual dose of soap, as an aperient, is half a dram or a dram: as a lithontriptic, half an ounce, or an ounce, or more, are taken in a day at proper intervals. It is given in the form of a bolus or pills; or made into an electuary with some grateful syrup, as that of orange peel; or dissolved in milk or other liquids. * It is excellently covered by chocolate: two drams in a pint are not in the least perceived;

SAPO.

perceived; the chocolate is thought by some better than without it. A little soap is always added in the composition of chocolate, to make it froth.

In watery liquors it dissolves only imperfectly, the solution being always turbid. Rectified spirit, though it has no action on the alkaline salt or oil separately, dissolves the soap into a limpid liquor. Proof spirit, free from acidity, dissolves it as perfectly, and in much larger quantity; rectified spirit not taking up one tenth its own weight, but proof spirit one third or more. The spirituous solutions bear to be largely diluted with pure water, without suffering any turbidness or separation of their parts: but on the addition of any acid, or of any combination of acids with earthy or metallic bodies, as the *sal catharticus amarus*, &c. the soap is resolved into its constituent ingredients; its alkaline salt being absorbed by the acid, and the oil rising to the surface. The oil, thus extricated from soap by acids, dissolves like essential oils, in rectified spirit.

Soap is employed externally for discussing rheumatic pains, arthritic tumours, the humours stagnating after sprains, &c. Some pretend that the indurated tophaceous concretions in arthritic joints have been resolved by the external use of soapy cataplasms. Several compositions for external purposes are prepared in the shops. One part of Spanish soap, shaved or cut in thin slices, is stirred into six parts of common plaster melted over the fire, and the mixture boiled till it acquires the consistence of a plaster; which is formed into rolls whilst hot, the soap disposing it to grow brittle as it cools† : some endeavour to promote the resolvent virtue of the soap, by adding to four parts

† Emplastrum saponis *Ph. Lond.*

parts of the common plafter, two of gum plafter, with one of foap‡. But foap acts to much better advantage in the form of a cataplafm or liniment, than in the ftiff one of a plafter. The officinal faponaceous liniments are made, by digefting three ounces of Spanifh foap in a pint of fpirit of rofemary till the fpirit is faturated, and diffolving in this folution an ounce of camphor ‖ : or by digefting two ounces of foap in a pint of rectified fpirit of wine, and afterwards adding an ounce of camphor, and two drams of oil of rofemary §. Sometimes opium is joined, by which the compound is fuppofed to be rendered more effectual for allaying violent pains: half an ounce of opium is digefted with the foap in the laft mentioned compofition. This is given alfo internally, in nervous colics, jaundices, &c.

Emp. faponac. † *Ph.Ed.*

Linimentum faponis ‖ *Ph. Lond.*

vulgo Balf. faponaceum § *Ph. Ed.*

Linimentum anodynum vulgo Balf. anodyn. *Ph. Ed.*

2. Sapo mollis. Soft foap. The common foft foap ufed about London, generally of a greenifh hue with fome white lumps, is prepared chiefly with tallow: a blackifh fort more common in fome other places, is faid to be made with whale oil. Both kinds are confiderably more acrid than the hard foaps, and are employed only for fome external purpofes: a mixture of equal parts of our common foft foap and quicklime is ufed as a mild cauftic.

3. Sapo volatilis. Volatile foap. Of this there are three kinds: one compofed of fixt alkalies and volatile oils; another, of volatile alkalies, and oils of the groffer or more fixt kind; and the third, in which both the alkali and the oil are volatile.

Fixt alkalies are very difficultly made to unite with diftilled oils. The moft commodious

ous method of obtaining the combination appears to be, by throwing the falt red-hot into a heated mortar, immediately reducing it into powder, then pouring on it, while it continues quite hot, by little at a time, an equal quantity or more of the oil, and continuing to grind them together, fo as to form a fmooth foft mafs. Stahl reports that the union may foon be obtained alfo, by agitating the falt with a fmall proportion of the oil, and a quantity of phlegmatic vinous fpirit; the fpirit feeming to ferve as a medium for joining them together. This medicine, prepared with oil of turpentine, was formerly celebrated as a diuretic, in nephritic complaints, and as a corrector of certain vegetables, particularly of opium: its virtues have not been fully determined by experience, nor does the prefent practice pay any regard to it. Beaumé obferves, that it confifts of only the refinous part of the oil united with the alkali; that the more fluid and well rectified the oil is, the lefs foap is obtained; and that by adding a little turpentine in fubftance, the preparation is confiderably expedited. *Sapo philofophicus tartareus, &c.*

Combinations of volatile alkalies with expreffed oils, and with the oily balfamic juices, are obtained more readily. One ounce of fpirit of fal ammoniac, and three of oil of olives, fhaken together in a wide-mouthed vial, unite perfectly, in a fhort time, into a white faponaceous liquid: or, for a more active preparation, one ounce of the volatile fpirit with quicklime is fhaken with two ounces of olive oil. Both thefe compofitions are very acrimonious, and are ufed only externally, as ftimulants, in rheumatic and ifchiadic pains. *Liniment. ammoniæ Ph. Lond.* *Linim. ammoniæ fortius Ph. Lond.*

Combinations of volatile alkalies with volatile oils, in a liquid form, have been already mentioned

tioned under the head of *sal alkalinus volatilis:* compositions of the same kind may be obtained in a solid state, by mixing the salt with the oil, and subliming them together. It may be observed, that in all these combinations made with volatile salts, though the pungency of the salt is more or less covered, it is never completely sheathed as that of the fixt alkalies is in the hard soaps; and that none of the compositions, in which either the alkali or the oil is volatile, are so perfectly saponaceous as those in which they are both of the more fixt kind.

SAPONARIA.

SAPONARIA major lævis C. B. *Saponaria officinalis Linn.* SOAPWORT or BRUISEWORT: a smooth herb, with plantain-like three-ribbed leaves set in pairs on short broad pedicles; producing, on the tops of the stalks, umbel-like clusters of red, purple, or whitish flowers, cut deeply into five segments nipt at the ends, standing in long cups, followed by pear-shaped capsules full of small seeds: the root is long, slender, spreading to a great distance, so as scarce to be extirpated, of a brownish colour on the outside, internally white, with a yellowish fibre in the middle. It grows wild, but not very common, in moist grounds, and flowers in July.

THE roots and leaves of saponaria discover to the taste a kind of glutinous softness or smoothness; accompanied, in the roots, with a sweetishness and slight pungency; in the leaves, with a degree of bitterness and roughness. The smoothness or soapiness, from which the plant received its name, is strongest in the leaves; which,

which, on being agitated with water, raife a flippery froth, and are faid to impart a detergent quality approaching to that of folutions of foap itfelf. This matter is diffolved alfo by rectified fpirit as well as water, and hence appears evidently of a different nature, from gummy or mucilaginous fubftances: on infpiffating the folutions, it remains entire in the extracts, and proves ftronger in the fpirituous extract than in the watery. This plant therefore, among us, difregarded, may be prefumed to have fome confiderable medicinal virtues: by the German phyficians, the roots are ufed in venereal maladies, and fuppofed to be fimilar, but fuperiour, to thofe of farfaparilla. * A phyfician in Paris is faid lately to have given the infpiffated juice of this plant to the quantity of half an ounce in a day, to perfons labouring under a gonorrhœa, with fuccefs.

SAPONARIÆ NUCULÆ.

NUCULÆ faponariæ non edules C. B. *Saponariæ fphærulæ arboris filicifoliæ* J. B. *Baccæ bermudenfes Marloe.* Soap-berries: a fpherical fruit, about the fize of a cherry; whofe cortical part is yellow, gloffy, and fo tranfparent, as to fhew the fpherical black nut, which rattles within, and which includes a white kernel. It is the produce of a fmall tree, growing in Jamaica and other parts of the Weft Indies, called by Sir Hans Sloane *prunifera racemofa, folio alato, cofta media membranulis utrinque extantibus donata, fructu faponario;* by Linnæus, *Sapindus Saponaria.*

It is faid that this fruit, at leaft its cortical part, has a very bitter tafte, and no fmell: that it raifes a foapy froth with water, and has fimilar effects with foap in wafhing: that it is a medicine of fingular and fpecific virtue in chlorofes: and that a tincture or extract are preferable to the berry in fubftance, from whence it may be prefumed that its foapy matter, like that of the faponaria, is diffoluble in fpirit. Its medicinal virtue was firft publifhed by Marloe in a letter to Mr. Boyle; but the fruit having been concealed under the fictitious name of Bermudas berries, its ufe died with the author. That Marloe's Bermudas berries were the fame with the foap-berries of America, had been fufpected by fome, and was confirmed by Dale in examining the Bermudas berries which Marloe had left under that title behind him. They are ftill, however, unknown in practice, and in the fhops.

SARCOCOLLA.

SARCOCOLLA Pharm. Lond. A concrete gummy-refinous juice, brought from Perfia and Arabia, in fmall, fpongy, crumbly, whitifh-yellow grains, with a few of a reddifh and fometimes of a deep red colour mixed with them: the tears, when entire, are about the fize of peas: the whiteft tears or fragments are preferred, as being the frefheft. The plant which produces this juice, and the place of its production, are unknown.

SARCOCOLLA has a bitterifh fubacrid tafte, followed by a naufeous kind of fweetifhnefs. It foftens in the mouth, bubbles and catches flame from a candle, diffolves almoft wholly in water, and greateft part of it in rectified fpirit. Its medicinal qualities are not well known: it is said,

SARSAPARILLA.

said, when taken internally, to act as a flow and dangerous purgative; externally, to cleanse and promote the cicatrization of ulcers: dissolved in breast-milk, to be an useful collyrium for defluxions on the eyes.

SARSAPARILLA.

SARSAPARILLA Pharm. Lond. & Edinb.
Zarza quibusdam: the root of a species of bindweed, *smilax. aspera peruviana sive sarsaparilla* C. B. *Smilax (sarsaparilla) caule aculeato angulato, foliis inermibus retuso-mucronatis Linn.* growing in the Spanish West Indies, and scarcely bearing the winters of our climate without shelter. The root consists of a number of strings, of great length, about the thickness of a goose-quill or thicker, flexible, free from knots, composed of fibres running their whole length, so that they may be stript in pieces from one end to the other. They are covered with a thin, brownish, or yellowish ash-coloured skin, under which lies a thicker, white, friable substance, and in the middle runs a woody pith.

This root has a farinaceous somewhat bitterish taste, and no smell. To water it communicates a reddish brown, to rectified spirit a yellowish red tincture, but gives no considerable taste to either menstruum. An extract, obtained by inspissating the spirituous tincture, has a weak, somewhat nauseous, balsamic bitterishness, which is followed by a slight but durable pungency: the watery extract is much weaker, and in larger quantity.

Sarsaparilla was first brought into Europe by the Spaniards, about the year 1563, with the character of a specific for the cure of the lues venerea, which made its appearance a little before that time. Whatever good effects it might have produced in the warmer climates, it was found to be insufficient in this, insomuch that many have denied it to have any virtue at all, and supposed that it could do no more than, by its farinaceous softness, to obtund the force of the gastric fluid, and thus weaken the appetite and digestion. It appears however, from experience, that though greatly unequal to the character which it bore at first, yet, in many cases, strong decoctions of it, drank plentifully and duly continued, are of very considerable service, for promoting perspiration, and what is called sweetening or purifying the blood and humours. In the medical observations published by a society of physicians in London, there are several instances of its efficacy in venereal maladies, as an assistant to mercury, or when mercury had preceded it use: it oftentimes answered, and that speedily, after mercurial unctions, and long continued courses of strong decoctions of guaiacum, had failed. Three ounces of the root are boiled in three quarts of river water, till the liquor when strained amounts to about one quart, which is taken at three or four doses, either warm or cold, every twenty-four hours. Dr. Harris says, that infants who have received the infection from the nurse, though full of pustules and ulcers, and sometimes troubled with nocturnal pains, are cured by sarsaparilla without mercurials: he directs the powder of the root to be mixed with their food.

Decoct. sarsapar. *Ph. Lond.*

* The London college have now admitted as officinals a simple decoction of sarsaparilla, made by boiling (after maceration) six ounces of the

root

root in eight pints of water to four; and a compound one, in which, to fix ounces of farfaparilla are added one ounce each of faffafras, guaiacum wood, and liquorice, and three drams of the bark of mezereon root, to be boiled in ten pints of water to five.   Decoct. farfapar. comp. *Ph. Lond.*

SASSAFRAS.

SASSAFRAS Pharm. Lond. & Edinb. The root of a large American tree of the bay kind, *(laurus (faffafras) foliis integris trilobifque Linn. Arbor ex florida ficulneo folio* C. B.*)* brought over in long ftraight pieces, very light and of a fpongy texture, covered with a rough fungous bark, outwardly of an afh-colour, inwardly of the colour of rufty iron.

This root has a fragrant fmell, and a fweetifh fubaftringent, aromatic tafte: the bark is much ftronger than the internal woody part; and the fmall twigs than the larger pieces. It gives out its virtues, together with a reddifh colour, totally to fpirit, lefs perfectly to water: the fpirituous tincture fmells weakly and taftes ftrongly, the watery fmells ftronger and taftes weaker of the root. Diftilled with water, it yields a fragrant effential oil, of a penetrating pungent tafte, fo ponderous as to fink in water, limpid and colourlefs when newly diftilled, by age growing yellowifh and at length of a reddifh brown colour: the remaining decoction, infpiffated, yields a bitterifh fubaftringent extract. Rectified fpirit, diftilled from the tincture made in that menftruum, brings over with it nothing confiderable: the infpiffated extract retains, along with the bitternefs and fubaftringency, nearly all the aromatic Ol. effentiale rad. faffafr. *Ph. Lond. & Ed.*

matter

matter of the root, though the smell is in great part suppressed in the extract as well as in the tincture.

Sassafras is used as a mild corroborant, diaphoretic, and sweetener, in scorbutic, venereal, cachectic, and catarrhal disorders. For these purposes, both the volatile and the fixt parts, the distilled oil and the watery extract, have been given with success: the spirituous tincture or extract, which contain both, appear to be the most elegant preparations. Infusions made in water, from the cortical or the woody part rasped or shaved, are commonly drank as tea: in some constitutions, these liquors, by their fragrance, are apt, on first taking them, to affect the head; an inconvenience, which is generally got the better of on continuing their use for a little time, and which neither the watery nor spirituous extracts are at all subject to.

SATUREIA.

SATUREIA hortensis sive cunila sativa plinii C. B. *Thymbra. Satureia hortensis* Linn. SUMMER SAVOURY; a low, shrubby, somewhat hairy plant: with small oblong narrow leaves, narrowest at the bottom, set in pairs; and small clusters, in the bosoms of the leaves, of pale purplish labiated flowers, whose upper lip is nipt at the extremity, the lower cut into three segments. It grows wild in some of the southern parts of Europe, and is sown annually in our culinary gardens.

THE leaves of savoury are a warm aromatic; of a grateful smell, like that of thyme but milder; and a penetrating pungent taste. To
rectified

SATYRION.

rectified spirit, they give out the whole of their active matter, together with a dark green tincture: water receives from them a reddish brown colour, and a considerable smell, but very little of their taste. In distillation with water, they yield a small quantity of a fragrant essential oil, very pungent, and of great subtility and volatility: the remaining decoction, infpissated, leaves a weakly bitterish, subastringent, ungrateful extract. Rectified spirit elevates in distillation much less than might be expected from the remarkable volatility of the oil: the extract smells agreeably, though weakly, of the favoury, and has a very warm, pungent, aromatic taste.

SATYRION.

SATYRION Pharm. Edinb. Orchis morio mas foliis maculatis C. B. Cynoforchis & testiculus caninus quibufdam. Orchis mafcula Linn. ORCHIS: a plant with six or seven long smooth narrow leaves, variegated with dark-coloured streaks or spots, iffuing from the root; and one or two embracing the stalk, which is single, roundish, and striated: on its top appears a long loose spike of irregular, naked, purplish red flowers, consisting each of six petala; one of which is large, cut into three sections, hanging downwards; the others smaller, forming a kind of hood above it, with a tail behind: the root confists of two roundish whitish tubercles, about the size of nutmegs, one plump and juicy, the other fungous and somewhat shrivelled, with a few large fibres at the top. It is perennial, grows wild in shady grounds and moist meadows, and flowers in the beginning of May or sooner.

THE

The plump roots or bulbs (the only part directed for medicinal use) have a faint somewhat unpleasant smell, and a viscid sweetish taste. They abound with a glutinous slimy juice, in virtue of which they have been found serviceable, like althea root and other mucilaginous vegetables, in a thin acrid state of the humours and erosions of the intestines. They have been celebrated also for aphrodisiac virtues, to which they appear to have little claim.

The substance brought from the eastern countries under the names of *Salep, salleb,* and *serapias,* and recommended, like our orchis root, in bilious dysenteries, defluxions on the breast, and as a restorative, appears to be no other than the prepared roots of some plants of the orchis kind, of which different species are said to be taken indiscriminately. The salep comes over in oval pieces, of a yellowish white colour, somewhat clear and pellucid, very hard and almost horny, of little or no smell, in taste like gum tragacanth. The common orchis root, boiled in water, freed from the skin, and afterwards suspended in the air to dry, gains exactly the same appearance: the roots thus prepared do not grow moist or mouldy in wet weather, which those, that have been barely dried, are very liable to: reduced into powder, they soften or dissolve as it were in boiling water into a kind of mucilage; which may be diluted, for use, with a larger quantity of water, or with milk.

* The following process for the preparation of salep from the English orchises, by Mr. Moult, of Rochdale, is published in the *Philos. Transact.* vol. lix. " The new root is to be washed in water, and the fine brown skin which covers it is to be separated by means of a small brush, or by dipping the root in hot water,

water, and rubbing it with a coarfe linen cloth. When a fufficient number of roots have been thus cleaned, they are to be fpread on a tin plate, and placed in an oven heated to the ufual degree, where they are to remain fix or ten minutes, in which time they will have loft their milky whitenefs, and acquired a tranfparency like horn, without any diminution of bulk. Being arrived at this ftate, they are to be removed, in order to dry and harden in the air, which will require feveral days to effect; or by ufing a very gentle heat, they may be finifhed in a few hours."

SAXIFRAGA.

SAXIFRAGA rotundifolia alba C. B. Saxifraga granulata Linn. WHITE SAXIFRAGE: a plant with kidney-fhaped crenated yellowifh-green leaves, and round flender purplifh branched ftalks, on the tops of which grow fhort loofe fpikes of pentapetalous white flowers, followed each by a two-horned capfule full of fmall feeds: the root is compofed of fmall fibres, with a number of little tubercles among them, about the fize of pepper-corns, containing under a chaffy covering, irregular whitifh bodies fomewhat brittle like the kernels of fruits. It is perennial, grows wild in fandy pafture-grounds, and flowers in May: the leaves and ftalks wither foon after flowering, and by degrees the tubercles of the roots alfo difappear.

THE leaves of this plant, of little or no fmell, and of a weak unpleafant tafte; and the tubercles of the roots, improperly called feeds, of no fmell, and in tafte fweetifh with a very flight acrimony; are recommended as aperients

and

and diuretics, in obstructions of the menses, stranguries, and nephritic cases. Among us, they have long been disused, and unknown in the shops;' a more common plant, of the same name, but of a different genus, and of more activity, having generally supplied their place, viz.

SAXIFRAGA VULGARIS.

SAXIFRAGA vulgaris sive anglica, Hippomarathrum anglicum, Fœniculum erraticum: Seseli pratense, silaus forte plinio C. B. Angelica pratensis apii folio Tourn. Peucedanum Silaus Linn. English or Meadow saxifrage: an umbelliferous plant, with winged leaves subdivided into oblong narrow sharp-pointed segments: the flowers are of a yellowish white colour, the umbel naked, but its subdivisions have several little leaves at their origin; the seeds are short, brownish or reddish, plano-convex, with three deep furrows so as to appear winged: the root is long, about the thickness of the finger, brownish or blackish on the outside, and white within. It is perennial, common in meadows and pasture grounds, and flowers in June.

The roots, leaves, and seeds of this plant have been commended as aperients, diuretics, and carminatives; and appear, from their aromatic smell, and moderately warm pungent bitterish taste, to have a better claim to these virtues than the preceding saxifrage. They are rarely or never used.

SCAMMONIUM.

SCAMMONIUM Pharm. Lond. & Edinb. Diagrydium. Scammony: the concrete gummy

SCAMMONIUM.

my-refinous juice of the roots of a species of convolvulus *(convolvulus (scammonia) foliis sagittatis postice truncatis, pedunculis bifloris Linn.)* distinguished by the leaves being shaped like an arrow-head and having two semicircular notches at the bottom on each side of the footstalk, the flowers being of a pale yellowish colour and standing two on one stem: it is a native of Syria, and has been lately found to bear the colds of our own climate. The scammony is extracted in Syria, by baring the upper part of the root in June, cutting off the top obliquely, and placing a shell or some other receptacle at the depending part to receive the milky juice, which on standing concretes into solid masses.

The best scammony is brought from Aleppo, in light spongy masses, easily friable, glossy, of different shades of colour from a grey or yellowish white almost to black, when reduced to powder of a brownish white colour. An inferior sort comes from Smyrna, in compact hard ponderous pieces, full of sand and other impurities. Such should be chosen as crumbles the most easily betwixt the fingers, grows instantly white on the contact of watery moisture, and leaves little or no feces on being dissolved. Its colour in the mass affords no criterion of its purity or goodness.

Scammony has a slight unpleasant smell, and a weak bitterish subacrid taste. It consists of about equal parts of resinous and gummy matter, and hence dissolves almost totally in a mixture of equal parts of rectified spirit and water, that is, in proof spirit. Rectified spirit takes up the resin, with some part of the gum: if the tincture be inspissated a little, and then mixed with water, the gum continues dissolved, and the pure resin precipitates. By trituration
with

with water, or by bare maceration, the scammony is refolved into a milky liquor verging to
greenifh; which on ftanding depofites fome
portion of the refin, but retains its milkinefs.

This gummy-refin is one of the ftrong ftimulating cathartics; more kindly in operation,
and hence in more general ufe, than moft of
the other fubftances of that clafs: the dofe is
from two or three grains to twelve. Sundry
ill qualities have been afcribed to it, which it is
not found to poffefs: and fundry correctors
have been devifed, which it does not appear to
want. In cold indolent ferous habits, fcammony itfelf procures generally a plentiful evacuation with great eafe and fafety: in inflammatory cafes, and the more irritable difpofi-
tions, it is indeed dangerous; but no otherwife fo than the reft of the ftrong purgatives;
and no otherwife than by virtue of that power
on which its efficacy in the oppofite circumftances depends.

By the fmallnefs of the dofe of this medicine,
its eafy folubility, and its having little tafte, it
is fitted for being commodioufly taken in almoft
any form. It is made in the fhops into a powder, *Pulvis e fcammonio comp. † Ph. Lond. ‡ Ph. Ed.*
with the addition of an equal weight of hard
extract of jalap, and a fourth of its weight of
ginger†; or with equal its weight of cryftals of
of tartar‡. It is likewife combined with aloes, *Pulv. e fcammon. cum aloe Ph. Lond.*
and alfo with calomel, in different officinal
powders. A fcammoniate electuary is com- *Pulv. e fcammon. cum calom. Ph. Lond.*
pofed of one ounce of fcammony, aromatifed
with half an ounce of cloves, half an ounce of
ginger, and a fcruple of the effential oil of
caraway-feeds made up with fyrup of rofes;
of which compofition, one dram and a half
contain fifteen grains of the fcammony§. Agree- *§ Electar. e fcammonio Ph. Lond.*
able purging troches, for thofe who are not
eafily

easily prevailed upon to take medicines of this kind in other forms, are prepared, by grinding together three drams of scammony, four drams of crystals of tartar, four drops of oil of cinnamon, and eight ounces of fine sugar, and moistening the mixture with so much rosewater as will render it of a due consistence for being formed: each tablet is made to weigh about a dram†, and consequently contains two grains and a half of scammony. One of the most elegant liquid preparations is a solution of the scammony in a strong infusion or decoction of liquorice, poured off from the feces, and aromatised with some grateful distilled water or aromatic tincture; as those of cardamom-seeds.

† Morsuli purgantes *Ph. Brandenburgh.*

The dried root of the plant, as well as its juice, may perhaps deserve some notice. Dr. Russel, to whom the public is obliged for an accurate history of this drug, relates that a decoction of half an ounce of the root procured five stools, without gripes, sickness, or any manner of uneasiness, and, on repeating the trial several times, had the same effect: and that the decoctions are entirely without smell, and in taste rather sweetish than disagreeable. Neither the stalks, leaves, flowers, or seeds, seemed to have any purgative virtue *(a)*.

SCILLA.

SCILLA *Pharm. Lond. & Edinb.* Scilla radice alba, & scilla vulgaris radice rubra C. B. Ornithogalum maritimum *Tourn.* Scilla maritima *Linn.* SQUILL or SEA-ONION: a plant with a

(a) Medical observations and inquiries, by a society of physicians in London.

large

large bulbous onion-like root; from which rife, firſt a naked ſtalk bearing ſeveral hexapetalous white flowers, and afterwards large green lily-like leaves with a remarkable rib in the middle of each. It grows ſpontaneouſly on ſandy ſhores in Spain and in the levant, from whence we are annually ſupplied with the roots. They ſhould be choſen large, plump, freſh, and full of a clammy juice: ſome are of a reddiſh colour, and others white, but no difference is obſerved in the qualities of the two ſorts, and hence the college allows both to be taken promiſcuouſly.

This root is to the taſte very nauſeous, intenſely bitter, and acrimonious: much handled, it exulcerates the ſkin. Taken internally, it acts as a powerful attenuant and aperient: in doſes of a few grains it promotes expectoration and urine: in ſomewhat larger ones, it proves emetic and ſometimes purgative. It is one of the moſt certain diuretics in hydropic caſes, and expectorants in aſthmatic ones, where the lungs or ſtomach are oppreſſed by tenacious phlegm, or injured by the imprudent uſe of opiates.

This medicine, on account of its ungrateful taſte, is moſt commodiouſly taken in the form of pills; into which the dried root may be reduced, by beating it with thrice its weight each of ammoniacum and leſſer cardamom-ſeeds in powder, and extract of liquorice, and a ſufficient quantity of ſimple ſyrup† or one dram of dried ſquills may be mixed with three drams each of powdered ginger and ſoap, and two drams of ammoniacum, making up the maſs with ſyrup of ginger‡. In whatever form ſquills are given, unleſs when deſigned to act as an

Pil. ſcilliticæ
Ph. Ed. †

Pil. e ſcilla
Ph. Lond. ‡

an emetic, the addition of some grateful aromatic material is of use, to prevent the nausea which of themselves they are very apt, even in small doses, to occasion.

The fresh root loses in drying about four fifths of its weight, without any considerable loss of its taste or virtue; the vapour which exhales appearing to be little other than merely aqueous. Hence four grains, which are the mean dose of the dry root in powder, are equivalent to near a scruple of the fresh squill. The most convenient way of drying it is, after peeling off the outer skin, to cut the roots transversely into thin slices (not to separate the coats, as has been usually directed) and expose them to a gentle warmth. *Scilla exsiccata Ph. Lond. & Ed.*

The ancients, in order to abate the acrimony of the squill for certain purposes, inclosed it (after separating the skin, and the fibres at the bottom with the hard part from which they issue) in a paste made of flour and water, and then baked it in an oven, till the paste became dry, and the squill soft and tender throughout. The squill, so prepared, was beaten with two-thirds its weight of flour, the mixture formed into troches, and dried with a gentle heat. These troches were supposed to be alexipharmac, and in this light were made an ingredient in theriaca. The virtues of the fresh squill may be preserved by beating it with sugar into a conserve, in the proportion of one ounce of the squill to five ounces of fine sugar. *Scilla cocta.* *Troch. e scilla.* *Conf. scillæ Ph. Lond.*

Water, wine, proof spirit, and rectified spirit, extract the virtues both of the fresh and the dry root. The London college have directed a tincture in which two ounces of fresh dried squills are digested for eight days in a pint of proof spirit. Nothing rises in distillation with *Tinct. scillæ Ph. Lond.*

Vol. II.   A a   any

any of these menstrua, the entire bitterness and pungency of the squill remaining concentrated in the inspissated extracts: the spirituous extract is in smaller quantity than the watery, and of a proportionably stronger almost fiery taste.

Alkalies considerably abate both the bitterness and acrimony of the squill: vegetable acids make little alteration in either, though the admixture of the acid taste renders that of the squill more supportable. These acids extract its virtue equally with watery or spirituous menstrua; and, as an expectorant in disorders of the breast, excellently coincide with it.

Acetum scillæ Ph. Lond. The college of London directs an acetous tincture to be prepared, by macerating a pound of the dry roots in six pints of vinegar, with a gentle heat: to the liquor pressed out, and after settling poured off from the feces, one twelfth its quantity of proof spirit is added, to prevent *Acetum scillit. Ph. Ed.* its growing soon foul. The college of Edinburgh for the same preparation directs the proportions of two ounces of dried squills, two pounds and a half of distilled vinegar, and three ounces of *Oxym. scil. lit. Ph. Lond.* rectified spirit. A scillitic oxymel is obtained by boiling a quart of the acetous tincture with three pounds of clarified honey, till the mixture *Syr. scillit. Ph. Ed.* acquires the consistence of a syrup: and a syrup of squills, by dissolving three pounds and a half of fine sugar in two pounds of the vinegar. These preparations are used, as expectorants, in doses of one, two, or three drams, along with cinnamon or some other grateful water: where the first passages are overloaded with viscid phlegm, an ounce or more is given at once, to procure a more speedy and effectual evacuation by vomit.

SCINCUS.

SCINCUS.

SCINCUS seu crocodilus terrestris Raii. The Skink: a small amphibious animal, of the lizard kind, clothed with greyish scales, caught about the Nile, &c. and thence brought, dried, to us, remarkably smooth and glossy as if varnished. The flesh of this animal, particularly of the belly, has been said to be diuretic, alexipharmac, aphrodisiac, and useful in leprous disorders. Whatever virtues it may have when used fresh, as food, it is not expected to be of any importance as it comes to us, and serves only to increase the number of the articles of which mithridate is composed.

SCORDIUM.

SCORDIUM Pharm. Lond. & Edinb. C. B. Chamædrys palustris & trissago palustris quibusdam. Teucrium Scordium Linn. Water-germander: a trailing plant, with oblong, oval, indented, soft hoary leaves, set in pairs, without pedicles: in their bosoms issue purplish monopetalous flowers, not above four or five together, each cut into five segments and followed by four small seeds lodged in the cup. It is sometimes found wild in watery places, but the shops are supplied chiefly from gardens: it is perennial, and flowers in June.

The leaves of scordium, rubbed betwixt the fingers, yield a moderately strong smell, somewhat of the garlic kind: to the taste they discover a considerable bitterness and some pungency; but the astringent power, which some ascribe to them, could not be distinguished, either by the taste, or by solution of chalybeate vitriol. They are recommended as alexipharmacs and corroborants, in malignant and putrid disorders, and in laxities of the intestines: they enter several officinal compositions in those intentions, and are sometimes employed externally in antiseptic cataplasms and fomentations.

On keeping the dry herb for some months, its smell is dissipated; and the bitterness, thus divested of the flavouring matter, proves considerably less ungrateful than at first. The leaves, moderately and newly dried, give out their smell and taste both to water and to rectified spirit; and tinge the former of a brownish, the latter of a deep green colour. In distillation, their peculiar flavour arises with water; but the impregnation of the distilled fluid is not strong, nor could any essential oil be obtained on submitting to the operation several pounds of the herb: the remaining decoction, inspissated, leaves a very bitter mucilaginous extract. Rectified spirit brings over little or nothing: the inspissated extract partakes in a considerable degree of the flavour of the scordium, and proves in bitterness also far stronger than the watery.

SCORZONERA.

SCORZONERA latifolia sinuata C. B. Viperaria & serpentaria hispanica quibusdam. Scorzonera hispanica Linn. VIPERS-GRASS: a plant with large sharp-pointed leaves, slightly sinuated about the edges, having a large prominent rib in the middle, joined to the stalks without pedicles: on the tops of the branches grow yellow flosculous flowers, set in scaly cups, followed by oblong roundish striated seeds winged with down: the root is long, single, from the size of a goose

SCROPHULARIA.

a goose-quill to that of the little finger, of a dark colour on the outside and white within. It is perennial, a native of Spain, and common in our culinary gardens.

THE roots of scorzonera have been employed medicinally as alexipharmacs, and in hypochondriacal disorders and obstructions of the viscera; but at present are more properly looked upon as alimentary articles, in general salubrious, and moderately nutritious. They abound with a milky juice, of a soft sweetish taste, but which in drying contracts a slight bitterness. Extracts made from them by water are considerably sweet and mucilaginous: extracts made by rectified spirit have a less degree of sweetishness, accompanied with a slight grateful warmth. In Cartheufer's experiments, the spirituous extract amounted to one third the weight of the root, and the watery to above one half: as his watery extract, though in larger quantity than the spirituous, was nevertheless, like mine, sweeter, it should seem that the sweet matter of scorzonera is somewhat different, in regard to its solubility, from that of most of the other vegetable sweets that have been examined, the spirituous extracts having generally much the greatest sweetness.

SCROPHULARIA.

FIGWORT: a plant with square stalks; the leaves set in pairs, at distances, in opposite directions; the branches terminated by loose spikes of irregular, purple, helmet-shaped flowers; each of which is followed by a roundish pointed capsule, containing numerous small seeds in two cells. It is perennial.

MATERIA MEDICA.

1. SCROPHULARIA *nodofa fœtida* C. B. *Mille-morbia quibufdam.* *Scrophularia nodofa Linn.* Common figwort or kernelwort: with the leaves fomewhat heart-fhaped and ferrated about the edges; the roots long, thick, and full of knots and tubercles. It grows wild in woods and hedges, and flowers in July.

The roots and leaves of this plant have been celebrated both internally and externally, againft inflammations, the piles, fcrophulous tumours, and old ulcers. Their fenfible qualities are, a rank fmell fomewhat like that of elder leaves but ftronger, and a difagreeable bitterifh tafte. The anodyne and anti-inflammatory virtues, which they are reckoned to exert in external applications, are attributed in great part to the odorous matter, which is fuppofed to be fomewhat of the narcotic kind: the root, which has lefs of this fmell than the leaves, has been generally preferred for internal ufe. At prefent, they are both among us difregarded.

2. SCROPHULARIA *aquatica major* C. B. *Betonica aquatica.* *Scrophularia aquatica Linn.* Greater water figwort, water betony: with the leaves oblong, nearly oval, crenated about the edges; the ftalks winged at the angles; the root compofed of numerous white ftrings iffuing from one head. It grows in watery places, and flowers in July.

The leaves of this fpecies are recommended for the fame purpofes as thofe of the preceding, to which they have by fome been preferred: in tafte and fmell, they are fimilar, but weaker. Mr. Marchant reports, in the Memoirs of the French Academy, that this plant is the fame with the *iquetaia* of the Brazilians, celebrated as a fpecific correftor of the ill flavour of fenna: on
his

SEDUM.

his authority, the Edinburgh college, in their common infusion of that drug, directed two thirds its weight of the water of figwort leaves to be joined; but as they have now discarded this ingredient, we may presume that it was not found to be of much use.

SEDUM.

SEDUM majus vulgare C. B. *Aizoon & barba jovis quibusdam. Sempervivum tectorum Linn.* HOUSELEEK or SENGREEN : a plant with numerous, thick, stiff, fleshy, pointed leaves, lying over one another in form of a roundish cluster; in the middle of which rises a stiff stalk, covered with smaller leaves, divided at the top into several branches, bearing purplish flowers with twelve petala, which are followed by the same number of capsules full of small seeds. It is perennial and evergreen, grows on old walls and the tops of houses, and flowers in June.

The leaves of houseleek, of no remarkable smell, discover to the taste a mild subacid austerity : their expressed-juice, of a pale yellowish hue when filtered, yields on inspissation a deep yellow, tenacious, mucilaginous mass, considerably acidulous and acerb: from whence it may be presumed, that this herb has some claim to the refrigerant and restringent virtues that have been ascribed to it. It is observable that the filtered juice, on the addition of an equal quantity of rectified spirit of wine, forms a light white coagulum, like creme of fine pomatum, of a weak but penetrating taste : this, freed from the fluid part, and exposed to the air, almost totally exhales. From this experiment it is concluded by some that houseleek contains

contains a volatile alkaline salt *(a)*: but the juice coagulates in the same manner with volatile alkalies themselves, as also with fixt alkalies: acids produce no coagulation.

SELENITES.

SELENITES: an earthy or stony concrete; not dissoluble in acids; calcining in a gentle heat into a soft powder†, which forms a tenacious paste with water: composed of calcareous earth and vitriolic acid.

† Plaster of Paris.

THE vitriolic acid, poured on crude calcareous earths, as chalk, limestone, marble, does not dissolve or unite with them, at least in any considerable degree: but if the earth be previously dissolved in any other acid, the vitriolic acid, superadded to this solution, absorbs the dissolved earth, and forms with it a concrete no longer soluble, which of course renders the liquor milky, and on standing settles to the bottom, either in a powdery or crystalline form, according as the liquor was less or more diluted with water. Native mineral concretes of this kind, when pellucid and crystalline, are called *selenites*; when composed of a number of thin transparent coats or leaves, *lapis specularis*, *Muscovy glass*, or *isinglass*; when in large stony masses, of a granulated texture, *gypsum*; and when the masses are of a fibrous texture, *striated gypsum* or *English talc*. All these substances are made to discover their composition, by strongly calcining them in contact with the burning fuel: the inflammable principle of the coals absorbs their vitriolic acid, from which combination is produced common

(a) Burghart, *Medicorum Silesiacorum satyræ, specim.* IV. *obs.* ii. *p.* 11.

sulphur,

SENNA.

fulphur, greateſt part of which exhales; and the remaining calcined earth, thus deprived of the acid, is found to be a perfect quicklime.

This concrete, in its different forms, has been recommended as an aſtringent in fluxes and hemorrhagies; a virtue which agrees but ill with its indiffolubility and want of taſte. It is often met with in the refidua of waters, both of the common and medicinal fprings.

SENNA.

SENNA. *Pharm. Lond. & Edinb. Folium orientale.* SENNA: the leaf of an annual, woody, pod-bearing plant *(ſenna alexandrina ſive foliis acutis C. B. Caſſia (ſenna) foliolis trijugatis quadrijugatiſque Linn.)* brought dry from Alexandria in Egypt. It is of a lively yellowiſh green colour, an oblong fomewhat oval figure, ſharp-pointed at the ends, about a quarter of an inch broad, and not a full inch in length. Some inferiour forts are brought from Tripoli and other places; theſe may be diſtinguiſhed by their being either narrower, longer, and ſharper pointed; or larger, broader, and round pointed, with ſmall pr m n n veins; or large, obtuſe, and of a freſh green colour without any yellow caſt.

SENNA is a moderately ſtrong, and in general a fafe cathartic: Geoffroy fpecifies hemorrhagies, inflammations of all kinds, and diſorders of the breaſt, as being almoſt the only exceptions to its uſe. The doſe in ſubſtance is from a ſcruple to a dram; in infuſion, from one dram to three or four. It gives out its virtue both to watery and ſpirituous menſtrua: to water and proof ſpirit it communicates a browniſh colour,

more

more or less deep according to the proportions; to rectified spirit, a fine green. There are two inconveniences often complained of in this medicine, its being liable in most constitutions, to occasion gripes; and its being accompanied with an ill flavour, which is apt to nauseate both the stomach and the palate. The first may be greatly obviated by dilution, the latter by aromatic and other additions; several compositions of this kind are prepared in the shops, both sufficiently palatable, and which operate for the most part with ease and mildness.

Inf. sennæ simpl. Ph. Lond.

The most simple infusion is that ordered by the London college, in which six drams of senna with half a dram of powdered ginger are directed to be macerated during an hour in half a pint of boiling water.

Infusum tamarindorum cum senna Ph. Ed.

For the acidulated infusions, six drams of tamarinds, one of cryftals of tartar, half a dram of coriander seeds, half an ounce of brown sugar, and one, two or three drams of senna are infused in eight ounces of boiling water, in an unglazed earthen vessel, for four hours, and then strained.

Infuf. sennæ tartarisat. Ph. Lond.

Or three drams of senna are infused in a quarter of a pint of boiling water, till the liquor has grown cold, with a dram of coriander seeds bruised, and half a dram of cryftals of tartar, which laft are previoufly boiled in the water till diffolved;

Infuf. sennæ limoniatum.

or with two drams of fresh lemon peel, and two drams by measure of lemon juice. The former committee of the London college observed, that this last was the most agreeable form, they had been able to contrive, for the exhibition of senna to those who are more than ordinarily offended with its flavour; and that though acids are generally supposed to impede the action of water on vegetables, yet the infusions of senna made with acids were found, from experience,

not

SENNA.

not to fail in their intention. Indeed if the acids really weaken the diffolving power of the water, which it is probable they do in fome degree, it fhould feem to be, on this account, rather of advantage than otherwife; for, as the committee further obferved, in a medicine very naufeous to many, it is of primary confequence that only the lighter and leaft difguftful parts be extracted. On this principle, fome macerate the fenna for a night in cold water, which becomes fufficiently impregnated with its purgative virtue, without extracting fo much, as boiling water does, of the naufeous matter: if the liquor, poured off from the fenna, be boiled a little by itfelf, great part of its ill flavour will be diffipated; and the remains of its offenfivenefs may be covered by infufing in it fome bohea tea. If the coction is continued for any confiderable time, the purgative virtue of the fenna will be diminifhed; for the infpiffated watery extracts are fcarcely found to purge fo much, as one fourth of the infufion or decoction they were made from, or fo much as an equal weight of the leaves in fubftance. The London college have now admitted an extract of this kind. *Extr. fennæ Ph. Lond.*

The officinal fpirituous tinctures of fenna are prepared by digeftion for fome days, in proof fpirit. The proportions, in the London pharmacopœia, are three ounces of fenna to a quart of the fpirit, to which are added four ounces of ftoned raifins, three drams of caraway feeds, and one dram of leffer cardamom feeds hufked: in the Edinburgh, two ounces of fenna to three pounds and an half of the fpirit, with the addition of one ounce of jalap, half an ounce of coriander feeds, and four ounces of white fugar-candy in powder, which laft is directed to be diffolved in the tincture after ftraining it from *Tinctura fennæ Ph. Lond. Tinct. fennæ comp. vulgo elixir falutis Ph. Ed.*

the

the other ingredients. Both thefe tinctures are agreeable and ufeful carminative purgatives, efpecially to thofe who have accuftomed themfelves to fpirituous liquors: the ill flavour of the fenna is in great meafure covered, and its offending the ftomach or producing gripes prevented, by the warm feeds and the fweets. Several compofitions of this kind have been offered to the public, under different names; the two above are inferiour to none; and fuperiour to moft of them.

SERPENTARIA.

SERPENTARIA VIRGINIANA Pharm. Lond. & Edinb. Serpentaria virginiana & Viperina & Colubrina virginiana Pharm. Parif. VIRGINIAN SNAKEROOT: the root of a fpecies of ariftolochia growing in Virginia and Carolina, *ariftolochia (ferpentaria) foliis cordatis oblongis planis, caulibus infirmis flexuofis teretibus, floribus folitariis Linn.* The root is fmall, light, bufhy, compofed of a number of ftrings or fibres iffuing from one head and matted together, of a brownifh colour on the outfide, and paler or yellowifh within.

SNAKEROOT has an aromatic fmell, approaching to that of valerian, but more agreeable, and a warm bitterifh pungent tafte, which is not eafily concealed or overpowered by a large admixture of other materials. It gives out its active matter both to water and rectified fpirit, and tinges the former of a deep brown, the latter of an orange colour. Greateft part of its fmell and flavour is carried off in evaporation or diftillation by both menftrua: along with water there arifes, if the quantity of the root
submitted

SESELI.

submitted to the operation be large, a small portion of a pale-coloured essential oil, of a considerable smell, but no very strong taste, greatest part of the camphorated pungency, as well as bitterishness of the root, remaining in the inspissated extract. The spirituous extract is stronger than the watery; not so much from its having lost less in the evaporation, as from its containing the active parts of the root concentrated into a smaller volume; its quantity amounting only to about one half of that of the other.

This root is a warm diaphoretic and diuretic. It is reckoned one of the principal medicines of the alexipharmac kind; and as such is in general use, in low malignant fevers and epidemic diseases, for raising the pulse, promoting a diaphoresis, and correcting a putrid disposition of the humours. It is given, in substance, from a few grains to a scruple or half a dram; in decoction or infusion, to a dram and upwards. Tinctures of it are prepared in the shops, by macerating three ounces of the root in a quart of proof spirit†, or two ounces, with one dram of cochineal, in two pounds and a half of the same spirit ‡.

Tinctura serpentariæ
† *Ph. Lond.*
‡ *Ph. Ed.*

SESELI.

LIGUSTICUM quod seseli officinarum C. B. *Laserpitium Siler Linn.* HARTWORT or SERMOUNTAIN: a tall umbelliferous plant, with large leaves, composed of oblong pointed sections set in pairs or three together: the entire umbel, and its subdivisions, have a circle of little leaves at their origin: the seeds are large, of a pale brown colour, oblong, flat on one side, convex and striated on the other,

other, and edged with a leafy margin: the root is large, thick, and branched. It is perennial, grows wild in some of the southern parts of Europe, is raised with us in gardens, and flowers in June.

Both the seeds of this plant, which are the part directed in our pharmacopœias, and the roots, appear to be useful aromatics, though not regarded in practice; of an agreeable smell, and a warm glowing sweetish taste. The roots have the greatest warmth and pungency: the seeds, the greatest sweetness and the most pleasant flavour. A spirituous extract of the seeds is a very elegant aromatic sweet.

SESELI MASSILIENSE.

SESELI MASSILIENSE fœniculi folio C. B. Fœniculum tortuosum J. B. Seseli tortuosum Linn. Hartwort of Marseilles: a large spreading branched umbelliferous plant; with the stalk and branches firm, woody, knotty, and variously bent; the leaves finely divided, like those of fennel, but somewhat thicker, shorter, stiffer, and more distant from one another; the seeds also in shape like those of fennel, and of a pale grey colour. It is perennial, and a native of the southern parts of Europe, from whence the seeds are sometimes brought to us.

The seeds of this plant have an agreeable aromatic smell, and a very warm biting taste: they are more pungent than those of the foregoing seseli, but want their sweetness.

SIGILLUM

SIGILLUM SALOMONIS.

CONVALLARIA *feu Sigillum Salomonis Ph. Edinb. (a)* Polygonatum *latifolium vulgare C. B.* Convallaria multiflora *Linn.* SOLOMONS-SEAL : a plant with unbranched ftalks, bearing oval narrow leaves ribbed like thofe of plantain, generally all on one fide : on the other fide hang oblong monopetalous white flowers, two or more together, on long pedicles, followed each by a black berry : the root is white, thick, flefhy, with feveral joints, and fome flat circular depreffions fuppofed to refemble the ftamp of a feal. It is perennial, grows wild in woods, and flowers in May.

THE roots of Solomons-feal are recommended externally as reftringents; and internally as incraffants and mild corroborants. They have little or no fmell; to the tafte they difcover a confiderable fweetnefs and vifcidity, followed by a very flight impreffion of bitterifhnefs and acrimony, which is diffipated by boiling. It is faid that they have been ufed with fuccefs in the hæmorrhoids *(b)*. The flowers, berries, and leaves, are acrid and poifonous *(c)*.

SIMAROUBA.

SIMAROUBA *Pharm. Lond. & Edinb.* the bark of the *Quaffia Simarouba, Linn. fuppl.* *Quaffia dioica, Pharm. Suec.* Evonymus fructu nigro tetragono *Barrer. Æquin.* It is brought from Guiana, in long pieces, of a yellowifh

(a) The Edinburgh college give the *Convallaria Polygonatum* of Linnæus as their fpecies.

(b) Cullen, *Mat. Med.* *(c)* Id.

white

white colour, light, tough, and of a fibrous texture.

Mr. de Juffieu reports, that this bark is of common use in Guiana, againft dyfenteric fluxes, and was brought from thence into Europe in the year 1713: that the fluxes which, in France, fucceeded the exceffively hot fummer of 1718, and which not only refifted, but were aggravated by, purgatives, aftringents, and ipecacoanha, happily yielded to fimarauba: that decoctions of an ounce or half an ounce in a fmall quantity of water, the dofe ufed by the natives of Guiana, occafioned often vomiting, almoft always uneafy fweats, and fometimes an increafe of the bloody and ferous difcharges by ftool; but that a decoction of two drams in a quart of water, boiled to the confumption of one third, divided into four dofes, and taken warm at intervals of three hours, abated the pain in one day, and when continued for a fhort time completed the cure, without producing any naufea or difturbance: that it is not accompanied with the ill effects of aftringents: that it abates fpafmodic and hyfteric fymptoms: that it anfwers beft in fluxes of the ferofo-bilious, bloody and mucous kind, fupported by a convulfive motion of the inteftines, where there is no fever, where the functions of the ftomach are unhurt, and in tenefmi *(a)*. Dr. Degner likewife made ufe of this bark in the above form, with good fuccefs, after proper evacuations, in an epidemic putrid dyfentery, which raged at Nimeguen during the fummer and autumn of 1736: he fays it acted mildly

(a) Mem. de l'acad. des fcienc. de Paris, 1729. Geoffroy, mat. med. ii. 211.

and

SINAPI.

and almoft infenfibly, and that its effects were fpeedier in bloody than in bilious difcharges: he takes notice alfo that the barks procured under the name of fimaruba, in different parts of Holland, from Leipfick, and from Paris, differed greatly in quality from one another; but does not mention what the differences were, nor the qualities of the genuine or beft fort *(a)*.

The fimaruba, which I have met with in our fhops, has a moderately ftrong, durable, not very ungrateful bitter tafte, without fmell, and without any manifeft aftringency. Macerated in water, or in rectified fpirit, it quickly impregnates both menftrua with its bitternefs, and with a yellow tincture. It feems to give out its virtue more perfectly to cold than to boiling water; the cold infufion being rather ftronger in tafte than the decoction; which laft, of a tranfparent yellow colour whilft hot, grows turbid and reddifh brown as it cools. The milky appearance, which Juffieu fays it communicates to boiling water, I have not obferved in the decoction of any of the fpecimens I examined.

SINAPI.

SINAPI Pharm. Lond. Sinapi album Pharm. Edinb. Sinapi rapi folio C. B. Sinapis nigra Linn. * *(b)*. MUSTARD: an annual plant; with long rough leaves divided to the rib into irregular fegments, of which the extreme one is

---

(a) Hift. dyfenteriæ biliofo-contagiofæ, in Append. ad act. nat. curiof. vol. v.

* *(b)* The *Sinapis alba Linn.* is the Edinburgh fpecies, which differs little from the black, or common, except in being lefs pungent and bitter. It fhould feem therefore to be lefs proper for external ufe, at leaft.

largeſt; producing, at the tops of the branches, tetrapetalous yellow flowers, followed, each, by a ſhort, ſmooth, quadrangular pod, divided longitudinally by a membrane which projects at the ends, containing ſmall, roundiſh, reddiſh-brown or dark-coloured ſeeds. It is a native of England, but commonly cultivated for medicinal and dietetic uſe.

Mustard seed is one of the ſtrongeſt of the pungent, ſtimulating, diuretic medicines that operate without exciting much heat. It is ſometimes taken, unbruiſed, to the quantity of a ſpoonful at a time; in paralytic, cachectic, and ſerous diſorders. In this manner of exhibition it generally opens the body; whereas the powder is apt to occaſion vomiting, in which intention it is ſometimes given diffuſed in warm water, of which repeated draughts muſt be drunk, to continue the effect. It is applied alſo, as an external ſtimulant, to benumbed or paralytic limbs; to parts affected with fixt rheumatic pains; and to the ſoles of the feet, in the low ſtage of acute diſeaſes, for raiſing the pulſe: in this intention, a mixture of equal parts of the powdered ſeeds and crumb of bread, with the addition, ſometimes, of a little bruiſed garlick, are made into a cataplaſm with a ſufficient quantity of vinegar.

Muſtard ſeed yields upon expreſſion a conſiderable quantity of oil, which is by ſome recommended externally againſt rheumatiſms and palſies, though it has nothing of that quality by which the ſeeds themſelves prove uſeful in thoſe diſorders; the oil being mild and inſipid as that of olives, and the pungency of the ſeed remaining entire in the cake left after the expreſſion.

expreſſion. Nor is any confiderable part of the pungent matter extracted by rectified fpirit; the tincture, which is of a pale amber colour, having very little taſte; and the extract, obtained by infpiſſating it, being only bitteriſh and oily: the quantity of extract is about one fixteenth the weight of the feeds. The bruifed feeds give out readily to water nearly the whole of their active matter: added to boiling milk, they curdle it, and communicate their pungency to the whey. Diſtilled with water, they yield a limpid eſſential oil, extremely pungent and penetrating both in fmell and taſte, and fo ponderous as to fink in the aqueous fluid: the remaining decoction, thus divefted of the principle in which alone the acrimony of the muſtard refides, leaves on being infpiſſated a fweetiſh mucilaginous extract.

* *S I U M.*

WATER-PARSNEP. A genus of umbelliferous plants, growing in watery fituations, with winged leaves, ſtriated feeds, and a polyphyllous involucrum. Of thefe, a fpecies has of late years come into frequent ufe under the name of *Sium aquaticum*, but this appellation equally fuiting the three Engliſh kinds, it was a matter of doubt which of them was intended, and different opinions were given by botaniſts. The London college have at length determined the point by admitting into their catalogue the

SIUM NODIFLORUM *Linn.* CREEPING WATERPARSNEP, which is diſtinguiſhed from the others by the reclining pofition of the leaves, and by the manner in which the umbels of flowers come out, chiefly from the axillæ of the leaves.

It is a very common plant, often entirely covering the bottom of ditches. In cutaneous eruptions, and the cafes termed fcorbutic, the expreſſed juice of the water-parfnep has been given in the dofe of three large ſpoonfuls twice a day to children, and three or four ounces every morning to adults, with great advantage. It is not naufeous, and is readily taken by children when mixed with milk. In thefe dofes, it has no fenfible effect on the head, ftomach, or bowels.*(a)*

SOLANUM.

NIGHTSHADE: a plant with a monopetalous flower, divided into five fegments, having its cup divided in the fame manner, with the fame number of ftamina in the middle, and followed by a juicy berry.

1. BELLADONNA *Pharm. Edinb. & Parif.* Solanum melanocerafos *C. B.* Solanum lethale. *Atropa Belladonna Linn.* Deadly nightfhade or dwale: with the leaves oval, pointed, fomewhat hairy; the flowers folitary in the bofoms of the leaves, of a dull purplifh colour, tubulous, flightly cut, with the ftamina feparate from one another; the berries of a gloffy black. It is perennial, grows wild in fome fhady wafte grounds, and flowers in July.

2. SOLANUM *Pharm. Parif.* Solanum officinarum *C. B.* Solanum nigrum *Linn.* Garden nightfhade: with the leaves oval, pointed, having generally fome irregular indentations; the flowers in clufters, white, not tubulous,

(a) Dr. Withering in the *Botan. Arrangement, fecond edit.* p. 293.

deeply

SOLANUM.

deeply cut, the fegments fpread out, and the tips of the ftamina united into one button; the berries black. It is annual, grows fpontaneoufly in cultivated grounds, and flowers in Auguft.

The leaves of thefe plants have a faint fmell, fomewhat of the narcotic kind, which in drying is diffipated: on the organs of tafte, whether frefh or dry, they make fcarcely any impreffion. Their effects are neverthelefs very powerful: in external applications, they are faid to act as refrigerants, refolvents, and difcutients: taken internally, in the quantity of not many grains, they are highly deleterious, the firft fomewhat the moft fo. In very fmall dofes, as an infufion in boiling water of half a grain or a grain of the dried leaves, they occafion a warmth over the whole body; which is often followed by a fweat, or an increafe of the urinary difcharge, or fome loofe ftools, or a ficknefs and vomiting; and often by a headach, giddinefs, dimnefs of the fight, and other paralytic fymptoms *(a)*. In fome cancerous, ulcerous, and hydropic cafes, thefe infufions have been repeated, at bed-time, every two or three nights or oftener, and the quantity of the leaves in each dofe increafed gradually to five or fix grains or more, with apparent benefit: but they are fo variable and irregular in their operation, and fo liable, not only

(a) Mr. Ray gives an account, from his own knowledge, of a pretty remarkable effect of a fmall piece *(particula)* of a frefh leaf of belladonna applied externally to a little ulcer, fuppofed cancerous, below the eye: the uvea became in one night fo relaxed, that it loft all power of contracting the pupil, which, though expofed to the ftrongeft light, continued dilated to four times its natural fize, till the leaf being removed the parts gradually recovered their tone. The application was repeated three feveral times, and produced always the fame effect. *Hift. plant.* 680.

to fail of giving relief, but to be productive of very alarming symptoms by strongly affecting the nervous system, that their use is deservedly laid aside. Their good effects, when they happen to prove medicinal, seem to depend, not on any alterative or peculiarly deobstruent power, but merely on the evacuations they produce: where they do not act as evacuants, they generally aggravate they complaints (a).

* In the *Med. Comment.* vol. i. p. 419, is a remarkable case of the efficacy of an external application of belladonna in discussing a scirrhous tumour in the rectum, near the anus, which almost totally blocked up the passage. The mode of application was a poultice of the root boiled in milk. This was applied to the anus and perinæum, and renewed morning and evening. In the space of a month it entirely dissolved the tumour, without any suppuration, or discharge of matter. The writer says that he could add more instances of the good effects of this plant externally applied.

The roots and berries appear to partake of the deleterious qualities of the leaves, though probably in different degrees: the berries in particular seem to be of much less activity. It is said that three or four of the berries of the deadly nightshade, which are reckoned more virulent than those of the other sort, have been sometimes eaten without injury: Gesner reports that there expressed juice, boiled with a little sugar to the consistence of a syrup, proves, in doses of a tea-spoonful, an effectual and safe anodyne, but gives a particular caution not to exceed this

(a) See Mr. Gataker's *observations* (and *the supplement* thereto) *on the internal use of the nightshade*, and Mr. Bromfield's *account of the English nightshades and their effects.*

dose

dose. The Edinburgh college has directed the inspissated juice of the leaves to be kept as an officinal. *Succus spissatus belladonnæ Ph. Ed.*

3. DULCAMARA *Pharm. Edinb. Solanum scandens seu dulcamara* C. B. *Amaradulcis & glycypicros quibusdam. Solanum Dulcamara Linn.* Woody nightshade or bittersweet: with several of the leaves, particularly the upper ones, cut deeply into three sections, or rather furnished with two smaller appendages at the bottom; the flowers in clusters, of a blue colour, with the segments spread out and the stamina united as in the second species; the berries red. It grows by the sides of ditches and in moist hedges, climbing upon the bushes, with winding, woody, but brittle stalks. It is perennial, and flowers in June and July.

The roots and stalks of this species impress, on first chewing them, a considerable bitterness, which is soon followed by an almost honey-like sweetness. They have been commended in different disorders, as high resolvents and deobstruents: their sensible operation is by sweat, urine, and stool; the dose from four to six ounces of a tincture made by digesting four ounces of the twigs in a quart of white wine. Experience has shewn, that they are by no means equally deleterious with the two preceding nightshades; that they act more regularly and uniformly: and that, without producing nervous complaints, they produce more considerable evacuations, especially by stool; but their virtues in particular cases have not yet been sufficiently ascertained.

* In a medical dissertation on this plant, printed at Upsal, a light decoction and infusion of the stalks is the preparation recommended, and is said to have been frequently employed with success in violent ischiadic and rheumatic pains. The efficacy of the dulcamara in the jaundice, scurvy, suppressed menses, and the lues venerea, is also mentioned from other authors.

SPERMA CETI.

SPERMA CETI Pharm. Lond. & Edinb.
SPERMACETI, improperly so called: a species of fat; found in certain wales, particularly in their heads; artificially purified, by boiling with alkaline lye, to a snowy whiteness; and afterwards broken into flakes. It differs from the other animal fats, in not being dissoluble by alkalies or combinable with them into soap; and in rising almost totally in distillation, not in form of a fluid oil, but in that of a butyraceous matter resembling, both in consistence and smell, the butter of wax. In long keeping, it is apt to turn yellow and rancid: the matter, very small in quantity, which has suffered this change, and which taints the rest, is found to have lost the discriminating characters of the spermaceti; being dissoluble both by alkaline lye and by vinous spirits, so as to leave the remainder white and sweet as at first.

THIS concrete, of a soft butyraceous taste and no remarkable smell, is given with advantage in tickling coughs, in dysenteric pains and erosions of the intestines, and in such cases in general as require the solids to be softened and relaxed, or acrimonious humours to be obtunded. It readily dissolves in oils, and unites by liquefaction with wax and resins; and in these forms is applied externally. For internal

SPIGELIA.

nal ufe, it may be diffolved in aqueous liquors into the form of an emulfion, by trituration with almonds, the yolk or white of an egg, and more elegantly by mucilages; or made into a lohoch with proper additions.

*SPIGELIA.

SPIGELIA Pharm. Lond. & Edinb. *Anthelmia* D^{ris}. Lining. *Spigelia marilandica* Linn. *Periclymeni virginiani flore coccineo planta marilandica, fpica erecta, foliis conjugatis* Catefby carol. INDIAN PINK: this plant has a perennial fibrous root, whence rife fingle ftems, befet with oppofite oval-lanceolate, entire leaves, and crowned with a fpike of tubular monopetalous red flowers, with five ftamina and one piftil. Each flower is fucceeded by two round united byvalvular capfules, containing feveral fmall feeds. It grows fpontaneoufly in South Carolina, and other fouthern provinces of North America.

The ufe of the root of this plant as an anthelmintic was communicated from the native Indians to the colonifts, and it has fince been much employed in that country. The firft account of its virtues is to be met with in a paper of Dr. Lining's, in vol. i. of the *Effays Phyfical and Literary*; and Dr. Garden has confirmed it in vol. iii. of the fame publication, and has given a figure and particular defcription of the plant.

The root is given both in powder and infufion; but the powder is efteemed moft efficacious. The dofe is not accurately afcertained, but extends from twelve to fixty or feventy grains of the powder. It is found to be moft efficacious when it purges, which it does not always do without fome additions. The exhibition

bition of a vomit previous to the use of the Indian pink has proved very serviceable. It sometimes produces disagreeable effects on the nervous system, such as giddiness, dimness of the sight, and convulsive motions of the muscles of the eye. These, according to Dr. Garden, are more likely to happen from a small dose than a large one, the latter more certainly proving purgative or emetic. Dr. Lining, on the other hand, represents these effects as consequent upon too large a dose. It is said to act powerfully as a sedative in abating the exacerbations of low remittent worm-fevers.

SPINA CERVINA.

SPINA CERVINA *Pharm. Lond.* Rhamnus catharticus five Spina cervina *Pharm. Edinb.* Rhamnus catharticus *C. B. & Linn.* Spina infectoria et cervispina quibusdam. BUCKTHORN: a prickly bush, or low tree, common in hedges: with oval pointed leaves; producing in June small greenish flowers; and about the beginning of October ripening its black berries, which contain a dark green juice, with four seeds in each. The berries of the black alder and dogberry tree, which are frequently, in our markets, mixed with or substituted for those of buckthorn, may be distinguished, by their juice having no greenness, and by their containing only one or two seeds.

BUCKTHORN BERRIES have a faint unpleasant smell, and a bitterish, acrid, nauseous taste. They operate briskly by stool; and occasion, at the same time, a thirst and dryness of the mouth and throat, and not unfrequently severe gripes, especially if water-gruel or other soft
diluents

diluents are not freely drank foon after taking them. The dofe is faid to be, about twenty of the frefh berries in fubftance; twice or thrice that number in decoction; a dram, or a dram and a half, of the dried berries; an ounce of the expreffed juice; or half an ounce of the rob or extract obtained by infpiffating the juice. Among us they have been employed only in the form of a fyrup, in which they feem to operate lefs unkindly than in any other, and which is given by itfelf in dofes of three or four fpoonfuls, or mixed in fmaller quantities with other cathartics. The college of Edinburgh directs the fyrup to be prepared by boiling the depurated juice with fugar to a due confiftence: that of London adds a little ginger, and pimento, with a view to cover in fome degree the ill flavour of the buckthorn: but notwithftanding this improvement of the medicine, it is ftill fo unpleafant and fo churlifh, that it has now almoft fallen into difufe.

Syrupus fpinæ cervinæ Ph. Lond. & Ed.

SPIRITUS VINOSUS.

VINOUS SPIRIT: an inflammable fluid, obtained by diftillation from wines or other fermented liquors. As firft diftilled, it partakes both of the phlegm or watery part, and of the oil, of the fermented liquor; which oil, in the liquors commonly ufed for this purpofe, is naufeous and fetid.

1. Spiritus vinosus rectificatus *Pharm. Lond. & Edinb.* Rectified fpirit of wine: a vinous fpirit purified as much as poffible both from its phlegm and ill fmell.

Spirits drawn from wine, fuch as French brandy, may be in great meafure purified by
fimple

simple diftillation, in tall veffels, with a gentle heat, the pure fpirituous part rifing before the phlegm: if French brandy be thus diftilled to one half, the diftilled fpirit proves tolerably pure. But wine or brandy being in this country too dear an article for diftillation; and all vinous fpirits, when perfectly purified, being one and the fame thing; this purification is chiefly practifed, among us, on the cheaper fpirits of melaffes and malt liquors. Thefe fpirits, when freed by diftillation from greateft part of their phlegm, are ftill found, particularly the latter, to abound with a very offenfive oil. To feparate this, they are mixed with equal their quantity of fpring water, and the fpirit gently drawn off again: a confiderable portion of the oil is thus left behind in the water, which now proves turbid, and milky, and very naufeous both in fmell and tafte. By repeating this ablution with frefh quantities of water, the fouleft and moft offenfive fpirits may be purified from all ill flavour.

Though fpirits, by this treatment, may be divefted of their oil, they cannot be freed wholly from phlegm; the gentleft heat, in which they can be diftilled, being fufficient to raife a little watery vapour. To complete the purification, therefore, a little fixt alkaline falt, thoroughly dried and powdered, is added; which, imbibing the phlegm, is thereby diffolved into a ponderous liquid, that does not mingle with the fpirit, but fettles at the bottom. If the fpirit is very phlegmatic, four pints will require a pound of the alkali: if the diftillation has been performed with due care, half this quantity, or lefs, will be fufficient: in either cafe, if all the falt diffolves, the fpirit is to be digefted with a little more, till at leaft a part remains

SPIRITUS VINOSUS.

remains undiffolved. The fpirit, now poured off, is to be again diftilled, in order to feparate from it a portion of the falt, which has united with it, and which, though extremely minute, is fufficient to vary, in fome refpects, its qualities. As fome particles of the alkali are apt to be carried up with it even in the diftillation, fo as to communicate an ill flavour, it is advifable to previoufly add a fmall portion of calcined vitriol or burnt alum, which will completely abforb the alkali, without giving any new impregnation to the fpirit. In this manner was prepared the fpirit ufed in the experiments of the prefent work under the name of rectified or pure fpirit of wine.

Vinous fpirits, thus rectified, have a very hot pungent tafte, without any particular flavour. They readily take fire, and burn totally away, without leaving any mark of an aqueous moifture behind; though on catching the vapour that exhales from the flame, a confiderable quantity of mere water is collected. On diftilling them with the gentleft heat, the laft runnings prove as colourlefs, flavourlefs, and inflammable, as the firft. They diffolve diftilled vegetable and animal oils, and all the pure refins, into an uniform tranfparent fluid. They are the lighteft of almoft all known liquids: expreffed oils, which fwim on water, fink freely in thefe to the bottom: a meafure which holds ten ounces by weight of water, will contain little more than eight and a quarter of pure fpirit.

2. SPIRITUS VINOSUS TENUIOR *Pharm. Lond. & Edinb.* Proof fpirit: the fame fpirit containing an admixture of an equal quantity [by meafure] of water. " The beft proof fpirit is
" that

"that diftilled from French wine; but for
"common ufes, may be employed the fpirit
"drawn from the fyrupy matter which feparates
"in the purification of fugar, commonly called
"melaffes†."—The fpirits ufually met with,
under the name of proof, are thofe diftilled
from different fermented liquors, freed from
their phlegm and their flavour only to a certain
degree. Their purity, with regard to flavour,
may be judged by the tafte, efpecially if the
fpirit be firft duly diluted: of their ftrength, or
the proportion of phlegm contained in them,
the leaft uncertain criterion feems to be their
gravity, which is eftimated moft commodioufly
by the hydrometer. For the nicer purpofes, a
pure flavourlefs proof fpirit may be obtained by
mixing the foregoing rectified fpirit with an
equal meafure of pure water.

† *Ph. Lond.*

Alkohol *Ph. Lond.*

* The laft London pharmacopœia has given
three forms of vinous fpirit. The pureft is
their *alkohol*, made by digefting rectified fpirit
of wine with hot fixed alkali, and then redif-
tilling. Of this, the fpecific gavity to water is
ftated at 815 to 1000. Their rectified fpirit has
95 parts of alkohol, and 5 of water, in 100
parts; and fhould have the fpecific gravity of
835 to 1000. Their proof fpirit contains 55
parts of alkohol, to 45 of water, and weighs
as 930 to 1000.

Rectified spirit coagulates all the fluids of
animal bodies, that have been tried, except bile
and urine. It hardens the folid or confiftent
parts, and preferves them from corruption.
Applied externally to living animals, it ftrength-
ens the veffels, contracts the extremities of the
nerves, and deprives them of fenfibility: hence
its power of reftraining hemorrhagies, abating

fuperficial

SPIRITUS VINOSUS.

superficial pains, &c. Received into the stomach, undiluted, it produces the like effects; thickening the fluid, and contracting all the solid parts which it touches, and destroying, at least for a time, their use and office: if the quantity taken is confiderable, a palfy, or apoplexy, follows, and speedily proves mortal.

Proof spirits, and such as are diluted below the proof strength, have the same effects in a lower degree. Externally they are of use in corroborant, anodyne, and antiseptic fomentations. Taken inwardly, in small quantity, they strengthen lax fibres, incraffate thin fluids, and warm the habit: in larger quantity, they disorder the senses, destroy voluntary motion, and produce, like the rectified spirit, a mortal apoplexy or palfy.—Vinous spirits, therefore, in small quantity and properly diluted, may be applied to useful purposes in the relieving of some disorders; whilst in larger ones, or imprudently continued, they act as a poison of a particular kind. Their moderate use is the most serviceable to those, who are exposed to heat and moisture, to corrupted air, or other causes of colliquative and putrid diseases; the most pernicious in the opposite circumstances, and to those who are afflicted with hysterical or hypochondriacal complaints; for whatever temporary relief these spirituous cordials may afford in the lownesses to which hysterical and hypochondriacal persons are subject, we entirely agree with Dr. Pemberton, that there are none who feel so soon the ill effects arising from their habitual use.

SPONGIA.

SPONGIA.

SPONGIA Pharm. Lond. & Edinb. SPONGE: a soft, light, very porous and compressible substance, readily imbibing water; found in the sea, adhering to rocks, particularly in the Mediterranean, about the islands of the Archipelago. It has been commonly supposed a vegetable production, but is more probably, like the corallines, of animal origin. Chemically analysed, it yields, like animal substances, a volatile alkaline salt, and this even in larger quantity than I have obtained from any of the other animal matters except the bags of the silkworm: the caput mortuum, incinerated, yields also a large proportion of fixt salt, not however an alkaline one like that of vegetables, but chiefly of the marine kind: a like salt is obtainable by boiling the sponge in water without burning.

Dry sponge, from its property of imbibing and swelling by moisture, is sometimes used as a tent for dilating wounds and ulcers: for this purpose, after being carefully freed from the small stones generally lodged in it, it is dipt in melted wax, and the wax squeezed out from it in a press †. * It has also been found to be the most efficacious of all those substances which have been employed to suppress hemorrhagies on the ground of their strong adhesion to the mouths of the wounded vessels; such as agaric, puff ball, &c. For this purpose, a very dry and solid piece, of a cubical or conical form, should be applied in close contact with the vessel, and retained by proper compression. It soon adheres with great force; and indeed its difficult removal is one of its chief inconveniencies.

† Spongia præparata Ph. Paris.

encies. Very large arteries have been prevented from bleeding by this application *(a)*.

Burnt in a clofe earthen veffel, till it becomes black and friable†, it has been given in dofes of a fcruple againft fcrophulous complaints and cutaneous defedations; in which it has fometimes been of fervice, in virtue, probably, of its faline matter, the proportion of which, after the great reduction which the other matter of the fponge has fuffered in the burning, is very large. By virtue of this faline matter alfo, the preparation, if ground in a brafs mortar, corrodes fo much of the metal, as to contract a difagreeable taint and fometimes an emetic quality: hence the college exprefsly orders it to be powdered in a mortar of glafs or marble. *The burned fponge is a principal article in what is called the Coventry method of cure in the bronchocele, and alfo in that publifhed by Mr. Proffer *(b)*.

† Spongia ufta *Ph. Lond. & Ed.*

STANNUM.

STANNUM Pharm. Lond. & Edinb. TIN: a filver-coloured metal, not liable to ruft, but lofing its brightnefs in the air; eafily flexible, and making a crackling noife in being bent; little more than feven times fpecifically heavier than water; fufible in a heat far below ignition, and fomewhat lefs than that in which lead melts. Heated till almoft ready to melt, it proves extremely brittle, fo as to fall in pieces from a blow. Melted, and nimbly agitated at the time of its beginning to congeal (as by fhaking in a wooden box rubbed on the infide with chalk) it

Stannum pulveratum *Ph. Lond.*
Stanni pulvis *Ph. Ed.*

(a) See White's *Cafes in Surgery.*

(b) See Wilmer's *Cafes in Surgery,* and Proffer *on the Bronchocele.*

is reduced partly, and by repetitions of the process totally, into powder. Continued in fusion for some time, and kept stirring with an iron rod, it changes into a dusky calx; which, urged longer in the fire, gains a perfect whiteness, a mark of the purity of the tin. It is corroded by vegetable acids, and renders them turbid and whitish: the nitrous acid pretty readily dissolves it, but soon depofites a part in form of a thick mucilage, especially if the acid has any admixture of the vitriolic: the vitriolic and marine acids are very difficultly made to act upon it: its most perfect menstruum is a mixture of the marine and nitrous.

The principal use of this metal in the present practice is as an anthelmintic: even the flat worms, which too often elude the force of other medicines, are said to be effectually destroyed by powdered tin. The common dose of the powder is from a scruple to a dram; but Dr. Alston affirms, in the Edinburgh medical essays, that its success depends chiefly on its being given in much larger quantities, as half an ounce or an ounce. It is possible, that the anthelmintic virtues of tin may proceed, not so much from the pure metal, as from a certain substance of a different nature, which there are grounds to suspect that the purest sorts of tin usually met with, participate of; filings of tin, held in the flame of a candle, emit a thick fume smelling like garlick: Mr. Marggraf reports *(a)*, that by gentle diffolution in aqua regis and flow evaporation, he obtained cryftals, which on being exposed to the fire, with the addition of some fixt alkaline salt to absorb their acid, sublimed into a white concrete; and that this ex-

(a) Memoires de l'acad. roy. des sciences de Berlin, tom. iii.

haled

haled in the fire in fumes of a strong garlick smell, formed with sulphur yellow and red compounds, and whitened copper (see *Arsenicum*). It must be observed, however, that notwithstanding these strong presumptions, not to say proofs, of an arsenical impregnation in tin, the metal taken in substance has not been observed to be noxious, though the fumes which it emits in a red heat are undoubtedly so.

A sparkling gold-coloured preparation of tin, called mosaic gold, is prepared by adding six ounces of quicksilver to twelve of melted tin, pulverizing the mass when grown cold, mixing with it seven ounces of flowers of sulphur and six of sal ammoniac, and subliming in a matras: the mosaic gold is found under the sublimed matter, with some dross at the bottom. This preparation is chiefly valued for its beautiful appearance: as a medicine it is at present little regarded, though formerly held in considerable esteem against hysterical and hypochondriacal complaints, malignant fevers, and venereal disorders. It appeared, upon experiment, to be little more than a calx of tin: tin, calcined by itself, gains nearly as much in weight, as it does by being made into mosaic gold; and the mosaic gold, melted with inflammable fluxes, is revived into tin again without suffering much more loss than the simple calx. The volatile ingredients, sal ammoniac, sulphur, and quicksilver, sublime in the process, partly escaping, and partly forming the scoriæ: great part of the sulphur and mercury are found united together into the form of cinnabar.

A salt of tin is directed to be prepared, from Sal Jovis. twelve ounces of calx of tin and four of aqua regia diluted with twenty-four of water: after digestion for two days, the vessel is to be shaken,

shaken, the more ponderous part of the undiſ-ſolved calx ſuffered to ſettle, the turbid liquor poured off and evaporated nearly to dryneſs, and the maſs further exſiccated on brown paper: to the remaining calx, half the quantity of freſh menſtruum is to be added, and the proceſs repeated. Of the virtues of this ſalt I can ſay nothing from experience, except that it is in taſte very ſharp and almoſt corroſive. Nor do I apprehend the uſe of calcining the metal, as tin uncalcined diſſolves much more eaſily and more plentifully: the ſolution is in both caſes the ſame, the fire in the calcination diſſipating only the inflammable principle of the tin, which the acid equally does in the ſolution and evaporation. Hoffman ſays, that ſolution of tin is a ſtrong purgative.

STAPHISAGRIA.

STAPHISAGRIA Pharm. Lond. & C. B. Staphys, pedicularia, & herba pedicularis quibuſdam. Delphinium Staphyſagria Linn. STAVESACRE: a plant with large leaves, ſet on long pedicles, deeply divided into ſeveral ſegments; producing irregular blue flowers with a tail behind like thoſe of larkſpur, followed by pods containing large, rough, triangular, darkcoloured ſeeds. It is annual, a native of the ſouthern parts of Europe, from whence the ſhops have been generally ſupplied with the ſeeds.

The ſeeds of ſtaveſacre have a diſagreeable ſmell, and a very nauſeous bitteriſh burning taſte. They were formerly employed ſometimes as a cathartic, in doſes of from twelve grains to a ſcruple: but they operate with ſo much violence

STOECHAS.

violence both upwards and downwards, and are so liable not only to diforder the bowels, but likewife to inflame the throat, that their internal ufe has been long laid afide. They are now ufed only in external applications, for fome kinds of cutaneous eruptions, and for deftroying infects. Their acrimony is extracted partially by water, totally by rectified fpirit, and not elevated in diftillation by either.

STOECHAS.

STOECHAS purpuræa C. B. *Lavandula Stœchas Linn.* FRENCH LAVENDER: a low fhrubby plant, with fmall oblong narrow leaves, bearing on the tops of the branches fhort thick fpikes or fcaly heads, from which iffue feveral fmall purple labiated flowers, followed each by four feeds inclofed in the cup. It is a native of the fouthern parts of Europe, common in our gardens, and flowers in May or June. The fhops have been generally fupplied, from Italy and the fouth of France, with the flowery tops, often mouldy, and never equal to thofe of our own growth.

THE beft ftechas which we receive from abroad has no great fmell or tafte; Pomet affirms, that fuch as is to be met with in the fhops of Paris is entirely deftitute of both; whereas ours, both whilft frefh and when carefully dried, has a pretty ftrong aromatic fmell, and a moderately warm pungent bitterifh tafte. Diftilled with water, it yields a confiderable quantity of a pale-coloured fragrant effential oil: the remaining decoction is unpleafantly bitterifh and fubaftringent. With rectified fpirit, it gives over nothing confiderable, greateft part of the active matter of the ſtechas being left in the extract. Both the herb itſelf and its preparations are much leſs grateful than lavender, with which it is ſuppoſed to have ſome agreement in virtue.

*STRAMONIUM.

STRAMONIUM Pharm. Edinb. Solanum fœtidum, pomo ſpinoſo oblongo C. B. Datura Stramonium Linn. Thorn-apple: an herbaceous plant, with a thick branched ſtalk, two or three feet high, large ſinuated indented leaves, and long tubular white or purpliſh flowers, ſucceeded by large, prickly, green, fleſhy ſeed-veſſels, which open at the end in four diviſions, and diſcloſe numerous black ſeeds. It is ſown in gardens, and ſometimes found wild among rubbiſh. It flowers in July.

This plant, which has been long known as a narcotic poiſon, has been introduced into the catalogue of medicines by Dr. Stœrck. An extract made from the expreſſed juice of the leaves is acrid and ſaline to the taſte, and yields cryſtals of nitre on ſtanding. This preparation given in doſes of from one to five grains twice or thrice a day, is ſaid to be a very powerful remedy in various convulſive and ſpaſmodic diſorders, epilepſy and mania. The accounts of other practitioners have confirmed that of the firſt introducer, and it has been received into ſome pharmacopœias.

An abridged account of its medicinal properties, with ſome inſtances of its efficacy, from a treatiſe printed at Upſal by Dr. Wedenberg, is to be met with in the *Med. Comment.* vol. iii. p. 18. An ointment prepared from the

the leaves has been found to give eafe in external inflammations and hæmorrhoids.

Several inftances are recorded of the bad effects of inadvertently eating the feeds of thornapple. Emetics and purgatives give the fpeedieft relief in thefe cafes, which it is fometimes neceffary frequently to repeat, as fome of the feeds have been found to lodge a confidetable time in the ftomach.

STYRAX.

STYRAP Pharm. Lond. Styrax calamita Pharm. Edinb. SOLID STORAX: an odoriferous refin, exuding in the warmer climates from a middling-fized tree *(ftyrax folio mali cotonei* C. B. *Styrax officinale Linn.)* with leaves like thofe quince, flowers like thofe of the orange tree, and fruit like filberds; a native of Afia, and, as is faid, of Italy. Two forts of this refin have been commonly diftin guifhed in the fhops.

1. *Storax in the tear:* not in feparate tears, or exceeding rarely, but in maffes, fometimes compofed of whitifh and pale reddifh brown tears, and fometimes of an uniform reddifh-yellow or brownifh appearance; unctuous and foft like wax, and free from vifible impurities. This is fuppofed to be the fort which the ancients received from Pamphylia in reeds or canes, and which was thence named calamita.

2. *Common ftorax:* in large maffes, confiderably lighter and lefs compact than the foregoing, and having a large admixture of woody matter like faw-duft. This appears to be the kind intended by the London college, as they direct their *ftyrax calamita* to be purified, for me- Styrax coladicinal tus *Ph. Lond.*

dicinal ufe, by foftening it with boiling water, and preffing it out from the feces betwixt warm iron plates; a procefs which the firft fort does not ftand in need of. And indeed there is rarely any other than this impure ftorax to be met with in the fhops.

The writers on the materia medica in general prefer the pure ftorax in the tear, and reject that which is mixed with woody matter. It appears however, upon comparifon, that this laft, notwithftanding its large proportion of impurities, is the moft fragrant of the two: nor is it difficult to affign a reafon for this fuperiority, as the pure juice muft have required, for its infpiffation to a firm confiftence, a longer expofure to the fun and air, and confequently loft more of its volatile parts, than when abforbed and thickened by the woody fubftance.

Common ftorax, infufed in water, imparts to the menftruum a gold yellow colour, fome fhare of its fmell, and a flight balfamic tafte. It gives a confiderable impregnation to water by diftillation, and ftrongly diffufes its fragrance when heated, though it fcarcely yields any effential oil. Hence, in the purification of it by ftraining, it is apt to fuffer a confiderable lofs of its finer matter, which is partly diffipated by the heat, and partly kept diffolved by the water: a part of the ftorax is likewife defended by the woody fubftance from the action of the prefs, and and left behind among the feces. It may be purified rather more elegantly by means of rectified fpirit, which readily diffolves the fine refin, leaving only the impurities and a little inert gummy matter: the fpirit gently diftilled off from the filtered reddifh-yellow folution, brings over with it very little of the fragrance of the ftorax; and the remaining refin is more fragrant

grant than the finest storax in the tear which I have met with. The pure resin, distilled without addition, yields, along with an empyreumatic oil, a portion of saline matter similar to the flowers of benzoine; I have sometimes also extracted from it a substance of the same nature by coction in water.

Storax is one of the most agreeable of the odoriferous resins, of a mild taste, of no great heat or pungency, nearly similar, in its medical as in its pharmaceutic qualities, to benzoine and balsam of Tolu. It is not, however, much used in common practice, unless as an ingredient in some of the old compositions.

STYRAX LIQUIDA.

LIQUID STORAX: a resinous juice; obtained from a large tree with angular leaves like those of the maple, and a round fruit composed of a number of pointed seed-vessels, called by Ray *styrax aceris folio*, by Linnæus *Liquidambar styraciflua*, a native of Virginia and Mexico, and lately naturalized to our own climate. The juice called liquidambar is said to exude from incisions made in the trunk of this tree, and the liquid storax to be obtained by boiling the bark or branches in water.

Two sorts of liquid storax are distinguished by authors: one, the purer part of the resinous matter that rises to the surface in boiling, separated by a strainer, of the consistence of honey, tenacious like turpentine, of a reddish or ash brown colour, moderately transparent, of an acrid unctuous taste, and a fragrant smell, faintly resembling that of the solid storax, but somewhat disagreeable: the other, the more impure

impure part, which remains on the ſtrainer, untranſparent, in ſmell and taſte much weaker, and containing a conſiderable proportion of the ſubſtance of the bark. What is moſt commonly met with under this name in the ſhops, is of a weak ſmell, and a grey colour, and is ſuppoſed to be an artificial compoſition.

Liquid ſtorax has been employed chiefly in external applications. Among us, it is at preſent almoſt wholly in difuſe.

SUCCINUM.

SUCCINUM Pharm. Lond. & Edinb. Ambarum citrinum & electrum quibuſdam. Amber: a ſolid, brittle, bituminous ſubſtance, dug out of the earth or found upon the ſea ſhores, moſt plentifully in Poliſh Pruſſia and Pomerania; of a white, yellow, or brown colour, ſometimes opake, and ſometimes very clear and tranſparent; of very little taſte; and ſcarcely any ſmell, unleſs heated or briſkly rubbed, in which circumſtances it yields a pretty ſtrong one, to moſt people agreeable.

Boiled in water, it neither ſoftens, nor undergoes any ſenſible alteration. Digeſted in rectified ſpirit, it imparts a yellowiſh or browniſh colour, a fragrant ſmell, and a bitteriſh aromatic taſte: by repetitions of the proceſs with freſh quantities of ſpirit, a conſiderable part of the amber by degrees diſſolves. The ſpirit, diſtilled off from the tinctures, is ſtrongly impregnated with their ſmell; nevertheleſs the remaining balſam, or ſoft extract, is found to be very ſtrong both in ſmell and taſte.

By alkalies, fixt and volatile, the vegetable, nitrous, and marine acids, it is ſcarcely at all acted

acted upon: the vitriolic acid diffolves it into a deep purple liquor, from which the amber is precipitated on the mixture of any other acid, or of water, or fpirit of wine *(a)*.

The fpirituous tincture and balfam are medicines of great efficacy in hyfterical diforders, cachexies, the fluor albus, fome rheumatic pains, and in debilities and relaxations in general: in fome cafes of this kind they have taken place, after bark and other corroborants of the vegetable kingdom had been given with little effect. The fpirit, which diftils in concentrating the tincture, may be referved for extracting a frefh tincture, either from another parcel of 'amber, or from that which remained after the former extraction. It is faid that if a little vitriolic acid be previoufly combined with the fpirit, it will diffolve more of the amber than pure vinous fpirits. The amber is fometimes given alfo in fubftance, levigated into an impalpable powder, but does not appear to act with fo much advantage in this form as in a diffolved ftate.

Succinum præparatum *Ph. Lond.*

This concrete, expofed to the fire in open veffels, melts into a black mafs, takes flame, emits a copious fmoke, with a fmell like that which arifes from the finer kinds of pitcoal, and burns almoft entirely away. Diftilled in a retort, it yields firft an acidulous phlegm intermingled with a thin limpid oil, which grows thicker and deeper coloured as the fire is increafed: at length a brownifh faline matter arifes into the neck of the retort, fucceeded by a groffer oil, and at laft, in a great heat, by a black thick pitchy matter. About the time that the

(a) See Stockar's *Specimen inaugurale de fuccino.*

firft

firſt oil begins to riſe, the amber melts in the retort, and, unleſs the heat be cautiouſly regulated, is apt to boil over into the receiver: to prevent this accident, ſome previouſly mix with the amber an equal quantity of clean ſand, which does not appear, however, to be of much uſe, for with due care, the proceſs ſucceeds equally without as with it.

The ſalt is purified from its adhering oil, either by ſublimation, or by repeated ſolution, filtration, and cryſtallization†. When perfectly pure, it is of a white colour, and of a penetrating gratefully acid ſubaſtringent taſte. It diſſolves in rectified ſpirit ſparingly by the aſſiſtance of heat, not at all in the cold. Of cold water, in the common temperate ſtate of the atmoſphere, it requires for its ſolution above twenty times its own weight; of boiling water, only about twice its weight: in ſlow cooling, it ſhoots into triangular priſmatic cryſtals, with the points obliquely truncated. In the heat of boiling water, it does not exhale, or ſuffer any viſible alteration: in a greater one, it firſt melts, then riſes in white fumes, and concretes again in the upper part of the glaſs into fine white flakes; leaving behind a ſmall quantity of a dark coaly matter. It effervesces with alkalies and abſorbent earths, and forms with them compound ſalts ſomewhat reſembling thoſe made with vegetable acids, its acid matter ſeeming to have a conſiderable analogy to the acids of the vegetable kingdom, and being eſſentially diſtinct from the three called mineral acids *(a)*: mixed with acids, it makes no ſenſible commotion. By theſe characters this ſalt may be diſtinguiſhed from all the other matters that have been mixed

† Sal ſuccini purif. *Ph. Lond. & Ed.*

(a) See Neumann's *chemical works, p.* 237, *and* Stockar's *ſpecimen.*

with

SULPHUR.

with or vended for it. With regard to its virtues, it is accounted aperient, diuretic, and antihyfteric: its great price has prevented its coming much into ufe, and probably its real virtues, though doubtlefs confiderable, fall greatly fhort of the opinion that has been generally entertained of them.

The oil†, diftilled again by itfelf, is divided into a thinner oil which arifes‡, and a thicker part, which remains behind, called balfam of amber: fome diftil it from brine of fea falt, or from plain water‖, by which it becomes purer than when diftilled without addition. This oil has a ftrong bituminous fmell, and a hot pungent tafte; and approaches more to the nature of the mineral petrolea than of the vegetable or animal diftilled oils, being very difficultly if at all, diffoluble in vinous fpirits. It is fometimes given internally, in dofes of ten or twelve drops, as an antihyfteric and emmenagogue; and fometimes employed externally in antihyfteric, paralytic, and rheumatic liniments or unguents.

† Ol. fuccini Ph. Ed.
‡ Ph. Lond.
‖ Ol. fuccini rectificatum Ph. Ed.

SULPHUR.

SULPHUR *Pharm. Lond.* BRIMSTONE: a yellow concrete, of no tafte, and fcarcely any fmell; melting in a fmall degree of heat into a vifcous red fluid, and totally exhaling on an increafe of the heat; readily inflammable, and burning with a blue flame and a fuffocating acid fume.

It confifts of the vitriolic acid combined with a fmall proportion of inflammable matter. If a combination of pure vitriolic acid with a pure fixt alkaline falt be melted in a clofe veffel with the addition of a little powdered charcoal, a true fulphur will be produced, and the compound

pound will be the fame (excepting for the earthy part of the charcoal) as if the alkali had been melted with common brimftone. And contrariwife, if a combination of alkaline falt with common brimftone be reduced into powder, and roafted with a gentle heat, the inflammable principle exhales, and the remainder proves the fame as if the alkali had been combined with the pure vitriolic acid: the diminution of weight, refulting from this avolation of the inflammable principle, does not exceed two drams upon fixteen ounces of the fulphur *(a)*.

Greateft part of the fulphur met with in the fhops is either extracted from certain ores by a kind of diftillation *(b)*; or prepared from minerals abounding with vitriolic acid, by ftratifying them with wood, which being fet on fire, the fulphur is collected in cavities made in the upper part of the pile *(c)*. The largeft quantities are brought from Saxony, in irregular maffes, which are afterwards melted and caft into cylindrical rolls. Sulphur is found likewife native in the earth; fometimes in tranfparent pieces, of a greenifh or bright yellow colour; more commonly in opake grey ones with only fome ftreaks of yellow: this laft is the fort which is underftood by the name *fulphur vivum*, though what is fold under this name in the fhops is no other than the drofs which remains after the fublimation of fulphur. The native fulphurs fhould never be employed for any internal ufe without purification: they almoft

(a) Vide Stahlii *Menfis Julius, Experimenta & animadverfiones ccc, &c.*

(b) Leopold, *Relatio de itinere fuo Suecico.*

(c) Hoffman, *Obfervationes phyfico-chemicæ*, lib. iii. obf. 9.

always

SULPHUR.

always participate of arsenic, which is discoverable in some by their having naturally more or less of a red colour, and in the others by their exhibiting this colour after a part of the pure sulphur has been separated by sublimation.

Sublimation is the most effectual method of purifying sulphur from arsenical as well as earthy admixtures; and by the same process it is reduced into a fine powder, somewhat of a softer kind than that obtained by triture. Those who prepare the flowers in the way of trade, use for the subliming vessel a large iron pot, capable of holding two or three hundred weight: this is placed under an arched chamber, lined with glazed tiles, which serves for the recipient. Some shall portion of sulphur that rises first, especially when the vessels are very large, or the air not sufficiently excluded, is apt to take fire, and give out its acid, which adhering to the flowers that sublime afterwards, communicates to them a sensible acidity or roughness; in consequence of which, they are sometimes found to coagulate milk, when taken internally to produce gripes, and to receive from some metalline vessels a disagreeable taint: hence the London college directed such of the flowers, as might happen to concrete or melt together from the vicinity of the receiver to the fire, to be reduced to powder, not with metalline instruments, but either in a wooden mill, or in a marble mortar with a wooden pestle. From this extraneous or superficial acid they are freed, by boiling them in water, and afterwards carefully washing them with cold water†.

Flor. sulph. *Ph. Lond. & Ed.*

† Flor. sulph. loti *Ph. Lond.*

PURE SULPHUR, taken in doses of from ten grains to a dram or more, gently loosens the belly and promotes perspiration. It seems to pass through

through the whole habit; and manifestly transpires through the skin, as appears from the sulphureous smell of persons who have taken it, and silver being stained in their pockets to a blackish hue as by the vapour of sulpureous solutions. In consequence of these properties, and of this subtility of parts, it promises to be of great medicinal powers; but what its particular virtues are, experience has not as yet clearly determined. It is principally recommended against the piles, in disorders of the breast, and in cutaneous eruptions: in the itch indeed it is a certain remedy, whether internally or externally used, but in other kinds of eruptions it has not equally succeeded, and perhaps its efficacy against the first depends not so much on its purifying the blood, as on its fumes being destructive to the cuticular animalcules to which the present theory ascribes that distemper. It remarkably corrects or restrains the power of certain mineral substances of the more active kind: by the admixture of sulphur, mercury becomes inert, the virulent antimonial regulus mild, and arsenic itself almost innocent: hence though sulphur should contain a small proportion of arsenic, it possibly may not receive from that poisonous ingredient any very hurtful quality.

This concrete is not acted on by water, by acids, or by vinous spirits; but dissolves, by the assistance of heat, in oils both expressed and distilled, and in the mineral petrolea: when dissolved, it yields a very offensive smell, and discovers to the taste a nauseous pungency and heat. Expressed oils and petroleum dissolve it more readily and more plentifully than the distilled, taking up so much as to become thick and almost consistent: the college of London
directs

SULPHUR.

directs one part of flowers of fulphur and four of oil-olive†, and the fame proportions of the flowers and of petroleum ‡, and that of Edinburgh one ounce of the flowers and eight ounces of oil-olive ||, to be boiled together till they unite into the confiftence of a balfam. Effential oils do not load themfelves fo much with the fulphur as to become thick. As foon as the fulphur begins to be ftrongly acted on either by expreffed or effential oils, which happens nearly about the point of ebullition, or in fuch a heat as the fulphur by itfelf would melt in, the matter rarefies and fwells up greatly, fo as to require the veffel to be very large and occafionally removed from the heat; and at the fame time throws out impetuoufly great quantities of an elaftic vapour, which, if a free exit is not allowed it, produces violent explofions. The volatile flavour of the effential oils is in great meafure diffipated in this procefs by the great heat requifite for effecting the folution: a more elegant compofition of this kind might be obtained by adding to the effential oil a proper quantity of the balfam made with expreffed oils, which will unite with it by a gentle warmth. The balfams of fulphur have been employed externally for cleanfing and healing foul running ulcers. They are recommended internally in fome cachectic and hydropic cafes; as alfo in coughs and confumptions, in which they promife, by their manifeft heat and acrimony, to be oftener injurious than beneficial: they have been frequently obferved to hurt the appetite, and excite febrile fymptoms.---It may be obferved, that in thefe folutions the component parts of the fulphur are in fome meafure difunited from one another; infomuch that a confi-

† Oleum fulphurat.
‡ Petroleum fulphur.
|| Balf. fulph. craffum.

derable quantity of sulphureous acid, but no actual sulphur, separates in distillation.

Kali sulphurat. Ph. Lond. †
Hepar sulphuris.

Fixt alkaline salt, stirred by little and little into twice [or rather half] [or a fifth part†] its weight of sulphur in fusion, unites with it into a red mass called from its colour *liver* of sulphur. This compound has a fetid smell, and a nauseous taste: it dissolves in water, and deliquiates in the air, into a yellow fluid: thrown, whilst hot from the fire, into rectified

Tinctura sulphuris.

spirit of wine, in the proportion of about four ounces to a pint, and digested about twenty-four hours, it communicates a rich gold colour, a particular not ungrateful smell, and a hot somewhat aromatic taste. Solutions of the

Syrupus sulphuris.

liver in water, made with sugar into a syrup; and a few drops of the tincture mixed with a glass of canary or other rich wine, to which it communicates a milky hue; have been sometimes given in the same intentions as the balsams, and seem to be accompanied with the same inconveniences.

Flowers of sulphur may be dissolved in water by boiling them with thrice their weight of quicklime, though not so readily as by alkaline salts. If the filtered solution be long exposed to the atmosphere, or if air from the lungs be blown into it for a short time through a glass pipe, the lime gradually separates, as it does from common lime-water; and the sulphur, which was dissolved by means of the lime, separates and precipitates along with it. Common alkalies, fixt or volatile, added to the solution of sulphur in lime-water, occasion a precipitation of the lime, the sulphur continuing dissolved; caustic alkalies make no precipitation.

SULPHUR.

On adding to the sulphureous solution, whether made by lime or by alkalies, some of the weak spirit of vitriol, (or any other acid) the liquor becomes milky, an extremely fetid and diffusive inflammable vapour arises, and on standing for some time the sulphur settles to the bottom in form of a white powder, which, when washed with fresh quantities of water, becomes insipid and inodorous, and is vulgarly called *lac* or milk of sulphur: the liquor after the precipitation retains still a sulphureous impregnation, which further additions of acid will not precipitate. The method of preparing the lac with fixt alkalies is the most expeditious and least troublesome, provided the sulphur has been thoroughly united with a sufficient quantity of the alkali†; and on the other hand, quicklime gives the preparation a more saleable whiteness. The medicine proves in either case nearly the same: it would be exactly the same if the precipitation was made with any other acid than the vitriolic; which forms with the dissolved lime a selenitic concrete, not separable from the lac by any ablution, but with the alkali a neutral salt, which by hot water may be totally dissolved and washed off; whereas the combinations of all the other acids, with lime as well as with alkalies, are easily dissoluble even in cold water. The pure lac is not different in quality from pure sulphur itself; to which it is preferred, in external applications, only on account of its colour. The whiteness does not proceed from the sulphur having lost any of its parts in the operation, nor from any new matter superadded: on being melted with a gentle fire, it resumes its yellow hue.

† Sulphur præcipitat. *Ph. Lond.*

A solution of sulphur in volatile alkaline spirits may be obtained, by boiling half a pound

Tinct. sulph. volatilis *vulgo*

of flowers of sulphur with a pound of quicklime, in a gallon of water, till half the liquor is wasted; then putting the remainder into a retort, with eight ounces of powdered sal ammoniac, and distilling with a gradual fire. The spirit comes over loaded with the sulphur, and has a strong offensive smell, somewhat resembling that which rises in the precipitation of the lac. Hoffman says, a mixture of it with thrice its quantity of spirit of wine, given in doses of thirty or forty drops, proves a powerful diaphoretic; and that applied externally as a fomentation, with the addition of camphor, it alleviates gouty pains.

The flowers of sulphur in substance seem to be preferable for internal use to any of the preparations: they are certainly more safe, and and perhaps not less effectual; as they do not heat or irritate the first passages, and yet are evidently dissolved in the body and carried through the habit. They are most commodiously taken in the form of troches: the college of London directs for this purpose two ounces of the washed flowers, and four of double refined sugar, to be beaten together, and made up with mucilage of quince seeds; that of Edinburgh, one ounce of the flowers of sulphur, ten grains of flowers of benzoine, fifteen grains of factitious cinnabar, and two ounces of fine sugar, to be formed with mucilage of gum tragacanth: by the addition of the flowers of benzoine in this last prescription, the medicine is supposed to be rendered more efficacious in some disorders of the breast.

Trochisci e sulphure Ph. Lond.

Trochisci e sulphure sive diasulphuris Ph. Ed.

Unguentum sulphuris † Ph. Lond. — e sulphure sive antiphoricum ‡ Ph. Ed.

A sulphureous ointment, for the itch, is prepared, by mixing two ounces of the unwashed flowers, with three † or eight ‡ ounces of the simple ointment called pomatum †, or of hogs-lard ‡, and a scruple of essence of lemons. Half this

SUMACH.

this quantity is, in moſt caſes, ſufficient for a cure; though it may be proper to renew the application, and touch the parts moſt affected, for ſome nights longer, till the whole quantity is exhauſted. Some have been of opinion, that this external uſe of ſulphur is unſafe; that as ſulphur taken inwardly promotes the expulſion of impure humours and the eruption of cutaneous eﬄoreſcences, it muſt act, when outwardly applied, by repreſſing them. This conſequence, however, does not follow; nor is it by affecting the humours that it performs the cure: for it equally removes the itch, whether uſed internally or externally, by its vapours diffuſed through the ſkin. All the danger, that is to be apprehended from ſulphureous unguents, is that which may ariſe from the obſtruction of the cutaneous pores by the unctuous matter; and to prevent any diſorders from this cauſe, only a part of the body is to be anointed at one time.

SUMACH

SUMACH ſive Rhus obſoniorum. Rhus folio ulmi C. B. Rhus coriaria Linn. SUMACH: a ſhrub or low tree; with oval, pointed, ſerrated, downy leaves, having each a red rib running along the middle, ſet in pairs without pedicles; producing cluſters of ſmall yellowiſh or greeniſh flowers, each of which is followed by a ſmall, red, flattiſh berry, including a roundiſh reddiſh-brown ſeed. It is a native of the ſouthern parts of Europe, and cultivated in ſome of our gardens.

THE berries of ſumach have an acid auſtere taſte: they were formerly uſed for reſtraining bilious

bilious fluxes, and hemorrhagies, and colliquative hectic fweats: fome direct an infufion of half an ounce of the berries, and others two or three drams of an extract made from them by water, for a dofe. The leaves and young twigs are ftrong aftringents, and have been directed in the fame intentions.

TACAMAHACA.

TACAMAHACA. A refin; obtained from a tree, refembling the poplar, (*Populus balfamifera Linn.*) bearing, at the extremities of the branches, fmall roundifh fruits, including a feed like a peach-kernel; a native of the temperate parts of the continent of America, and in a fheltered fituation enduring the winters of our own country.

Two forts of this refin are fometimes to be met with. The beft, called, from its being collected in a kind of gourd-fhells, tacamahaca in fhells, is fomewhat unctuous and foft, of a pale yellowifh or greenifh colour, a bitterifh aromatic tafte, and a fragrant delightful fmell approaching to that of lavender and ambergris. This fort is very rare. That commonly found in the fhops is in femitranfparent grains or glebes, of a whitifh, yellowifh, brownifh, or greenifh colour, and of a lefs grateful fmell than the foregoing. The firft is faid to exude from the fruit of the tree, the other from incifions made in the trunk. The tree, as raifed among us, affords in its young buds, or the rudiments of the leaves, a refinous juice of the fame kind of fragrance (*a*).

(*a*) See article *Populus.*

Tacamahaca is used chiefly as an ingredient in warm nervine plasters; though the fragrance and taste of the finer sort points out its being applicable to other purposes, as an internal balsamic corroborant. Both kinds dissolve in rectified spirit into a gold-coloured liquor, only a small quantity of impurities being left: they impregnate water also considerably with their smell and taste, but give out very little of their substance to this menstruum.

TALCUM.

TALC: an earthy concrete; of a fibrous or leafy texture; more or less pellucid, bright or glittering; smooth and slippery to the touch; in some degree flexible and elastic, so as scarce to be pulverable; soft, so as to be easily cut; suffering no change in an intense fire, or no other than a diminution of its brightness, flexibility, and unctuosity; not acted upon by acids, either in its crude state, or after vehement calcination. There are several different appearances of this earth; among which, the greenish foliaceous *Venice talc* has been selected for medicinal use; though it does not appear capable of answering any medicinal intention, as not being dissolved, or sensibly affected, by any known humid menstruum: on account of its unctuous softness, and the silver hue, which it exhibits when reduced by rasping or otherwise into powder, it has been employed externally as a cosmetic. The fibrous flexible *amianthus* or *asbestos*, and the more rigid fibrous *alumen plumosum*, seem to approach to the nature of the talcs, and made an equally insignificant addition to the articles of the materia medica in our former pharmacopœias.

TAMARINDUS.

TAMARINDUS Pharm. Lond. & Edinb. Oxyphœnicon. TAMARIND: the fruit of a pretty large tree, *(siliqua arabica quæ tamarindus C. B. Tamarindus indica Linn.)*; growing in Arabia and in the East and West Indies. The fruit is a pod, somewhat resembling a bean-cod, including several hard seeds, together with a dark-coloured viscid pulp: the East India tamarinds are longer than those of the West, the former containing six or seven seeds each, the latter rarely above three or four: they nevertheless seem both to be the produce of one species of plant. The pulp, with the seeds, connected together by numerous tough strings or fibres, are brought to us freed from the outer shell: the oriental sort is drier and darker coloured than the occidental, and has more pulp: the former is sometimes preserved without addition, the latter has always an admixture of sugar.

The pulp of tamarinds is an agreeable laxative acid; of common use in inflammatory and putrid disorders, for abating thirst and heat, correcting putrefaction, and loosening the belly. The dose, as a laxative, is two or three drams: an ounce or two prove moderately cathartic. It is an useful addition to the purgative sweets, cassia and manna, increasing their action, and rendering them less liable to produce flatulencies: the resinous cathartics are said to be somewhat weakened by it. Tournefort relates that an essential salt may be obtained from tamarinds, by dissolving the pulp in water, and setting the filtered solution, with some oil upon the

the furface, in a cellar for feveral months; that the falt is of a fourifh tafte, and difficultly diffoluble in water; and that a like falt is fometimes found alfo naturally concreted on the branches of the tree. The falt, as Beaumé obferves, may be obtained more expeditioufly, by clarifying the decoction of the tamarinds with whites of eggs, then filtering and evaporating it to a proper confiftence, and fetting it to cool: the falt fhoots into cryftals, of a brown colour, and very acid tafte; but in diffolving and cryftallizing them again, or barely wafhing them with water, they lofe almoft all their acidity, the acid principle of the tamarinds feeming not to be truly cryftallizable.

TANACETUM.

TANACETUM *Pharm. Lond. and Edinb.* Tanacetum vulgare luteum C. B. Tanafia, athanafia, & parthenium mas quibufdam. Tanacetum vulgare Linn. TANSY: a plant with large leaves, divided to the rib, on both fides, into oblong deeply indented fegments; producing, on the tops of the ftalks, feveral gold-coloured difcous flowers, in umbel-like clufters, followed by fmall oblong blackifh feeds. It is perennial, grows wild by road-fides and about the borders of fields; and flowers in June and July.

THE leaves and flowers of tanfy have a ftrong, not very difagreeable fmell, and a bitter fomewhat aromatic tafte: the flowers are ftronger though rather lefs unpleafant than the leaves. They give out their virtue both to water and fpirit, moft perfectly to the latter: the tincture, made from the leaves, is of a fine green, from the flowers of a bright pale yellow colour.

Diftilled

Diſtilled with water, they yield a greeniſh-yellow eſſential oil, ſmelling ſtrongly of the herb: the remaining decoction, inſpiſſated, affords a ſtrong bitter ſubſaline extract. The ſpirituous tinctures give over alſo, in inſpiſſation, a conſiderable part of their flavour *(a); a part of it remaining, along with the bitter matter, in the extract.

This plant is uſed as a warm deobſtruent bitter, in weakneſs of the ſtomach and in cachectic, and hyſteric diſorders; and likewiſe as an anthelmintic. The ſeeds have been chiefly recommended in this laſt intention, and ſuppoſed by ſome to be the *ſantonicum* of the ſhops, from which they differ not a little in quality as well as in appearance, being much leſs bitter, and of a more aromatic flavour.

TARAXACUM v. DENS LEONIS.

TARTARUM.

TARTARUM vini albi vel rubri. TARTAR: an acid concrete ſalt thrown off from wines, after complete fermentation, to the ſides and bottoms of the caſks; of a red or white colour, and more or leſs droſſy, according to the colour and quality of the wine. The white is generally purer than the red: both kinds, when purified, are exactly the ſame.

THIS ſalt is one of thoſe which are moſt difficultly diſſoluble in water, being ſcarcely affected by it in the cold, and requiring ten or

---

*(a) Alcohol, diſtilled from tanſy, proved, after ſtanding for upwards of fifteen years, richly impregnated with the flavour of the plant, and ſufficiently grateful. *M. S. of Dr. Lewis.*

twelve

twelve times its own weight when affisted by a boiling heat. From this faturated folution the tartar begins to feparate almoft as foon as the boiling ceafes: if the quantity of water is greater, as about twenty times the weight of the falt, it continues long enough fufpended to be paffed, with due care, through a woollen ftrainer or a filter. The filtered liquor appears nearly colourlefs, whether the tartar made ufe of was red or white: if haftily cooled, the falt feparates in fmall grains like fand, but if the veffel is clofely covered, and the heat very leifurely diminifhed, it fhoots into femitranfparent whitifh *cryftals*: if the filtered liquor be kept boiling, a thick fkin forms on the furface, which, being taken off with a perforated wooden fkimmer, is fucceeded by frefh cuticles, till the whole of the falt is thus formed into what is called *creme* of tartar. The refining of tartar is practifed, in the way of trade, chiefly about Montpellier, from whence the fhops are generally fupplied both with the cryftals and creme; the procefs being fo troublefome, and requiring fo large conveniences, that it is fcarcely ever attempted here. A certain earth, of the argillaceous kind, is added in the procefs, the chief ufe of which feems to be, to promote the feparation of the colouring matter; for the falt extracted from the coloured tartars by water only is feldom of perfect whitenefs. It is faid that the earth generally contains fome fmall portion of chalky matter, foluble in acids, which of confequence will be taken up by the tartar; I have fometimes obferved folutions of the cryftals to depofite an earthy precipitate on adding alkaline lye. The purer fort of white tartar, unrefined, efpecially that of Rhenifh wine, is, for many purpofes, particularly for combinations

Cryftalli tartari *Ph. Lond. & Ed.*

Cremor tartari.

tions with other bodies, not inferiour either to the creme or cryftals.

Pure tartar, in dofes of half a dram or a dram, is a mild cooling aperient: two or three drams gently loofen the belly; and fix or eight prove moderately cathartic. Its acidity and laxative power are its medical characters.

* The difficult folubility of creme of tartar being an objection to its medical ufe, fome experiments were made by Dr. Peter Jonas Berg, for rendering it more foluble by certain additions, without altering its medicinal qualities. Borax was found to anfwer beft for this purpofe. To four parts of creme of tartar, one of borax was added. Thefe were diffolved in a fufficient quantity of water, and the liquor ftrained. About a fixteenth part of impurities were left behind. The pure folution evaporated yielded an acid and extremely foluble white falt (a).

Tartar, diffolved in water, effervefces with fixt alkaline falts, and faturates, of the vegetable alkalies, near one third its own weight. The compound falt, refulting from their union, is a neutral one, more purgative than the tartar itfelf, and far eafier of folution; whence its name foluble tartar. This falt is prepared, either by boiling the refined tartar in a fufficient quantity of water till it is diffolved, and then dropping in ftrong alkaline lye; or by diffolving the alkali in boiling water in the proportion of a pound to a gallon†, or to fifteen pounds‡, and then adding the tartar, till a frefh addition occafions no further effervefcence; which generally happens before triple the weight of the

Kali tartarifat. † *Ph. Lond.*
Alcali fixum vegetabile tartarifatum *vulgo* tart. folub. ‡ *Ph. Ed.*
Sal. vegetabilis *guibufd.*

(a) *Nova Acta Phyfico-Medica Academiæ Cæfareæ Leopoldino-Carolinæ Naturæ Curioforum, tom. quart.*

alkali

alkali is thrown in: the liquor is then filtered while hot, and either cryftallized or evaporated to drynefs. As this falt difficultly cryftallizes, infpiffation to drynefs is the moft convenient method; and in this cafe, to fecure the neutralization of the falt, the tartar may be made to prevail at firft, and the liquor fuffered to cool a little before filtration, that the redundant tartar may concrete and feparate from it; or the neutralization may be more perfectly obtained by means of ftained papers, as mentioned at the end of the article *Acetum*.

Of the mineral fixt alkali or foda, this acid faturates, according to the faculty of Paris, four fifths its own weight. The London college directs fix parts of the acid to five of the alkali. The neutral falt refulting from its coalition with this alkali, is fomewhat lefs diffoluble than that with the vegetable; and fhoots much more eafily, into pretty large, hard, multangular cryftals, fome columnar and flattifh, others more irregular. It is milder in tafte, and faid to be lefs purgative, requiring to be given to the quantity of an ounce or an ounce and a half to purge effectually: eight drams are reckoned by fome to be equivalent, in cathartic power, to fix of the foluble tartar. *Natron tartarifatum Ph. Lond. Soda tartarifata vulgo fal rupellenfis Ph. Ed. Sel de Seignette, Rochelle falt.*

Tartar forms likewife foluble compounds with all the abforbent earths, and with fome metallic bodies, but with thefe laft it is difficultly made to fatiate itfelf completely, the part that is firft faturated feeming to impede the action of the reft; for after long boiling, a very confiderable part of the tartar feparates on cryftallization unchanged.

It is obfervable, that if any of thefe combinations of tartar, with alkalies, with earths, or with metals, be diffolved in water, and any other

other acid added, the pure tartar separates and falls to the bottom, as acid, and as difficult of solution, as at first; the substance, that was combined with it, being absorbed by the acid superadded. As the acids of the vegetable kingdom, whether native or fermented, vinegar, lemon juice, &c. have this effect of disuniting tartar from all the bodies that are combinable with it, equally with those of the mineral kingdom; it follows, that the tartareous acid, is of a kind essentially different from all the other known vegetable ones, and that no acid, unless it be tartar itself, can be joined in prescription to the *tartarum solubile*, the *sel de Seignette*, or the combinations of tartar with earths of metals.

*In the *Swedish Transactions*, part iii. for the year 1770, was published an analysis of creme of tartar by Mr. Scheele. By this it appears, that creme of tartar is not a pure acid, but a compound salt, containing the fixed vegetable alkali united with a superabundance of the tartareous acid. It differs, therefore, from soluble tartar, only in the proportion of acid it contains.

TELEPHIUM.

TELEPHIUM: a plant with unbranched stalks, clothed with thick fleshy oval leaves, but producing no leaves immediately from the root: the flowers stand in form of umbels on the top of the stalk, and are followed, each, by from three to six pods full of small seeds: the root is irregular and knobby. It is indigenous in England, and perennial.

1. CRASSULA. *Telephium vulgare* C. B. *Anacampseros, fabaria, & faba crassa quibusdam.*

Sedum

TEREBINTHINÆ.

Sedum. Telephium Linn. Orpine: with the leaves very flightly or not at all ferrated: growing in hedges and moift fhady grounds, and producing reddifh or whitifh pentapetalous flowers in June. The leaves have been fuppofed to be poffeffed of an anti-inflammatory power; but their virtues appear to be very inconfiderable, as they have no fmell, and only an herbaceous mucilaginous tafte.

2. RHODIOLA *five rofea: Rhodia radix* C. B. Rofewort: with ferrated leaves; growing in mountainous places, and producing yellow tetrapetalous flowers in the fpring. The root of this fpecies, of little fmell when frefh, has when dry a very pleafant one, refembling, when the root is in perfection, that of the damafk rofe: in this odorous matter confifts the medical virtue of the rhodiola, and its principal medical difference from the preceding fpecies. Linnæus obferves, that when raifed in gardens, it has not one hundredth part of the fmell or virtue of that which is produced on its native mountains.

TEREBINTHINÆ.

TURPENTINES: the native balfams or refinous juices of certain trees. Four kinds are diftinguifhed by medical writers.

1. TEREBINTHINA CHIA *Pharm. Lond.* Chio or Cyprus turpentine: generally about the confiftence of thick honey, very tenacious, clear and almoft tranfparent, of a white colour with a caft of yellow and frequently of blue, of a warm pungent bitterifh tafte, and a fragrant fmell more agreeable than that of any of the other turpentines. It is the produce of the common

common terebinth *(terebinthus vulgaris .C. B. Piftachia terebinthus Linn.)*, an evergreen bacciferous tree or fhrub, growing fpontaneoufly in the eaftern countries and in fome of the fouthern parts of Europe. The turpentine brought to us is extracted in the iflands whofe name it bears, by wounding the trunk and branches a little after the buds have come forth: the juice iffues thin and clear as water, and by degrees thickens into the confiftence in which we meet with it. A like juice, exuding from this tree in the eaft, infpiffated by a flow fire, is faid by Kæmpfer to be ufed as a mafticatory by the Turkifh women, for preferving the teeth, fweetening the breath, and promoting appetite.

2. TEREBINTHINA VENETA *Pharm. Edinb.* Venice turpentine: ufually thinner than any of the other forts, of a clear whitifh or pale yellowifh colour, a hot pungent bitterifh difagreeable tafte, and a ftrong fmell, without any thing of the fine aromatic flavour of the Chian kind. The true Venice turpentine is faid to be obtained from the larch *(larix C. B. Pinus larix Linn.)*, a coniferous tree, with fmall cones, and fhort leaves ftanding in tufts, which fall off in the winter, growing in great abundance on the Alps and Pyreneans, and not uncommon in the Englifh gardens. Though this kind of turpentine bears the name of Venice, it is not the produce of the Venetian territories: it is brought from fome parts of Germany, and one greatly refembling it, as is faid, from New England. In the fhops this turpentine is often fupplied by a compofition of rofin and the diftilled oil of common turpentine.

3. TEREBIN-

TEREBINTHINÆ. 417

3. TEREBINTHINA ARGENTORATENSIS. Straſburg turpentine: generally of a middle conſiſtence between the two foregoing, more tranſparent and leſs tenacious than either, in colour yellowiſh brown, in ſmell more agreeable than any of the other turpentines, except the Chian, in taſte the bitteriſh yet leaſt acrid. This juice is extracted, in different parts of Germany, from the ſilver and red fir, by cutting out, ſucceſſively, narrow ſtrips of the bark, from as high as a man can reach to within two feet of the ground. In ſome places, a reſinous juice is collected from certain knots under the bark: this, called *lacryma abiegna* and *oleum abietinum*, is accounted ſuperiour to the turpentine. Neither this turpentine, nor any thing under its name, is at preſent common in the ſhops.

4. TEREBINTHINA VULGARIS *Pharm. Lond.* Common turpentine: about the conſiſtence of honey, of an opake browniſh white colour, the coarſeſt, heavieſt, in ſmell and taſte the moſt diſagreeable, of all the kinds of turpentine. It is obtained from the wild pine (*pinus ſilveſtris C. B. & Linn.*), a low coniferous tree, with the leaves longer than thoſe of the firs and iſſuing two together from one tubercle, growing wild in the different parts of Europe. This tree is extremely reſinous, inſomuch that, if not evacuated of its juice, it often ſwells and burſts. The juice, as it iſſues from the tree, is received in trenches made in the earth, and afterwards freed from its groſſer impurities by colature through wicker baſkets. The cones of the tree appear to contain a reſinous matter, of a more grateful kind than that of the trunk: diſtilled while freſh, they are ſaid to yield a fine eſſential oil greatly ſuperiour to that of the turpentines. Carpathicum oleum Germanis.

Vol. II.   E e   All

All these juices dissolve totally in rectified spirit, but give out little to watery menstrua: they become miscible with water, into a milky liquor, by the mediation of the yolk or white of an egg, and more elegantly by mucilages. Distilled with water, they yield a notable quantity of a subtile penetrating essential oil†, vulgarly called spirit; a yellow‡ or blackish resin remaining in the still: this is the common rosin of the shops. It is supposed that the officinal Burgundy pitch ‖, which is brought from Saxony, is a preparation of the same kind, only less divested of the oil, made by boiling the common turpentine till it acquires a due consistence. The essential oil, redistilled by itself in a retort, with a very gentle heat, becomes more subtile, and in this state is called ethereal, or rectified §; a thick matter remaining behind, called balsam of turpentine ¶. A like balsam is obtained also by distilling, with a stronger fire, the common resin; from which there arises, first a thin yellow oil, and afterwards the thicker dark-reddish balsam, a blackish resin† remaining in the retort.

All the turpentines are hot stimulating corroborants and detergents. They are given, where inflammatory symptoms do not forbid their use, from half a scruple to half a dram and upwards, for cleansing the urinary passages and internal ulcerations in general, and in laxities of the seminal and uterine vessels. They seem to act in a peculiar manner on the urinary organs, impregnating the water with a violet smell, even when applied externally, particularly the Venice sort. This last is accounted the most powerful as a diuretic and detergent, and the Chio and Strasburgh as corroborants: they all loosen the belly, and Venice most; and on this account

† Ol. terebinth. *Ph. Lond. & Ed.*
‡ Resina flava *Ph. Lond.*
— alba *Ph. Ed.*
‖ Pix Burgundica *Ph. Lond. & Ed.*

§ Ol. terebinth. rectific. *Ph. Lond.*
¶ Bals. terebinth.

† Resina nigra seu colophonia.

account they are fuppofed by Riverius and others to be lefs hurtful than fuch irritating diuretics, as are not accompanied with that advantage. Terebinthinate glyfters, in obftinate coftivenefs, are faid to be much preferable to faline, as being more certain and durable *(a)*. The common turpentine, as being the moft offenfive, is rarely given internally: its principal ufe is in fome external applications, among the farriers, and for the diftillation of the oil.

The oil is a moft potent ftimulating detergent diuretic. It is fometimes given, in dofes of a few drops, in rheumatifms and fixt pains of the joints; and fome have ventured on much larger quantities. Cheyne recommends, as a perfect cure for fciaticas, though of many years ftanding, from one to four drams of the ethereal oil, to be taken with thrice its quantity of honey, in a morning fafting, with large draughts of fack whey after it, and an opiate at bed-time: this medicine is to be repeated, with the intermiffion of a day now and then, if daily repetitions cannot be borne, for four or five days, or eight at furtheft *(b)*. It appears, however, highly imprudent, to venture on fuch large dofes at once of a medicine fo very hot and ftimulating. Boerhaave, after recounting, not without fome exaggeration, its ftyptic, anodyne, healing, antifeptic, and difcutient virtues when applied hot externally, and its aperient, warming, fudorific and diuretic qualities when taken internally, adds, that it muft be ufed with great caution; that when taken too freely, it affects the head, excites heat and pain therein, and, violently urging a diabetes, brings on a flux of

(a) Cullen, Mat. Med.
(b) Effay on the gout, edit. 10. § lxxi. p. 119.

the

the femen and of the liquor of the proftates; and that in venereal runnings, in which it has by fome been commended, it tends to inflame the parts and increafe the diforder.

The balfam and the infpiffated refins are ufed chiefly externally: the balfam is lefs pungent than the oil, and the refins much lefs fo than the turpentines in fubftance. The common yellow refin, in tafte confiderably bitter, is fometimes given as an internal corroborant, in preference to the turpentines themfelves, as being divefted of the ftimulating oil.

TERRA JAPONICA.

CATECHU vulgo Terra Japonica Pharm. Lond. Terra japonica dicta Pharm. Edinb. JAPAN EARTH, improperly fo called. *The plant which yields the terra japonica, grows in the Eaft Indies, and is called *coira*, or *caira*, by the natives of Bahar province. It appears to be the fame with that mentioned by Cleyerus and Herbert de Jagur, from which the natives of Pegu prepare this extract: they name the tree *Kheir* or *Khadira*.

It is a fpecies of the genus *mimofa* of Linnæus, and called by him *mimofa catechu*. Its ftem grows to about a foot in thicknefs, and from three to five feet high. It branches out into a thick fpreading top, feldom above twelve feet high. The bark is thick and rough. The wood is extremely hard and heavy; its exteriour part varies from a pale brown to a dark red, fometimes approaching to black, but always covered with one or two inches thick of white wood. The leaves are doubly pinnated, and have two prickles at the bafe. From the axillæ of the leaves, arife denfe fpikes of fmall flowers, fucceeded by pods.

The

TERRA JAPONICA.

The *terra japonica* is an extract of the wood of this plant, prepared in the following manner. After the trees are felled, the exteriour white wood is carefully cut off, and the interiour coloured part is cut into chips, with which narrow-mouthed unglazed earthen pots are filled, and water poured upon them till it appears among the upper chips. When this is half evaporated by boiling, the decoction, without straining, is poured into a flat earthen pot, and boiled to one third part; this is set in a cool place for one day, and afterwards evaporated by the heat of the sun, stirring it several times in the day; when it is reduced to a considerable thickness, it is spread upon a mat or cloth which has previously been covered with the ashes of cow-dung; this mass is divided into square or quadrangular pieces by a string, and completely dried by turning them frequently in the sun.

This extract is called by the natives *cutt*; by the English *cutch*. In making it, the pale brown wood is preferred, as it produces the fine whitish extract. The darker the wood is, the blacker the extract, and of less value. From the slovenly manner in which the preparation is made, it generally contains a considerable quantity of earth, besides what may be designedly put in it for the purpose of adulteration *(a)*.

This concrete is a mild astringent, more agreeable in taste than most of the other substances of that class, being accompanied with a considerable degree of sweetness. It is often suffered to dissolve leisurely in the mouth; both as a topical restringent for laxities and exulcera-

(a) From Mr. Kerr's account, in vol. v. of the *London Medical Observations*.

tions of the gums; and in alvine and uterine fluxes, and catarrhal coughs and hoarsenefs; medicines of this kind acting in general to much better advantage when thus gradually swallowed, than when taken in full doses at once. With this view the terra japonica is made in the shops into troches; beaten with equal its weight of gum-arabic, and four times the weight of both of sugar of roses, and so much water to be dropt in as will reduce them into a mass of a due consistence for being formed. The Edinburgh college directs a compound electuary, of which terra japonica is the basis, joined with other astringents and aromatics, and a small proportion of opium, which is a very elegant and efficacious medicine of the kind.

Trochifci e terra japon.

Electuar. japonic. vulgo confect, japon. Ph. Ed.

Japan-earth dissolves almost totally in water, excepting the impurities; which are usually of the sandy kind, and in considerable quantity, amounting, in the specimens I examined, to about one eighth of the mass. Of the pure matter, rectified spirit dissolves about seven eighths, into a deep red liquor: the part, which it leaves undissolved, is an almost insipid mucilaginous substance. In the shops a solution of it is made in proof spirit, with the addition of cinnamon, a spice the best adapted of any to the intention of this medicine; three ounces of the japan earth and two of cinnamon are digested in a quart† or two pounds and a half‡ of the spirit, and the strained tincture given commonly in doses of two or three tea-spoonfuls. It dissolves also in volatile alkaline spirits, in alkaline lye, in the mineral acids, partially and more difficultly in the vegetable acids, and not at all in oils: all the solutions are of a red or purplish colour.

Tinctura catechu † Ph. Lond.
— japonic.
‡ Ph. Ed.

* By

*By the natives where this extract is made, it is employed medicinally as a cooler in the difeafes confidered by them as of a hot nature. It is faid, when profufely ufed, to deftroy the venereal appetite. It is given at the rate of two ounces a day to tame vicious horfes. It is a principal ingredient in one of their ointments of great repute, compofed of blue vitriol four drams, Japan-earth four ounces, alum nine drams, white refin four ounces; thefe are reduced to a fine powder, and mixed by the hand with ten ouncs of oil-olive, and water enough to give the mafs a proper confiftence. This ointment is ufed in every fore, from a frefh wound to a venereal ulcer; and has been found remarkably ferviceable by European practitioners *(a)*.

TERREA ABSORBENTIA.

ABSORBENT EARTHS: diftinguifhable from other earthy and ftony fubftances by their folubility in acids. Such are, the mineral calcareous earths, as chalk: the animal calcareous earths, as crabs-claws, oyfter-fhells, egg-fhells, pearl, coral, coralline: animal earths not calcareous, as crabs-eyes and burnt hartfhorn. See the refpective fubftances; which have been feparately treated of, fo far as concerned each in particular; and whofe general and common qualities were referved for this article.

The obvious and immediate virtue of thefe bodies is, to obtund acid humours in the firft paffages, and thus to relieve the cardialgic and other complaints occafioned by them: the relief,

(a) Kerr's Account, above quoted.

however, which they afford, is oftentimes only temporary; from their acting only upon the acid already generated, without correcting the indifpofition which tends to produce more. If no acid humours are contained in the firft paffages, thefe earthy bodies, not foluble by any other kind of fluid, can have no falutary operation; and, by concreting with the vifcous contents of the ftomach into indigeftible maffes, may prove injurious in a high degree (a).

Abforbents are of more general ufe in infancy than in adult age; acidities being very familiar to young children, being often in that tender age productive of alarming fymptoms, and having a greater or lefs fhare in moft of their difeafes; whereas, in adults, they are much lefs frequent, accompanying chiefly hypochondriacal affections, cardialgiæ, and fuch diforders as happen in the firft paffages from the immoderate ufe of acid and fermentable food.

An hypothefis formerly obtained, which afcribed the acute difeafes of adults to a morbific acid: againft which the abforbent earths were introduced as the moft direct alexipharmacs. This theory is now juftly exploded; thefe difeafes, inftead of being produced, being in general moft fuccefsfully controlled, by acids. The ufe of abforbents, in different kinds of fevers, is neverthelefs ftill continued, and sometimes perhaps with advantage: for, though the earths of themfelves are apparently rather injurious than beneficial, yet as acids are often given freely at the fame time, the folution of the earth in the acid may prove a medicine more ferviceable in particular cafes than the

(a) Vide Tralles *Virium terreis remediis afcriptorum examen rigorofius.*

acid unobtunded. It is however, doubtlefs, more advifable, to ufe the earth previoufly diffolved in the acid, than to give them feparately.

The college of Berlin, fenfible of the advantage of having the earths, in thefe cafes, previoufly diffolved, or reduced to a *foluble faline form; as well as of the abfurdity, retained in other German pharmacopœias, of precipitating them from their folutions by fixt alkaline falts, and thus rendering them wholly inert; directs them to be digefted in diftilled vinegar, with a gentle heat, till the menftruum ceafes to act, and the filtered folution to be infpiffated to drynefs. This preparation is greatly preferable to the fimple imbibition with vinegar or lemon juice recommended by fome; as by this laft management the earth is made foluble only in part, and in an undeterminable proportion.

Magifterium folubile, corallorum, perlarum,&c. P. Brandenb.

Solutions of thefe earths in vegetable acids are in tafte fomewhat auftere. The different earths differ fomewhat from one another, both in the degree and in the fpecies of the tafte, and probably alfo in the medical effects, of the folutions: but whether thefe differences are fuch, that fome of them, as crabs-claws, pearl, coral, and bezoar, are moft difpofed to promote a diaphorefis in fevers, while others, as egg-fhells and oyfter-fhells, act rather by promoting urine, as feems to be generally fuppofed, has not been determined by fair experience, the earths having rarely been given in a diffolved or in a foluble ftate. It is moft probable that they all act, when diffolved, as mild cooling reftringents; for when given in fubftance, as abforbents, in cafes of acidities, they all tend to reftrain fluxes of the belly, or to bring on coftivenefs, an

effect

effect to which regard ought to be had in the use of them.

There are two soluble earths, not commonly ranked among the absorbents, whose effects, when combined with acids, are known with more certainty, as they have been used oftener, so combined, than otherwise; to wit, the aluminous earth and magnesia; of which the one is strongly styptic, and the other moderately purgative.

Combinations of the absorbent earths with the nitrous and marine acids are bitterish and of great pungency, particularly those with the marine: the medical effects of these solutions are little known. The vitriolic acid does not dissolve them into a liquid form, but precipitates them from all the others, and is thus combined with them into concretes nearly insipid.

Experiments have been made for determining the comparative strength of different absorbents, or the quantities of acid they are capable of satiating. Langius reports, that ten grains of crabs-claws destroyed the acidity of forty drops of spirit of salt; that egg-shells, crabs-eyes, and mother of pearl, taken in the same quantity, saturated fifty drops each; red coral, white coral, and fixt alkaline salt, sixty drops each; volatile alkaline salt and pearl, eighty drops each; chalk, an hundred drops; oyster-shells, an hundred and twenty; and some lime stones no less than an hundred and sixty *(a)*. These experiments however (admitting their accuracy, and the acid to have been equally neutralized in all, which may be reasonably questioned) do not answer the end so perfectly as could be wished; for, to different acids, the earths have different

(a) Vide Langii *Opera omnia medica, Lipsiæ* 1704, *p.* 452 *& seq.*

habitudes;

THEA.

habitudes: from a set of experiments made by Homberg, it appears that oyster-shells, for example, require for their solution more of the marine acid than coral does; whereas of the nitrous acid, contrariwise, the coral requires more than 'the oyster-shells *(a)*. Neither the nitrous nor the marine acids are those which absorbents are destined to satiate in the human stomach, and by which their strength should be examined: the acids of the vegetable kingdom, and the acid of milk, may be presumed to be the most analogous to such as are generated in the bodies of animals. On trying, with these, the several substances enumerated at the beginning of this article, the differences in their absorbent powers appeared not to be very great: they all saturated pretty nearly the same quantities of the acids; and there remained, from all, quantities very considerable, but not very greatly different, of a matter which further additions of the acid would not dissolve.

THEA.

TEA: the leaf of a Chinese shrub, *evonymo affinis arbor orientalis nucifera flore roseo Pluk. alm. Thea bohea & viridis Linn.* The leaves, carefully picked, are dried hastily on warm iron plates; whereby they are said to lose in great measure some noxious qualities which they have when fresh, and to preserve their admired flavour which by slow exsiccation would be lost. The several sorts of tea brought to us are supposed to be the leaves of the same plant, collected at different times, and cured in a somewhat

(a) Vide Memoires de l'acad. roy. des sciences de Paris, pour l'ann. 1700.

somewhat different manner: Neumann suspects that the brown colour, and the flavour, of the bohea sorts, are introduced by art.

Both the green and bohea teas have an agreeable smell, and a lightly bitterish sub-astringent taste: with solution of chalybeate vitriol, they strike an inky blackness. They give out their smell and taste both to watery and to spirituous menstrua: to water, the green sorts communicate their own green tincture, and the bohea their brown: to rectified spirit they both impart a fine deep green. On gently drawing off the menstrua from the filtered tinctures, the water is found to elevate nearly all the peculiar flavour of the tea, while rectified spirit brings over little or nothing, leaving the smell as well as the taste concentrated in the extract: both extracts are very considerably astringent, and not a little ungrateful; the spirituous most so.

Infusions of tea, as dietetic articles, have been extravagantly commended by some and condemned by others; and notwithstanding the frequency of their use, they real effects are scarcely as yet clear. They seem, when moderately used, to be for the most part innocent: in some cases, they seem to be salutary: in some, they are apparently prejudicial. They dilute thick juices and quench thirst more effectually, and pass off by the natural emunctories more freely, than mere watery fluids: they refresh the spirits in heaviness and sleepiness, and seem to counteract the operation of inebriating liquors. From their manifest astringency, they have been supposed to strengthen and brace up the solids, but this effect experience does not countenance: it is in disorders, and in constitutions, wherein

corroborants

THLASPI.

corroborants are moft ſerviceable, that the' immoderate uſe of tea is peculiarly hurtful; in cold indolent habits, cachexies, chloroſes, dropſies, and debilities of the nervous ſyſtem.

THLASPI.

THLASPI: a plant with oblong narrow undivided leaves joined immediately to the ſtalks, on the tops of which grow numerous tetrapetalous flowers, each of which is followed by a ſhort flat ſeed-veſſel divided tranſverſely into two cells.

1. *Thlaſpi arvenſe ſiliquis latis C. B. Thlaſpi arvenſe Linn.* Treacle-muſtard: with roundiſh-pointed jagged leaves, and broad capſules containing about four ſeeds in each cell. It is annual, and grows wild in corn-fields.

2. *Thlaſpi arvenſe, vaccariæ incano folio majus C. B. Thlaſpi campeſtre Linn.* Mithridate muſtard: with hoary ſharp-pointed leaves ſhaped like an arrow-head; and only one ſeed in each cell. It is biennial, and grows in fields and open clayie grounds.

The ſeeds of theſe plants have an acrid biting taſte, approaching to that of the common muſtard; with which they agree nearly in their pharmaceutic properties, their pungent matter being totally extracted by water, only partially by rectified ſpirit, and being elevated by water in diſtillation. They have, joined to their acrimony, an unpleaſant flavour, ſomewhat of the garlick or onion kind; and this they give out to ſpirituous as well as watery menſtrua. They are rarely made uſe of any otherwiſe than

as

as ingredients in the compofitions whofe names they bear: though fome recommend them in different diforders, preferably to the common muftard.

THUS.

THUS Pharm. Lond. Frankincenfe: a folid brittle refin, brought to us in little glebes or maffes, of a brownifh or yellowifh colour on the outfide, internally whitifh or variegated with whitifh fpecks. It is fuppofed to be the produce of the pine that yields the common turpentine, and to concrete upon the furface of the terebinthinate juice foon after it has iffued from the tree.

This refin has a bitterifh acrid unpleafant tafte, and no confiderable fmell: it diffolves totally in rectified fpirit, but is fcarcely acted upon by watery menftrua. It may be looked upon as a mild corroborant; though at prefent it is little otherwife made ufe of than as an ingredient in theriaca, and externally in plafters. An officinal plafter, made with half a pound of frankincenfe and three ounces of dragons blood in powder ftirred into two pounds of the common plafter melted, now takes its name which was formerly *roborans*, from this ingredient.

Empl. thuris Ph. Lond.

THYMELÆA.

THYMELÆA Pharm. Parif. A fhrubby plant; with fmooth uncut leaves; and monopetalous flowers fet thick together: each flower is cut into four acute fections, and followed by an oblong, red, yellow, or black berry, containing one feed, which refembles a hemp-feed.

1. THYMELÆA

THYMELÆA.

1. THYMELÆA: *Thymelæa foliis lini* C. B. *Daphne Gnidium Linn.* Spurge-flax: with the ſtalks and branches clothed with evergreen leaves like thoſe of flax; and white flowers in cluſters on the tops.

2. LAUREOLA *ſeu Chamælæa: Laureola ſempervirens flore viridi, quibuſdam laureola mas* C. B. *Daphne laureola Linn.* Spurge laurel: with evergreen ſhining bay-like leaves, ſtanding ſeveral together, only at the tops of the branches; and greeniſh flowers on pedicles in their boſoms.

3. MEZEREUM *Pharm. Lond.* Mezereon *Pharm. Edinb. Laureola folio deciduo, flore purpureo, officinis laureola femina* C. B. *Daphne Mezereum Linn.* Spurge-olive, widow-wail: with pale purpliſh or white flowers clothing the branches; on the tops of which appear, after the flowers have fallen, bay-ſhaped leaves not ſhining.

The firſt of theſe plants grows on mountainous places in the ſouthern parts of Europe: the ſecond in moiſt woods in ſome parts of England: the third, a native of Germany *(a)*, is cultivated in our gardens, on account of the elegance and earlineſs of its flowers, which ſometimes appear in the end of January: the berries of all the ſorts ripen in Auguſt or September.

The leaves of theſe plants have little or no ſmell, but a nauſeous, acrid, very durable taſte: taken internally, in ſmall doſes, as ten or twelve grains, they are ſaid to operate with

---

(a) The mezereon has of late been obſerved to be a native of England alſo, being found plentifully in ſome woods near Andover in Hampſhire.

violence,

violence, by ſtool, and ſometimes by vomit, ſo as not to be ventured on with ſafety unleſs their virulence be previouſly abated by long boiling, and even then they are much too precarious to be truſted to.

The flowers are of a different nature, being in taſte little other than mucilaginous and ſweetiſh, and of a light pleaſant ſmell.

The pulpy part of the berries appears alſo to be harmleſs; but the ſeeds, called *coccognidia* or *grana cnidia*, are as acrid, and as virulently purgative as the leaves.

*The bark of the ſpurge laurel, macerated in water, has of late been much employed in France as a topical application to the ſkin, for the purpoſe of excoriating and exciting a diſcharge. That of the mezereon has been recommended for the ſame purpoſe.

The root of the mezereon has lately been uſed with ſucceſs in caſes of venereal nodes. Dr. Ruſſel, to whom the public is obliged for the communication of its efficacy in this frequently obſtinate complaint, obſerves, in the medical inquiries above-mentioned, that the cortical part of the root, on firſt chewing, is not pungent, but after a little time proves greatly ſo; and that the diſagreeable ſtimulus in the fauces laſts for many hours: that a decoction of an ounce of the freſh cortical part in a gallon and a half of water (the boiling being continued till half a gallon is waſted, and an ounce of ſliced liquorice added towards the end) may be taken to the quantity of half a pint four times a day, is not nauſeous to the taſte, and has not been found to diſagree with any ſtomach or conſtitution, or to remarkably increaſe any of the ſecretions; but that on doubling the quantity of the mezereon, the decoction proved ſo pungent, that

THYMIAMA.

that no ſtomach could bear it. He recommends the above decoction principally in thoſe venereal nodes that, proceed from a thickening of the membrane of the bones, which appears to be the cauſe of greateſt part of theſe tumours, at leaſt when recent: when there is an exoſtoſis, nothing is to be hoped for from this medicine; and when the bone is carious, no cure is to be expected without an exfoliation, though even here it ſometimes diſperſes the tumour, as appears from ſome of the caſes which he relates. In a thickening of the perioſteum from other cauſes, it has likewiſe had good effects.

THYMIAMA.

THYMIAMATIS cortex *Officinarum Germaniæ: Thus judæorum quorundam.* A bark, in ſmall browniſh-grey pieces, intermixed with bits of leaves, ſeeming as if the bark and leaves had been bruiſed and preſſed together; brought from Syria, Cilicia, &c. and ſuppoſed to be the produce of the liquid-ſtorax tree.

This bark has an agreeable balſamic ſmell, approaching to that of liquid-ſtorax, and a ſubacrid bitteriſh taſte accompanied with ſome flight aſtringency. Infuſions of it in water are of an orange colour, in taſte and ſmell ungratefully balſamic: inſpiſſated, they leave a dark reddiſh brown extract, retaining ſome of the ſmell of the bark, in taſte auſtere, ſlightly bitter, and of a mild aromatic acrimony. To rectified ſpirit it communicates a dark colour like that of a ſolution of balſam of Peru: the ſpirit, diſtilled off from this tincture, is highly fragrant, inſomuch that a dram communicates an agreeable odour to ſome quarts of water: the remaining extract is likewiſe of a pleaſant ſmell, and amounts to at leaſt one eighth of the weight of the bark.

This bark, said to be common in the German shops, is in this country very rarely to be met with. Cartheuser and Hoffman, from whom the above account is extracted, report, that it affords an excellent fumigation for œdemas, rheumatisms, and catarrhs; and that the spirituous tincture and extract, and the distilled spirit, are useful anodynes or antispasmodics in convulsive coughs and other disorders.

THYMUS.

THYME: a low shrubby plant; consisting of numerous slender tough stalks, with little roundish leaves in pairs, and loose spikes, on the tops, of purplish or whitish labiated flowers, whose upper lip is nipt at the extremity, the lower divided into three nearly equal segments.

1. THYMUS *Pharm. Edinb. Thymum vulgare folio tenuiore C. B. Thymus vulgaris Linn.* Common thyme: with upright stalks, and dark brownish green somewhat pointed leaves; a native of the southern parts of Europe, common in our gardens, and flowering in June and July.

This herb is a moderately warm pungent aromatic. To water it imparts, by infusion, its agreeable smell, but only a weak taste, with a yellowish or brown colour: in distillation, it gives over an essential oil, in quantity about an ounce from thirty pounds of the herb in flower, of a gold yellow colour if distilled by a gentle fire, of a deep brownish red if by a strong one, of a penetrating smell resembling that of the thyme itself, but less grateful, in taste excessively hot and fiery: the remaining decoction, inspissated, leaves a bitterish, roughish, subsaline extract. The active matter, which by water is only partially dissolved, is by rectified spirit

THYMUS.

spirit dissolved completely; though the tincture, in colour blackish-green, discovers less of the smell of the thyme than the watery infusion: the spirit brings over in distillation a part of its flavour, leaving an extract of a weak smell and of a penetrating camphorated pungency.

2. SERPYLLUM *Pharm.-Edinb. Serpyllum vulgare minus* C. B. *Thymus Serpyllum Linn.* Mother-of-thyme: with trailing stalks, and obtuse leaves: growing wild on heaths and dry pasture grounds. This also is an elegant aromatic plant, similar to the foregoing species, but milder, and in flavour rather more grateful. Its essential oil is both in smaller quantity and less acrid, and its spirituous extract comes greatly short of the penetrating warmth and pungency of that of the other. It is said to afford an agreeable distilled water, more durable, but less active and penetrating than peppermint*(a)*. Both the leaves themselves, and their spirituous tincture, are of a bright green colour, without any thing of the brown or blackish hue of those of common thyme.

3. THYMUS CITRATUS *Serpyllum foliis citri odore* C. B. Lemon-thyme: in appearance differing little from the second sort, of which Linnæus makes it a variety, except that it is more upright and more bushy; a native of dry mountainous places, common in gardens, and flowering as the others in July. This species is less pungent than the first sort, more so than the second, and much more grateful than either: its smell in particular is remarkably different, approaching to that of lemons. Dis-

(a) Cullen, *Mat. Med.*

tilled

tilled with water, it yields a larger quantity than the other forts, of a yellowifh very fragrant oil of the lemon flavour, containing nearly all the medicinal parts of the plant, for the remaining decoction is almoft infipid as well as inodorous. It gives over alfo with rectified fpirit its finer odorous matter; a lefs agreeable flavour, and a moderate warmth, remaining in the fpirituous extract.

TILIA.

TILIA femina folio majore C. B. Tilia europæ Linn. LIME or LINDEN: a tall fpreading-branched tree, with large heart-fhaped, ferrated, foft, fomewhat hairy leaves: in the bofoms of thefe rife long narrow leafy productions, from the middle rib of which iffue one or three pedicles bearing three flowers apiece, or one pedicle bearing nine: the flower is whitifh, pentapetalous, and followed by a kind of dry berry about the fize of a filberd. It is a native of England, flowers in July, and begins to lofe its leaves in Auguft.

THE flowers of the lime-tree are fuppofed to have an anodyne and antifpafmodic virtue: Hoffman feems to entertain a great opinion of them in thefe intentions, and as his theory deduces moft difeafes from fpafms and fpafmodic ftrictures, they are accordingly very frequent in his prefcriptions: he fays he knew a chronical epilepfy cured by the ufe of an infufion of them drank as tea. The frefh flowers have a moderately ftrong fmell, in which their virtue (whatever it may be) feems to confift, and which in keeping is foon diffipated: when divefted of this odorous principle, they difcover to the tafte only

TITHYMALUS.

only a ftrong mucilage, from which may be extracted, by rectified fpirit, a flightly bitterifh fubaftringent matter.

TITHYMALUS.

SPURGE: a plant with fmall fmooth leaves, round ftalks full of a milky juice, and umbel-like clufters of tetrapetalous flowers, whofe cups are divided into four fegments fet alternately with the petala: the flower is followed by a roundifh or three-fquare capfule containing three feeds.

1. TITHYMALUS PARALIOS: *Tithymalus mari‍timus* C. B. *Euphorbia Paralias Linn.* Sea fpurge: with oblong narrow flax-like leaves, broadeft in the middle, clothing the ftalks, and lying over one another in an upward direction, like fcales; and two roundifh, heart-fhaped, or kidney-fhaped leaves encompaffing each of the fubdivifions of the umbel: found wild on fandy fhores, and flowering in June. All the parts of this plant are extremely acrid irritating cathartics; apt to inflame the mouth, fauces, and ftomach; operating with fo great violence, that though fome may perhaps have borne their operation without much injury to the conftitution, yet common prudence forbids their being ever ventured on. Several correctors have been employed for them, but none with commendable fuccefs: maceration of the middle bark of the root in vinegar, directed by the faculty of Paris, renders it indeed lefs virulent, but of precarious operation: digeftion of the milky juice with alkaline falts, recommended by others, leaves it ftill too acrid. For alleviating inflammatory fymptoms produced by imprudently

dently fwallowing or tafting thefe acrid fubftances, milk, plentifully drank, feems the moft effectual remedy. Gerard relates, that on taking but one drop of the milk of the fea fpurge, it did fo fwell and inflame in his throat, that he hardly efcaped with his life, and that on drinking milk, the extremity of the heat ceafed.

2. TITHYMALUS CYPARISSIUS *C. B.* *Euphorbia Cyparissias Linn.* Cyprefs fpurge: with numerous oblong flender leaves, not wider in the middle than at the ends; the umbel divided into numerous ramifications, each of which is divided and fubdivided into two; the divifions perforating as it were the two roundifh leaves which encompafs them; a native of Germany, Switzerland, and fome other parts of Europe. This fpecies, though allowed by the faculty of Paris to be ufed indifcriminately with the preceding, is in all its parts lefs acrimonious. Poterius fays he has found half a dram or a dram of the powdered root to act as a mild cathartic; and that the juice obtained from the bruifed herb and root, depurated and exficcated in the fun, is of the fame operation with fcammony *(a).*

SEVERAL other fpurges are enumerated in catalogues of the materia medica, under the names of *efula, pityufa, cataputia, lathyrus, alypum, peplus, apios,* &c. among which there does not appear to be any one more virulent than the firft above defcribed, or lefs virulent than the fecond. None of them are among us ventured on for any internal ufe: the milky juice

(a) Pharmacopœia fpagyrica, lib. iii. *fect.* 3.

of the wild fpurges is fometimes applied externally by the common people for confuming warts.

TORMENTILLA.

TORMENTILLA Pharm. Lond. & Edinb. Tormentilla fylveftris C. B. Heptaphyllum. Tormentilla erecta Linn. TORMENTIL or SEPTFOIL: a plant with flender, weak, upright ftalks; oblong leaves, indented towards the extremity, and converging from the indented part to their juncture with the ftalk, ftanding generally feven at a joint; and fmall yellow tetrapetalous flowers on the tops of the branches, followed by naked feeds; the root is generally crooked and knotty, of a dark brown or blackifh colour on the outfide, and reddifh within. It is perennial, grows wild in woods and on commons, and flowers in June.

TORMENTIL ROOT is a ftrong and almoft flavourlefs aftringent, and gives out its aftringency both to water and rectified fpirit, moft perfectly to the latter: the watery decoction, of a tranfparent brownifh red colour whilft hot, becomes turbid in cooling like that of the Peruvian bark, and depofites a portion of refinous matter: the fpirituous tincture, of a brighter reddifh colour, retains its pellucidity. The extracts, obtained by infpiffation, are intenfely ftyptic, the fpirituous moft fo. It is generally given in decoction: an ounce and a half of the powdered root may be boiled in three pints of water to a quart, adding, towards the end of the boiling, a dram of cinnamon; of the ftrained liquor, fweetened with an ounce of any agreeable fyrup, two ounces or more may be taken four or five times a day.

TRICHOMANES.

TRICHOMANES Pharm. Edinb. Polytrichum five trichomanes C. B. *Callitrichum. Asplenium Trichomanes Linn.* ENGLISH MAIDENHAIR: a small plant, without stalks: the leaves are long, narrow, composed of little roundish dark-green segments set in pairs along a shining black rib: the seeds are a fine dust lying on the backs of the leaves. It is perennial, and grows wild on shady grounds and old walls.

THIS herb has a mucilaginous somewhat sweetish and roughish taste, and little or no particular flavour. It is accounted serviceable in disorders of the breast, particularly in tickling coughs and hoarseness from thin acrid defluxions, and in these intentions has been long substituted among us to the *adianthum*, from which it appears to be very little, if at all, different in quality. It is usually directed in infusion or decoction, with the addition of a little liquorice: a pectoral syrup is prepared in the shops, from an infusion of five ounces of the dry leaves and four of liquorice root in five pints of boiling water.

<small>Syrup. pectoralis.</small>

TRIFOLIUM PALUDOSUM.

TRIFOLIUM PALUDOSUM Pharm. Lond. Menyanthes Pharm. Edinb. Trifolium palustre C. B. *Menyanthes Trifoliata Linn.* BUCKBEAN: a plant with large oval leaves, pointed at each end like those of the garden bean, set three together on long pedicles, which embrace the stalk to some height, and there parting leave it naked to near the top, where issues a short spike

TURPETHUM.

of pretty large reddifh white monopetalous flowers, each of which is cut into five fegments, hairy on the infide, and followed by an oval feed-veffel. It is perennial, grows wild in marfhy places, and flowers in May.

The leaves of buckbean have a bitter penetrating tafte, which they impart both to watery and fpirituous menftrua; without any remarkable fmell or flavour. They have of late years come into common ufe, as an alterative and aperient, in impurities of the humours, and fome hydropic and rheumatic cafes. They are ufually taken in the form of infufion, with the addition of fome of the acrid antifcorbutic herbs, which in moft cafes improve their virtue, and of orange peel or fome other grateful aromatic to alleviate their ill tafte: they are fometimes, among the common people, fermented with malt liquors, for an antifcorbutic diet drink. Their fenfible operation is by promoting urine and fomewhat loofening the belly.

TURPETHUM.

TURPETHUM five Turbith. Turbith: the cortical part of the root of a fpecies of convolvulus (*Convolvulus Turpethum Linn.*), brought from the Eaft Indies, in oblong pieces, of a brown or afh colour on the outfide and whitifh within: the beft is ponderous, not wrinkled, eafy to break, and difcovers to the eye a large quantity of refinous matter.

This root, on the organs of tafte, makes at firft an impreffion of fweetnefs; but when chewed for fome time, betrays a naufeous acrimony. It is accounted a moderately ftrong cathartic,

cathartic, but does not appear to be of the safest or most certain kind; the resinous matter, in which its virtue resides, being very unequally distributed; insomuch that, as is said, some pieces taken from a scruple to a dram, purge violently, whilst others, in larger doses, have very little effect.

TUSSILAGO.

TUSSILAGO Pharm. Lond. & Edinb. Tussilago vulgaris C B. *Farfara Bechium & ungula caballina quibusdam. Tussilaga Farfara Linn.* COLTSFOOT: a low plant, producing early in the spring single stalks, each of which bears a yellow flosculous flower followed by several seeds winged with down: the leaves which succeed the flowers, are short, broad, somewhat angular, slightly indented, green above, and hoary underneath. It is perennial, and grows wild in moist grounds.

The leaves and flowers of coltsfoot, in taste somewhat mucilaginous, bitterish, and roughish, and of no remarkable smell, are ranked among the principal pectoral herbs. Infusions of them, with a little liquorice or with the other herbs of similar intention, are drank as tea, and sometimes with considerable benefit, in catarrhous disorders and coughs threatning consumptions. They have been found serviceable in hectics and colliquative diarrhœas *(a)*.

TUTIA.

TUTIA Pharm. Lond. & Edinb. Tutia alexandrina. TUTTY: an argillaceous ore of

(a) Percival, *Ess. Med. and Exp.* II. 224.

zinc,

zinc, found in Perfia; formed on cylindrical moulds into tubulous pieces like the bark of a tree, and baked to a moderate hardnefs *(a)*; generally of a brownifh colour and full of fmall protuberances on the outfide, fmooth and yellowifh within, fometimes whitifh, and fometimes with a bluifh caft. Like other argillaceous bodies, it becomes harder in a ftrong fire; and after the zinc has been revived and diffipated by inflammable additions, or extracted by acids, the remaining earthy matter affords, with oil of vitriol, an aluminous falt (fee *Bolus* and *Calaminaris*).

TUTTY, levigated into an impalpable powder, is, like the lapis calaminaris and calces of zinc, an ufeful ophthalmic, and frequently ufed as fuch in ointments and collyria. Ointments for this intention are prepared in the fhops, by mixing the levigated tutty with fo much fpermaceti ointment as is fufficient to reduce it to a due confiftence†; or by adding one part to five parts of a fimple liniment made of oil and wax ‡.

Tutia prǽparata *Ph. Lond. & Ed.*

† Ung. tutiæ *Ph. Lond.*
‡ *Ph. Ed.*

VALERIANA.

VALERIANA filveftris Ph. Lond. & Edinb.
Valeriana filveftris major montana C. B. Valeri-

(a) The above account of the origin of tutty is fupported by the authority of Teixeira and Douglas, and by its chemical properties. That the common opinion, of its being a fublimate produced in the European founderies where zinc is melted with other metals, is erroneous, appears from hence; that tutty is not found, upon ftrict inquiry, to be known at thofe founderies; and by its confifting in great part of an earth not capable of rifing in fublimation. Thus much, however, is probable, that fublimates or the common ores of zinc are often mixed with argillaceous earths and baked hard, in imitation of the genuine oriental tutty.

ana

ana officinalis Linn. WILD VALERIAN: a plant with channelled ftalks; the leaves in pairs; each leaf compofed of a number of long narrow fharp-pointed fegments, indented about the edges, of a dull green colour, fet along a middle rib, which is terminated by an odd one; producing, on the tops of the ftalks, umbel-like clufters of fmall monopetalous flowers, each of which is divided into five fegments, fet in a very little cup, and followed by a fingle naked feed winged with down: the root confifts of tough ftrings with numerous fmaller threads, matted together, iffuing from one head, of a dufky brownifh colour approaching to olive. It is perennial, and grows wild in dry mountainous places.

Another fpecies, or variety, of wild valerian, is met with in moift watery grounds, diftinguifhable by the leaves being broader and of a deep gloffy green colour. Both forts have been ufed indifcriminately; but the mountain fort is by far the moft efficacious, and is therefore exprefsly ordered for the officinal fpecies by the London college.

THE mountain valerian root has a ftrong not agreeable fmell, and an unpleafant warm bitterifh fubacrid tafte: the ftrength of the fmell and tafte is the only mark to be depended on of its genuinenefs and goodnefs. It is a medicine of great efteem in the prefent practice againft obftinate hemicraniæ, hyfterical, and the different kinds of nervous diforders, and is commonly looked upon as one of the principal antifpafmodics. Columna reports, that he was cured by it of an inveterate epilepfy after many other medicines had been ufed in vain: on more extenfive trials it has been found, in fome
epileptic

VALERIANA.

epileptic cafes, to effect a cure, in feveral to abate the violence or frequency of the fits, and in many to prove entirely ineffectual: oftentimes, it either purges, or operates by fweat or by urine, or brings away worms, before it prevents a fit. The dofe of the root in powder is from a fcruple to a dram or two, which may be repeated, if the ftomach will bear it, two or three times a day. * A remarkable inftance of its efficacy in a catalepfy is given by Mr. Mudge (a): dofes of half an ounce of the powder were exhibited twice a day, and a lefs quantity was found ineffectual.

The powdered root, infufed in water or digefted in rectified fpirit, impregnates both menftrua ftrongly with its fmell and tafte, and tinges the former of a dark brown, the latter of a brownifh red colour. Water diftilled from it fmells confiderably of the root, but no effential oil feparates, though feveral pounds be fubmitted to the operation at once: the extract obtained by infpiffating the watery infufion, has a pretty ftrong tafte, difagreeably fweetifh and fomewhat bitterifh: the fpirituous extract is lefs difagreeable, and more perfectly refembles the root itfelf: the quantity of watery extract is about one fourth the weight of the root; of the fpirituous, about one eighth. Tinctures of it are prepared in the fhops, by digefting four ounces of the powdered valerian in a quart of proof fpirit†; in the fame quantity of the volatile aromatic fpirit‡; or of the dulcified fpirit of fal ammoniac‖. The root in fubftance, however, is generally found to be more effectual than any preparation of it. Among the mate-

Tinct. valer. fimp. † Ph. Lond.
—volat. ‡Ph. Lond.
‖ Ph. Ed.

(a) On the *Vis Vitæ*, &c.

rials

rials I have made trial of for covering its flavour, mace seemed to answer the best.

VANILLA.

VANILLA *seu Banilia* Pharm. Parif. *Aracus aromaticus.* VANELLOE: the fruit of a climbing plant *(volubilis filiquofa mexicana foliis plantaginis Raii hift. Epidendrum Vanilla Linn.)* growing in the Spanish West Indies. It is a long flattish pod, containing, under a wrinkled brittle shell, a reddish brown pulp, with small shining black seeds.

VANELLOES have an unctuous aromatic taste, and a fragrant smell like that of some of the finer balsams heightened with musk. They are used chiefly in perfumes; scarcely ever, among us at least, in any medical intention; though they should seem to deserve a place among the principal medicines of the nervous class. By distillation, they impregnate water strongly with their fragrance, but give over little or nothing with pure spirit: by digestion, spirit totally extracts their smell and taste, and in great measure covers or suppresses the smell.

VERBASCUM.

VERBASCUM Pharm. Edinb. *Verbafcum mas latifolium luteum* C. B. *Tapfus barbatus Candelaria & lanaria quibufdam. Verbafcum Thapfus Linn.* MULLEIN: a large plant, all over white and woolly; with a single woody stalk, clothed with oblong oval leaves joined to it without pedicles, bearing on the top a long spike of large yellow monopetalous flowers cut into five segments, and followed by conical seed feed-veffels. It is biennial, grows wild by road-fides, and flowers in July.

The leaves of mullein, recommended as mild aftringents, have a roughifh drying kind of tafte, and very little fmell. The flowers have alfo a flight roughifhnefs, with a confiderable fweetnefs. A decoction of the leaves has lately been ufed with fome fuccefs in diarrhœas *(a)*.

VERBENA.

VERBENA communis flore cæruleo C. B. *Hierobotane, herba facra, herba cephalalgica & perifterium quibufdam. Verbena officinalis Linn.* Vervain: a plant with wrinkled, oblong, obtufe leaves, deeply jagged and indented, fet in pairs on the ftalks, the upper ones divided into three fegments: on the tops of the branches appear irregularly labiated blue flowers, in long fpikes, without any leaves among them, followed each by four feeds inclofed in the cup. It is annual, grows wild in uncultivated places, and flowers in July or Auguft.

This herb has been celebrated for abundance of virtues, for which its fenfible qualities afford little or no foundation. It has no remarkable fmell, and hardly any tafte.

VERONICA.

VERONICA: a low, fomewhat hairy, trailing plant, with firm leaves fet in pairs: from the joints arife flender pedicles, bearing fpikes of blue monopetalous flowers, each of which is divided, as is the cup, into four fegments, and

(a) Home's *Clin. Caf. and Exp.*

followed

followed by a flat bicellular capfule, which opens at the upper broad part and fheds fmall brown feeds.

1. VERONICA MAS *five Betonica pauli. Veronica mas fupina & vulgatiffima* C. B. *Thea Germanica quibufdam. Veronica officinalis Linn.* Male fpeedwell: with crenated leaves of a roundifh oval figure; thofe on the flowering twigs, long, narrow, and not crenated. It is perennial, and grows wild on fandy grounds and dry commons.

The leaves of veronica have a weak not difagreeable fmell, which in drying is diffipated, and which they give over in diftillation with water, but without yielding any feparable oil. To the tafte they are bitterifh and roughifh: an extract made from them by rectified fpirit is moderately bitter and fubaftringent: the watery extract is weaker, though the quantity of both is nearly the fame; whence fpirit feems to extract their virtue more completely than water. This herb is of great efteem among the Germans; in diforders of the breaft both catarrhous and ulcerous, and for purifying the blood and humours: infufions of the leaves, which are not unpalatable, are drank as tea, and are found to operate fenfibly by urine.

2. TEUCRIUM *Act. med. berolinenf. Chamædrys fpuria major anguftifolia* C. B. *Veronica Teucrium Linn.* Mountain fpeedwell: with fharply ferrated leaves of a long oval figure; the lower embracing the ftalk by a broad bafis. It is a native of Germany.

The leaves and flowers of this fpecies have been greatly commended for dietetic infufions; and faid to promote perfpiration and urine, to be

VINCETOXICUM.

be in general falubrious, and medicinal in feveral diforders(a). Cartheufer obferves, that they impart to boiling water a greenifh colour, a pleafant balfamic fmell, and a much more agreeable tafte than the preceding veronica. Among us they have not yet been introduced, nor is the plant common : what has ufually been called *teucrium* is a plant of another genus, a large fpecies of germander.

VINCETOXICUM.

VINCETOXICUM, Afclepias, Hirundinaria. Afclepias albo flore C. B. *Afclepias Vincetoxicum Linn.* Swallow-wort, tame-poifon: a plant with unbranched ftalks; fmooth oblong acuminated leaves fet in pairs, and clufters of white monopetalous flowers, each of which is divided into five fections, and followed by two long pods full of a white cottony matter with fmall brownifh feeds: the root is large, compofed of a great number of flender ftrings hanging from a tranfverfe head, externally brownifh, internally white. It is perennial, grows wild in gravelly grounds in fome parts of England, and flowers in July.

The root of vincetoxicum has, when frefh, a moderately ftrong not agreeable fmell, approaching to that of wild valerian, which in drying is in great part diffipated; chewed, it impreffes firft a confiderable fweetnefs, which is foon fucceeded by an unpleafant fubacrid bitterifhnefs: an extract made from it by water, is moderately fweetifh, balfamic, and bitterifh; the fpirituous extract is ftronger in tafte, proportionably fmaller in quantity, and retains a

(a) Gohl, *Acta medica Berolinenf.* dec. I. vol. ii. n. 5.

part of the specific flavour of the root. It is recommended as resolvent, sudorific, and diuretic; in catarrhal, cachectic, and scrophulous disorders, and in uterine obstructions; in doses of from a scruple to a dram or more in substance, and three or four drams in infusion. It has been employed by some of the Germans as an alexipharmac, and hence received the name of *contrayerva Germanorum*. Some have however suspected it to possess noxious qualities, and observe that when fresh it excites vomiting. Among us it is scarcely ever made use of in any intention.

VINUM.

WINE: the fermented juice of the grape. It differs in colour, flavour, and strength, partly from differences in the grape itself, but chiefly from different managements or additions. Five sorts are employed in the shops as menstrua for medicinal substances: *Vinum album*, Mountain: *Vinum album. gallicum*, French white wine: *Vinum canarinum*, Canary or sack: *Vinum rhenanum*, Rhenish: *Vinum rubrum*, Red Port.

ALL wines consist of an inflammable spirit, and water, separable by distillation; an unctuous viscid substance, which abounds particularly in the sweet wines, as Canary, and impedes their dissolving power; and an acid, obvious in some to the taste, as in Rhenish, which hence becomes an useful menstruum for some bodies of the metallic kind, particularly iron and the antimonial regulus. In distillation, after the inflammable spirit has arisen, they all yield more or less of a peculiar grateful acid; a grosser tartareous acid remaining in the still, along with

Media substantia vini Beccheri.

with the unctuous and mucilaginous matter. In long keeping, a part of the tartar is thrown off from the wine, and incruftates the fides of the cafk.

Wine, confidered as a medicine, is a valuable cordial in languors and debilities; more grateful and reviving than the common aromatic infufions and diftilled waters, particularly ufeful in the low ftage of malignant or other fevers, for raifing the pulfe and fupporting the vis vitæ, promoting a diaphorefis, and refifting putrefaction. Dietetically, its moderate ufe is of fervice to the aged, the weak, and the relaxed, and to thofe who are expofed to warm and moift, or to corrupted air: in the oppofite circumftances, it is lefs proper, or prejudicial. Externally, it is ufed as a corroborant, antifeptic, and antiphlogiftic fomentation.

The acid obtained from wine by diftillation, apparently of a different nature from the acetous as well as from the native vegetable acids, feems to deferve fome regard, both as a medicine, and as a more elegant menftruum, for iron and fome other bodies, than the common acids.

With regard to the medical differences of wines, it may be obferved, that the effects of the full-bodied are much more durable than thofe of the thinner: that all fweet wines are in fome degree nutritious; the others not at all, or only accidentally fo, by promoting appetite and ftrengthening the organs concerned in digeftion: that fweet wines in general do not pafs freely by urine, and that they heat the conftitution more than an equal quantity of any other, though containing full as much fpirit: that thofe which are manifeftly acid pafs freely by the kidneys, and gently loofen the belly; and that moſt of the red ones are ſubaſtringent, and tend to reſtrain immoderate excretions.

VIOLA.

VIOLA Pharm. Lond. & Edinb. Viola martia purpurea flore ſimplici odoro C. B. Violaria. Viola odorata Linn. Violet: a low creeping plant, without any other ſtalk than the pedicles of the leaves and flowers: the leaves are roundiſh, ſomewhat heart-ſhaped, obtuſely crenated about the edges: the flower conſiſts of five irregular petala, of the deep purpliſh blue called, from the name of the plant, violet colour: the fruit is a little capſule divided into three cells, full of ſmall roundiſh ſeeds. It is perennial, grows wild in hedges and ſhady places, and flowers in March.

The flowers of a different ſpecies, greatly inferiour to the above, are frequently ſubſtituted in our markets. This ſort may be readily diſtinguiſhed; the herb, by its having ſtalks, which trail on the ground, and bear both leaves and flowers, and by the young leaves being hairy; the flower, by the three lower petala being ſpotted with white, and by their want of ſmell.

The officinal violet flowers have a very agreeable ſmell, and a weak mucilaginous bitteriſh taſte. Taken to the quantity of a dram or two, they are ſaid to be gently laxative or purgative; and the ſeeds, which have more taſte than the flowers, to be more purgative, and ſometimes emetic. The flowers give out to water both their virtue and their fine colour, but ſcarcely impart any tincture to rectified ſpirit, though they impregnate the ſpirit with their fine flavour, and probably alſo with their purgative quality.

VIOLA. 453

quality. An infusion of two pounds of the fresh flowers in five† or eight‡ pints of boiling water, passed through a fine linen cloth without pressure, is made in the shops into a syrup, which proves an agreeable laxative for children. Both the flowers themselves and the syrup lose their colour in being long kept: acids change them instantly into a red; alkalies, and sundry combinations of acids with earthy and metallic bodies, to a green: perfect neutral salts, or those compounded of an acid and alkali, make no alteration. Some have been accustomed to communicate to syrups a violet colour with materials of greater durability than the violet itself, or than any other blue flower: these sophisticated preparations may be distinguished by their colour withstanding alkalies and acids, or being affected by them in a different manner.

Syr. violar.
† *Ph. Lond.*
‡ *Ph. Ed.*

*VIOLA TRICOLOR *Linn.* Pansies, or heartsease: this well known plant has lately been recommended by a German physician, Dr. Strack, as a specific in the *crusta lactea* of children. He directs a handful of the fresh, or half a dram of the dried leaves to be boiled in half a pint of milk, which is to be strained for use. This dose is repeated morning and evening. He observes, that when it has been administered eight days, the eruption usually increases considerably, and the patient's urine acquires a smell like that of cats. When the medicine has been taken a fortnight, the scurf begins to fall off in large scales, leaving the skin clean. The remedy is to be persisted in, till the skin has resumed its natural appearance, and the urine ceases to have any particular smell.

VIPERA.

VIPERA.

VIPERA Pharm. Edinb. Coluber Berus Linn. The Viper or Adder, a viviparous reptile, about an inch or less in thickness, and twenty or thirty in length, with a small sharp-pointed tail. It is found in the heat of summer, under hedges in unfrequented places; and in winter retires into holes in the earth.

The poison of this serpent is confined to its mouth. At the basis of the phangs, or long teeth which it wounds with, is lodged a little bag containing the poisonous liquid; a very minute portion of which, if mixed immediately with the blood, proves fatal; though it does not appear to be pernicious when swallowed, provided there is no solution of continuity in the parts which it comes in contact with *(a)*. Our viper-catchers are said to prevent the mischiefs otherwise following from the bite, by rubbing oil-olive warm upon the part.

The flesh of the viper is perfectly innocent, and has been greatly commended as a medicine in sundry disorders. It appears to be very nutritious, and hence an useful restorative in some kinds of weaknesses and emaciated habits: but in scrophulous, leprous, and other like distempers, the good effects, which have been ascribed to it, are more uncertain: I have known a viper taken every day for above a month, in disorders of the leprous kind, without any apparent benefit. The form in which they are used to best advantage, is that of broth, that the wines (made commonly by macerating for a week, with a gentle heat, two ounces of

(a) See Dr. Mead's *Mechanical account of poisons,* essay i.

VIRGA AUREA.

the dried flesh in three pints of mountain) have any great virtue, cannot perhaps be affirmed from fair experience.

The fat of the viper is accounted particularly useful in disorders of the eyes; but what advantages it has above other soft fats, is by no means clear: see *Pinguedo*. It was formerly supposed to have some specific power of resisting the poison of the viper's bite, by being rubbed immediately on the part; but experience has now shewn that common oil is in this intention of equal efficacy.

VIRGA AUREA.

VIRGA AUREA angustifolia minus serrata C. B. *Herba doria & Consolida saracenica quibusdam. Solidago Virga aurea Linn.* GOLDEN ROD: a plant with long somewhat oval leaves, pointed at both ends, slightly or not at all indented; and upright spikes, along the stalks, of small yellow flowers, composed of several flosculi set in scaly cups, followed by small seeds winged with down. It is perennial, grows wild in woods and on heaths, and flowers in August.

THE leaves and flowers of golden rod are recommended as corroborants and aperients; in urinary obstructions, nephritic cases, ulcerations of the bladder, cachexies, and beginning dropsies. Their sensible qualities promise considerable medical activity: their taste, which they readily impart both to water and rectified spirit, and which remains entire in the inspissated extracts, is of a subtile penetrating durable kind, not very ungrateful, weak in the herb in substance, strong in the watery extract, and stronger in the spirituous.

VISCUS.

VISCUS.

VISCUS QUERNUS Viscum baccis albis C. B. Viscum album Linn. Misseltoe: a bushy evergreen plant, with woody branches variously interwoven; firm narrow leaves, narrowest at the bottom, set in pairs; and imperfect white flowers in their bosoms, followed each by a transparent white berry containing a single seed. It grows only on the trunks and branches of trees, and may be propagated by rubbing the glutinous berries on the bark that the seeds may adhere.

The leaves and branches of misseltoe, formerly recommended as specifics in convulsive and other nervous disorders, and now fallen into general neglect, do not appear to have any considerable medicinal power. Instances have indeed been produced of their seeming to prove beneficial: but as there are, perhaps, no disorders, whose nature is so little understood, whose causes are so various, and whose mitigations and exasperations have less dependence upon sensible things; there are none in which medicines operate more precariously, and in which the observer is more liable to deception.

Half a dram or a dram of the wood or leaves in substance, or an infusion of half an ounce, the doses commonly directed, have no sensible effect. Both the leaves and branches have very little smell, and a very weak taste, of the nauseous kind. In distillation they impregnate water with their faint unpleasant smell, but yield no essential oil. Extracts made from them by water are bitterish, roughish, and subsaline: the spirituous extracts, in quantity smaller than the watery,

watery, are in tafte ftronger, naufeous, bitterifh, and fubauftere: the fpirituous extract of the wood has the greateft aufterity, and that of the leaves the greateft bitterifhnefs. The berries abound, with an extremely tenacious, not ungrateful, fweet mucilage.

VITRIOLUM.

VITRIOLUM & Calcanthum Pharm. Parif.
VITRIOL: a faline cryftalline concrete, compofed of metal united with a certain acid called the vitriolic acid. There are three metals with which this acid is found naturally combined, zinc, copper, and iron: with the firft it forms a white, with the fecond a blue, and with the third a green falt.

1. VITRIOLUM ALBUM *Pharm. Lond. Vitriolum album five Zinci Pharm. Edinb.* White vitriol, or vitriol of zinc; found in the mines of Goflar, fometimes in tranfparent pieces, more commonly in white efflorefcences; which are diffolved in water, and cryftallized into large irregular maffes fomewhat refembling fine fugar; in tafte fweetifh, naufeous, and ftyptic.

The common white vitriol of the fhops contains a quantity of ferrugineous matter; of which, in keeping, a part is extricated from the acid, in an ochery form, fo as to tinge the mafs of a yellow hue. On diffolving the whiteft pieces in water, a confiderable portion of ochre immediately feparates: the filtered folution, tranfparent and colourlefs, becomes again turbid and yellow on being made to boil, and depofites a frefh ochery fediment; and a like feparation happens, though much more flowly, on ftanding without heat. Hence, when the

the solution is evaporated to the usual pitch, and set to crystallize, the crystals generally prove foul; unless some fresh acid be added (as an ounce of the strong spirit or oil of vitriol to a pound of the salt †) to keep the ferrugineous matter dissolved: this addition both secures the whiteness of the crystals, and prevents their growing soon yellow in the air. White vitriol generally contains also a small portion of copper, distinguishable by the cupreous stain which it communicates to polished iron immersed in solutions of it, or rubbed with it in a moist state. The quantity of this metal is so exceedingly minute, that it is not, perhaps, of any inconvenience in the intentions for which white vitriol is commonly employed: the separation, if it should be thought necessary, may be effected, by boiling the solution for some time, along with bright pieces of iron, which will extricate all the copper: by continued or repeated coction, greatest part of the ferrugineous matter also may be separated.

† *Zinc. vitriol. purif. Ph. Lond.*

White vitriol is sometimes given, from five or six grains to half a dram and more, as an emetic; and appears to be one of the quickest in operation of those that can be employed with safety. Its chief use is for external purposes, as a cooling restringent and desiccative: a dilute solution of it, as sixteen grains in eight ounces of water, with the addition of sixteen drops of weak vitriolic acid, is an excellent collyrium in defluxions and slight inflammations of the eyes, and, after bleeding and purging, in the more violent ones: a solution of it with alum, in the proportion of two drams of each to a pint of water, is used as a repellent fomentation for some cutaneous eruptions, for cleansing foul ulcers, and as an ejection in the fluor albus

Aqua vitriolica Ph. Ed.

Aqua aluminis comp. Ph. Lond.

albus and gonorrhœa when not accompanied with virulence. This vitriol is sometimes likewise employed as an errhine, and said to be a very effectual dissolvent of mucous matters; in which intention it is recommended, in the German ephemerides, against obstructions of the nostrils in new-born infants.

2. VITRIOLUM CÆRULEUM *Pharm. Lond. Vitriolum cæruleum sive cupri Pharm. Edinb.* Blue vitriol, or vitriol of copper, commonly called Roman or Cyprian vitriol, or blue-stone. This kind of vitriol is in many places produced from sulphureous ores of copper: the acid of sulphur is no other than the vitriolic; and the inflammable principle of the sulphur being dissipated either by fire or by a spontaneous resolution of the mineral, the acid remains combined with the copper (see *Pyrites):* the vitriol, now formed, is either extracted by the application of water, or washed out by rain or subterraneous waters: hence in some copper mines are found blue waters, which are true vitriolic solutions of copper, and which deposite that metal on the addition of iron or of any other substance which the acid more strongly attracts. The greatest part of the blue vitriol, now met with in the shops, is prepared in England, by artificially combining copper with sulphur or its acid.

The vitriol of copper is of an elegant sapphire blue colour; hard, compact, and semitransparent; when perfectly crystallized, of a flattish, rhomboidal, decahedral figure; in taste extremely nauseous, styptic, and acrid. Exposed to a gentle heat, it first turns white, and then of a yellowish red or orange colour: on increasing the fire, it parts, difficultly, with its

acid

acid, and changes at length to a very dark red calx, reducible, by fusion with inflammable fluxes, into copper.

This salt, like the other preparations of copper, acts, in doses of a few grains, as a most virulent emetic. Its use is chiefly external, as a detergent, escharotic, and for restraining hemorrhagies: for which last intention, a strong styptic liquor is prepared in the shops, by dissolving three ounces of blue vitriol and three of alum in two pounds of water, then adding one ounce and a half of oil of vitriol, and filtering the mixture for use.

Aqua styptica Ph. Ed.

*Blue vitriol has of late been considerably employed as an emetic by some practitioners; and is said to be by no means an unsafe one, as it operates the instant it reaches the stomach, before it has time to injure by its corrosive quality. The peculiar advantage in using it is represented to be, that it has no tendency to become also purgative, and that its astringent power prevents the tone of the stomach from being impaired after vomiting with it. It is much recommended in the early state of tubercles in the lungs; and the following method of exhibition is directed(a). Let the patient first swallow about half a pint of water, and immediately afterwards, the vitriol dissolved in a cup-full of water. The dose may be varied according to age, constitution, &c. from two grains to ten, or even twenty; always taking care to begin with small ones. After the emetic is rejected, another half pint of water is to be drunk, which is likewise speedily thrown up, and this is commonly sufficient to remove the nausea.

(a) Simmons, *on the Treatment of Consumptions,* p. 70.

VITRIOLUM.

In still smaller doses, the blue vitriol has been much used by some as a tonic in intermittents, and other diseases.

3. VITRIOLUM VIRIDE *Pharm. Lond. Vitriolum viride, sive ferri Pharm. Edinb.* Green vitriol, or vitriol of iron; commonly called English vitriol or copperas; the Roman vitriol of the Italian writers. This sort of vitriol is produced from sulphureo-ferrugineous pyritæ, as the blue from sulphureo-cupreous ones; and as the ferrugineous minerals are much easier of resolution than the others, the ferrugineous vitriol is much oftener found native. In this native state, neither sort is free from an admixture of the other; the native green vitriols having always more or less of a bluish cast, and the blue of a greenish. The common green vitriol is prepared in large quantity at Deptford and Blackwall near London, and at Newcastle, by boiling iron with the acid liquor, which runs from certain pyritæ after long exposure to the air: this vitriol appears to be purely martial, for if it should receive any cupreous particles from the mineral, the superadded iron would precipitate them. All vitriols may be freed perfectly from copper by adding iron to solutions of them: those, which contain even a small portion of that metal, readily discover it by staining the iron of a copper hue.

Pure vitriol of iron is considerably transparent, of a fine bright, though not very deep, grass green colour; of a nauseous astringent taste accompanied with a kind of sweetishness. Dissolved, and set to crystallize, it shoots into thick rhomboidal masses; a part generally rising at the same time in efflorescences about the sides of the vessel. The solution deposites in standing a considerable a considerable quantity, and in boiling a much larger one, of the metallic basis of the vitriol, in form of a rusty calx or ochre: iron seems to be the only metallic body that thus separates spontaneously, in any considerable quantity, from the vitriolic acid. On exposing the vitriol itself to a moist air, a similar resolution happens on its surface; which, sooner or later, according as the acid is more or less saturated with the metal, changes its green to a rusty hue. In a warm dry air, it loses a part of the phlegm or water necessary to its crystalline form, and falls by degrees into a white powder. Exposed to a gentle fire, it liquefies and boils up; but soon changes, on the exhalation of the watery part that rendered it fluid, to a solid, opake, whitish or grey mass; this, pulverized and urged with a stronger fire, continues to emit fumes, becomes yellow†, afterwards red, and at length, having parted with most of its acid as well as its phlegm, turns to a deep purplish-red calx‡, revivable by inflammable substances into iron.

† Vitriolum calcinatum *Ph. Ed.*
‡ Colcothar vitrioli *Ph. Ed.*
Chalcitis factitia *Ph.Paris.*

Pure green vitriol is in no respect different from the artificial *sal martis*. It is one of the most certain of the chalybeate medicines, scarcely ever failing to take effect where the calces and other indissoluble preparations pass inactive through the intestinal tube. It may be conveniently given in a liquid form, largely diluted with aqueous fluids: two or three grains or more, dissolved in a pint or a quart of water (which from this quantity receives no disagreeable taste) may be taken in a day, divided into different doses. This vitriol is used also, especially when calcined, as an external styptic: the styptic of Helvetius, and as is said that of Eaton, is no other than French brandy very slightly impregnated with the calcined vitriol: a dram

dram of the vitriol is commonly directed to a
quart of the fpirit, but only a minute portion Tinctura
of the dram diffolves in it. As French brandy ſtyptica.
has generally an aftringent impregnation from
the oaken cafks in which it has been kept, the
vitriol changes it, as it does the watery infufions
of vegetable aftringents, to a black colour; but
makes no fuch change in fpirituous liquors that
have not received fome aftringent tincture.

It is from the green vitriol that the acid
called vitriolic has been generally extracted; by
diftilling the calcined vitriol in earthen long-
necks, with a ftrong fire continued for two days
or longer. The diftilled fpirit appears of a
dark blackifh colour; and contains a quantity
of phlegm, greater or lefs according as the
vitriol has been lefs or more calcined. On
committing it a fecond time to diftillation, in a
glafs retort placed in a fand-heat, the phlegmatic parts rife firft, together with a portion of
the acid, and are kept apart under the name of
fpirit or *weak fpirit* of vitriol †: at the fame † Spir. vitri-
time the remaining *ftrong fpirit*, or *oil* as it is oli tenuis.
called, lofes its black colour and becomes triolicum *Ph.*
clear ‡, and this is the ufual mark for dif- *Lond. & Ed.*
continuing the rectification. The colleges of ‖ Acidum
London and Edinburgh now directs a weak vitriolicum
vitriolic acid of more certain ftrength, made by *Lond.*
mixing one part of the ftrong acid with feven Acid. vitriol.
or eight parts of water ‖. tenue, *vulgo*
The ftrong acid or oil of vitriol is the moft Spiritus vi-
ponderous of unmetallic fluids, and the moft *Ph. Ed.*
fixed of faline ones, yielding no fmell in the
greateft heat of the atmofphere, and requiring,
to make it boil or diftil, a heat confiderably
greater than that in which lead melts. Expofed
to the air, it imbibes its humidity, fo as to gain
by degrees an increafe of about twice its own
weight.

weight. Mixed directly with water, it produces a heat fo great as to render the veffel infupportable to the hand: glafs veffels are apt to crack from the fuddennefs of the heat, unlefs the commixture is very flowly performed. The moft ready method of diftinguifhing it, in a dilute ftate, or when mixed with other acids, is by adding a folution of fome calcareous earth, as chalk, made in any kind of acid liquor: this folution is by a minute portion of the vitriolic acid rendered milky, but fuffers no change from any other fpecies of acid; fee *Selenites*.

If the long-neck, in the extrication of the acid from vitriol, happens to crack in the fire, the acid that rifes after this period is found remarkably changed. It emits in the air fuffocating vapours like the fumes of burning brimftone, and rifes in diftillation with a heat not much greater than that which the hand can bear: to the tafte it difcovers little corrofivenefs or acidity. Combined with alkaline falts, it lofes its pungent odour; but on the addition of any other acid, it is difengaged from the alkali, fo as to rife again in diftillation as volatile and fuffocating as-before. It deftroys or whitens the blue and red colours of the flowers of plants; whereas, in its fixt ftate, like the other acids, it changes the blue to red, and heightens thofe which are naturally red. This volatile fpirit lofes its fuffocating odour, and refumes its corrofivenefs, fixednefs, and other qualities, by expofure to the air, which feems to carry off the inflammable principle whereon its volatility depended.

Spir. vitrioli volat. *Stahl*.

The fumes of burning brimftone are no other than the vitriolic acid in its volatile ftate; fee *Sulphur*. If a little burning fulphur be fufpended over fome water in a clofe veffel till the fumes

Aqua fulphurata, Gas fulphuris *vulgo*.

VITRIOLUM.

fumes fubfide, and this repeated with frefh portions of fulphur, till about half a pound has been ufed to a quart of water, the liquor will be found ftrongly impregnated with the volatile fuffocating acid, and in keeping for fome time, if the veffel is not clofely ftopt, it will become exactly fimilar to water acidulated with the fixt acid. If a very large glafs, open at bottom, be hung over the burning fulphur, in a damp place fcreened from wind, a part of the fumes will condenfe upon the fides of the glafs, and run down in drops, which may be collected by placing a glafs difh underneath: the acid thus obtained is called, from the fhape of the veffel that has been generally ufed for condenfing the fumes, fpirit or oil of fulphur by the *bell*. The quantity of acid collected by this procefs is very fmall, greateft part of the fumes efcaping: fixteen ounces of fulphur, in the moft favourable circumftances, yield fcarcely one ounce of phlegmatic fpirit; though it is certain, that out of this quantity of fulphur, more than fifteen ounces are pure acid, of fuch ftrength, as to require being diluted with above an equal quantity of water to reduce it to the pitch of common fpirit of fulphur; fo that if fulphur could be burnt without the lofs of any of its fumes, we might obtain double its weight of an acid of the ordinary ftrength. The procefs has lately been improved, by fome particular perfons, though not perhaps to this degree, yet fo far as to afford at a very low price almoft all the acid now fold under the name of oil of vitriol. The improvement confifts chiefly in burning the fulphur in very large glafs veffels, in the bottoms of which fome warm water is placed, whofe fteam ferves to collect and condenfe the fumes.

Spir. fulph. per campanam.

The acid of vitriol or sulphur, largely diluted so as to be supportable or but gratefully tart to the palate, is the most salubrious of all the mineral acids. It is mixed with watery infusions, spirituous tinctures and other liquids, as an antiphlogistic; as a restringent in hemorrhagies; and as a stomachic and corroborant in weaknesses, loss of appetite, and decays of constitution, accompanied with slow febrile symptoms, brought on by irregularities, or succeeding the suppression of intermittents by Peruvian bark. In several cases of this kind, after bitters and aromatics of themselves had availed nothing, a mixture of them with the vitriolic acid has happily taken place: the form commonly made use of is that of a spirituous tincture: six ounces of oil of vitriol are dropt by degrees into a quart of rectified spirit of wine, the mixture digested for three days in a very gentle heat, and afterwards digested for three days longer with an ounce and a half of cinnamon, and an ounce of ginger†; or a pint of an aromatic tincture drawn with proof spirit is mixed with four ounces of the strong acid‡: these liquors are given from ten to thirty or forty drops, in any convenient vehicle, at such times as the stomach is most empty. A mixture of oil of vitriol with spirit of wine alone, in the proportion of one part of the former to three of the latter, digested together for some time, is used in France as a restringent in gonorrhœas, female fluors, and spittings of blood.

† Elixir vitrioli *Ph. Ed.*

‡ Elix. vitrioli acidum.

Aqua rabelliana *vulgo* Eau de Rabel. *Ph. Paris.*

* This acid, diluted with water, has been given internally with great success in the itch. It was first used for this purpose in the Prussian army in 1756, and has since been much employed in several parts in Germany. The dose recommended

VITRIOLUM.

recommended is, from an eighth to a fourth of a dram of the pure acid twice or thrice a day. It is said to fucceed equally in the dry and moift itch; and when given to nurfes, to cure both themfelves and their children.

When oil of vitriol and rectified fpirit of wine are long digefted together or diftilled, a part of the acid unites with the vinous fpirit into a new compound, very volatile and inflammable, of no perceptible acidity, of a ftrong and very fragrant fmell, and an aromatic kind of tafte: this dulcified part, more volatile than the reft, feparates and rifes firft in diftillation, and may thus be collected by itfelf. The college of London directs a pound of oil of vitriol and a pint of rectified fpirit of wine to be cautioufly and gradually mixed (a great conflict and heat enfuing if they are mixed haftily) and fet to diftil with a very gentle heat till fulphureous vapours begin to arife: that of Edinburgh orders the fame quantity of the oil of vitriol to be dropt into four times as much of the vinous fpirit, and the mixture to be digefted in a clofe veffel, for eight days, previoufly to the diftillation, with a view to promote the coalition of the two ingredients. The different proportions of the acid fpirit to the vinous, in thefe prefcriptions, make no material variation in the qualities of the product, provided the diftillation is duly conducted; for the fmalleft of the above proportions of acid is much more than the vinous fpirit can dulcify, and all the redundant acid remains in either cafe behind. The true dulcified fpirit rifes in thin fubtile vapours, which condenfe upon the fides of the recipient in ftraight ftriæ: thefe are fucceeded by white fumes, which form either irregular

Spiritus ætheris vitriol. Ph. Lond.

striæ or large round drops like oil; on the firſt appearance of which, the procefs is either to be ſtopt, or the receiver changed. The ſpirit which theſe fumes afford, very different from the dulcified one, has a pungent acid ſmell like the fumes of burning ſulphur: on its ſurface is found a ſmall quantity of oil, of a ſtrong penetrating and very agreeable ſmell, readily diſſoluble in ſpirit of wine, to a large proportion of which it communicates the ſmell and taſte of the aromatic or dulcified ſpirit. The college of Edinburgh, in order to ſecure againſt any acidity in the dulcified ſpirit, order it to be rectified, by mixing it with an equal meaſure of water, in every pint of which a dram of ſalt of tartar has been diſſolved, and drawing off the ſpirit again by a gentle heat*(a).

Ol. vitrioli dulce Hoffm.

This ſpirit, taken from ten to eighty or ninety drops, ſtrengthens the ſtomach and digeſtive powers, relieves flatulencies, promotes urine, and in many caſes abates ſpaſmodic ſtrictures, and procures reſt. It is not eſſentially different from the celebrated mineral anodyne liquor of Hoffman; to which it is frequently, by the author himſelf, directed as a ſubſtitute. It is evident, from Hoffman's writings, that his anodyne was compoſed of the dulcified ſpirit and the aromatic oil which comes over after it, but the particular proportions of the two he has no where ſpecified: the faculty of Paris directs, under the title of his preparation, twelve drops

Liquor anodynus mineralis Hoffm.

---

* (a) The Edinburgh college, in their laſt pharmacopœia, have manifeſtly ſhewn how little they conceive the acid to enter as a conſtituent part of this preparation, and at the ſame time have directed an effectual method of preventing its preſence in it. They order the *acidum vitriolicum vinoſum*, vulgo *ſpiritus vitrioli dulcis*, to be made by ſimply mixing one part of vitriolic ether with two of rectified ſpirit.

of

VITRIOLUM.

of the oil to be diffolved in two ounces of the fpirit; the college of Wirtemberg feems to think, that all the oil, and all the fpirit, obtained in one operation, were mixed together, without regard to the precife quantities.

* The London college have now given a formula for making this oil, which they call *oleum vini*. A pint each of alcohol and vitriolic acid are gradually mixed, and diftilled, with a caution that the black froth which arifes do not pafs into the receiver. Of the diftilled liquor, the oily part is to be feparated from the volatile vitriolic acid. A fufficient quantity of cauftic alkaline lixivium is to be added to the oily part to correct its fulphureous odour, and then the æther is to be diftilled from it by a gentle heat. The *oleum vini* will remain in the retort, fwimming above a watery liquor, from which it is to be feparated.

In place of Hoffman's anodyne liquor they direct three drams of this oil to be mixed with two pounds of their *fpirit of vitriolic æther*, or *dulcified fpirit of vitriol*, as it was before called. Spirit.ætheris vitriol. comp. *Ph. Lond.*

The dulcified fpirit is fometimes ufed as a menftruum for certain refinous and bituminous bodies, which are more difficultly and languidly acted upon by pure vinous fpirits. It is often mixed with aromatic and ftomachic tinctures, in cafes where the ftomach is too weak to bear the acid elixirs above-mentioned: eight ounces are commonly added to a pint of the officinal aromatic tincture†, in which it does not, like the acid undulcified, occafion any precipitation; or the ingredients of the aromatic tincture are infufed in the dulcified acid, inftead of common rectified fpirit‡. A medicine of this kind was formerly in great efteem under the name of Vigani's volatile elixir of vitriol, the preparation † Elix. vitrioli dulce ‡ *Ph. Ed.*

of which was long kept a secret, and first made public in the *pharmacopœia reformata*: it is prepared by macerating, in some dulcified spirit of vitriol free from acidity, a small quantity of mint leaves curiously dried, till the spirit has acquired a fine green colour: to prevent the necessity of filtration, during which the more volatile parts would exhale, the mint may be suspended in the spirit in a fine linen cloth.

If the dulcified spirit, rectified as above prescribed from a solution of fixt alkaline salt, be shaken with equal its quantity of a like solution, and the mixture suffered to rest; an ethereal fluid rises to the surface, and great part of the dulcified spirit may be recovered again from the remainder by distillation. I am informed by Dr. Hadley, that he has observed the largest proportion of ether to be obtained, by using the strongest vitriolic acid of the shops with equal its quantity by measure of spirit of wine, and distilling immediately by a heat sufficient to make the mixture boil; and that by this management, from three pints of oil of vitriol, and six pints of rectified spirit of wine, he obtained two pints and a half of the ether.
* The following is the method prescribed for making ether, in the last Edinburgh pharmacopœia. To thirty-two ounces of rectified spirit of wine in a glass retort, add at once an equal weight of strong spirit of vitriol. Mix them gradually by gentle agitation, and immediately set them to distil in sand previously heated, so that the mixture may be brought as soon as possible to boil, in which heat it is to be continued till sixteen ounces are come over. The receiver must be cooled by water or snow. To this distilled liquor, two drams of the strong alkaline caustic are to be added, and the distillation

VITRIOLUM.

tillation repeated in a very high retort, with a very gentle heat, till ten ounces are come over. To the refiduum after the firſt diſtillation may be added ſixteen ounces of freſh rectified ſpirit, when more ether will be procured; and this may be ſeveral times repeated. The laſt London pharmacopœia directs ether to be made by mixing two pounds of dulcified ſpirit of vitriol with one ounce (by meaſure) of cauſtic alkaline lixivium, and diſtilling over with a gentle heat fourteen ounces by meaſure.

Liquor æthereus vitriolicus Ph. Ed.

Æther vitriolic. Ph.Lond.

The ether or ethereal ſpirit is the lighteſt, moſt ſubtile, volatile, and inflammable, of all known liquids: it quickly exhales in the air, diffuſing an odour of great fragrance: it does not mingle with water, with acid liquors, with alkaline liquors, or with vinous ſpirits, at leaſt not in any conſiderable quantity, only a ſmall portion of the ether being imbibed by them: it unites with oils in all proportions, diſſolves balſams, and reſins, and extracts the oily and reſinous parts of vegetables. It has been hitherto regarded chiefly as a matter of curioſity, nor are its medicinal qualities as yet much known*(a). Malouin looks upon it as one of the moſt perfect tonics, friendly to the nerves, cordial and anodyne; and ſays he has found it to be a good remedy in rheums, for abating coughs, eſpecially thoſe of the convulſive kind. Its great volatility renders the taking of it very incommodious: the author above-mentioned orders, as the moſt convenient form, from three to twelve drops to be dropt on ſugar or pow-

---

*(a) It has ſince come more into uſe in flatulent and ſpaſmodic complaints, the gout in the ſtomach, nervous aſthmas, and the like. Though it will not mix with water, it may be diffuſed in a ſufficient quantity of it ſo as to be taken without much difficulty.

dered lipuorice, a little warm water or some warm infusion to be immediately added, and the whole swallowed directly. It has been reported to give immediate ease in violent headachs, by being rubbed on the temples.

The vitriolic acid saturates a larger quantity of fixt alkaline salts than any of the other acids, and dislodges therefrom such other acids as have been previously combined with them: of the strong spirit or oil of vitriol, about five parts are sufficient for eight of the common vegetable fixt alkalies. The neutral salt resulting from its coalition with this kind of alkali, is of a bitterish taste, very difficultly soluble in water, and scarcely susible in the fire: in small doses, as a scruple or half a dram, it is an useful aperient; in larger ones, as four or five drams, a mild cathartic, which does not pass off so hastily as the *sal catharticus*, and seems to perform its office more thoroughly. This salt has been commonly prepared with the alkali obtained from tartar, and is hence called vitriolated tartar: some dilute the oil of vitriol with six times the quantity of warm water, and drop into it a solution of the alkaline salt till a fresh addition occasions no further effervescence: others direct it to be made from the residuum after extracting the nitrous acid from nitre by means of the vitriolic (see *Nitrum*), but in order to get rid of the superfluous acid the matter is first to be exposed to a strong heat, and then dissolved in boiling water, and the salt crystallised.

With the mineral fixt alkali, and the earth called magnesia, this acid forms compound salts of a bitterer taste, somewhat less purgative, and much easier of solution, than that with vegetable alkalies: with volatile alkalies a very pungent ammoniacal salt, whose medicinal effects are

---

Sal enixum & Arcanum duplicatum *quibusdam.*

Alcal. fix. veget. vitriol. *vulgo* tartarum vitriolatum *Ph. Ed.*

Kali vitriolat. *Ph. Lond.*

VITRIOLUM.

are not well known. The ftrong acid, boiled on argillaceous earths to drynefs, corrodes a portion of them, and concretes therewith into an auftere ftyptic falt. Calcareous earths it does not diffolve into a liquid ftate, but may be combined with them, by precipitation from other acids, into an indiffoluble concrete feemingly of no medicinal activity. Among metallic bodies, it diffolves zinc and iron readily; copper, filver, quick-filver, lead, and tin, very difficultly: it is fitted for acting on the two firft by dilution with three or four times its quantity of water: the others require the undiluted acid, and a heat fufficient to make it boil; when, the more phlegmatic parts exhaling, fo much of the pure acid matter remains combined with the metals, as to render them, in part at leaft, diffoluble in water; fee the refpective metals.

The medical qualities of the acid in its volatile ftate are very little known, and thofe of the combinations thereof with alkalies not at all, though they fhould feem to deferve inquiry. The volatile acid of burning brimftone may be commodioufly transferred into fixt alkalies, by dipping linen cloths in a ftrong folution of the alkali and fufpending them over the fumes, of which they will quickly imbibe fo much as to neutralize the alkali: this neutral falt being rubbed off, the cloths may be again moiftened with the alkaline lye, expofed to the acid fumes, and thefe proceffes alternately repeated(a). The neutral falt thus obtained differs greatly in its tafte and other properties, and doubtlefs alfo in its medical virtues, from that which is produced by the coalition of the fixt acid with the

(a) Vide Stahlii Experimenta & animadverfiones ccc.

fame

same alkali, that is, from vitriolated tartar. It dissolves more easily in water, and shoots, not into octangular crystals, but into small slender ones like short needles. On adding to it the fixt vitriolic acid (or even the weaker acids of nitre or sea-salt) the volatile acid is disengaged from the alkali; and though, in the compound salt, its pungent smell was wholly suppressed, it now rises in distillation as pungent and suffocating as the original fumes of the brimstone. The neutral salt, in a dry form, may be kept unchanged for years: dissolved in water, and exposed for some time to the air, or if roasted with a gentle heat, it becomes the same with vitriolated tartar.

ULMARIA.

ULMARIA sive Regina prati. Barba capræ floribus compactis C. B. *Spiræa Ulmaria Linn.* MEADOWSWEET or QUEEN-OF-THE MEADOWS: a plant with tall, smooth, reddish, brittle stalks; and oval, sharp-pointed, indented leaves, set in pairs along a middle rib, with smaller pieces between, and at the end a larger odd one divided into three sections, wrinkled and green above, white underneath: on the tops come forth large thick clusters of little whitish flowers, followed each by several crooked seeds set in a roundish head. It is perennial, common in moist meadows, and flowers in June.

THE leaves of ulmaria recommended as mild astringents, discover to the taste or smell very little foundation for any medical virtues. The flowers have a strong and pleasant smell, in virtue of which they are supposed to be antispasmodic and diaphoretic, and which in keeping

ing is soon diffipated, leaving in the flowers only an infipid mucilaginous matter. As thefe flowers are more rarely ufed in medicine than their fragrant fmell might rationally perfuade, Linnæus fufpects that the neglect of them has arifen from the plant being poffeffed of fome noxious qualities, which it feemed to betray by its being left untouched by cattle: it may be obferved, however, that the cattle, which refufed the ulmaria, refufed alfo angelica, and other herbs, whofe innocence is apparent from daily experience.

ULMUS.

ULMUS Pharm. Lond. & Edinb. Ulmus campeftris & theophrafti C. B. Ulmus campeftris Linn. Elm: a tall common tree; covered with a rough, chapt, brownifh, brittle bark, under which lies a white, fmooth, tough, coriaceous one; producing in the fpring, before the leaves appear, imperfect flowers, followed by flat roundifh capfules, containing each a fingle feed.

THE inner tough bark of the elm tree, of no manifeft fmell, difcovers, on being chewed, a copious flimy mucilage, of no particular tafte: the outer brittle bark is much lefs flimy, but equally void of fmell and tafte. It may therefore be prefumed, that if elm bark has been found of ufe in nephritic cafes, in which it is recommended by authors; or externally againft burns, for which it is applied by the common people; it was of ufe no otherwife than as a fimple emollient. Neither the purgative virtue afcribed to it by fome, nor the aftringent by others, appear to have any foundation.

* A decoction

*A decoction of the inner bark of elm has been employed in cutaneous difeafes in fome of our hofpitals; and an account of its efficacy has been publifhed by Dr. Lyfons in vol. ii. of *Medical Tranfactions*, and fince, in a feparate work. In making this decoction, four ounces of the bark frefh from the tree are boiled in two quarts of water to one. It is of a beautiful light purple colour, when the elm is in flower; but browner at other times. Its tafte is mildly aftringent; and an extract from it is very auftere. It has no purgative effects, as fome have alledged, but rather the contrary. Where it fucceeds, it generally at firft increafes the efflorefcence. Patients are ufually directed to drink half a pint twice a day, and to perfift in the ufe of it fome months. It is now received into the London pharmacopœia.

Decoct. ulmi *Ph. Lond.*

URINA.

URINA Pharm. Parif. URINE: The recent urine of healthy fubjects is naufeoufly bitter, very faline, fcarcely manifeftly alkaline or acid. As foon as it begins to putrefy, it emits volatile alkaline vapours; and if diftilled, when moderately putrefied, by a gentle heat, it yields a concrete volatile alkaline falt: as volatile alkalies have a ftrong antifeptic power, the vapours of putrefied urine are not obferved, like thofe of cadaverous animal fubftances, to be productive of putrid difeafes. A pungent cauftic volatile fpirit may likewife be obtained from recent urine, by infpiffating, and then diftilling it with the addition of quicklime.

If the putrefied urine be flowly infpiffated, in glafs or ftone-ware veffels, to the confiftence of a thin fyrup, and fet for fome weeks in a cold

cold place, brown cryftals will fhoot from it, confifting partly of marine falt, and partly of a falt of a peculiar kind, which fhoots before the marine, and which, by repeated folutions, filtrations, and cryftallizations, may be purified both from that falt and from the adhering oil. In this ftate†, it appears perfectly neutral, and imprefles on the tongue a fenfe of coolnefs with a flight bitterifhnefs: laid on a red-hot iron, it bubbles, emits volatile alkaline vapours, and runs into a colourlefs pellucid fubftance refembling fine glafs: this apparent glafs is manifeftly acid, though but weakly fo, diffolves in water, neutralizes alkaline falts, and with volatile alkalies regenerates the original neutral falt. One of its moft diftinguifhing characters is, that a mixture of it with inflammable matters, as foot or powdered charcoal, on being heated to ignition in an open veffel, emits flafhes like lightening, and, on being diftilled in a retort with a moderately ftrong fire, yields the highly inflammable concrete called phofphorus.

† Sal microcofmicum, *five* fal effentiale urinæ.

Urine is fometimes applied externally, boiled with bran, as a refolvent and difcutient, in which intentions it is faid to be very efficacious. Recent cows urine has been drank in the fpring, to the quantity of a pint or more every morning, for feveral days, as an attenuant and deobftruent in different diforders: the naufeous draught purges plentifully by ftool, and fometimes vomits. The peculiar falt of urine is but of late difcovery, and its medicinal qualities are as yet unknown.

URTICA.

URTICA.

URTICA *Pharm. Lond. & Edinb. Urtica urens maxima* C. B. *Urtica dioica Linn.* COMMON STINGING NETTLE. Infusions and decoctions of this herb, or its expressed juice, are recommended in different disorders as aperients, and said to loosen the belly: the juice, depurated and gently inspissated, discovers a considerable taste, of the subsaline kind.

UVÆ PASSÆ.

RAISINS: rich sweet grapes, dried by the sun's heat in the warmer parts of Europe. Two sorts are directed for medicinal use. 1. UVÆ PASSÆ MAJORES *Pharm. Lond. Passulæ majores. Pharm. Edinb.* Raisins of the sun; the fruit of the *vitis damascena* dried upon the tree; the stem of each cluster, when the grapes are ripe, being cut almost through, so as to prevent the afflux of any fresh juice. 2. UVÆ PASSÆ MINGRES *seu* CORINTHIACÆ. Currants; the fruit of the *vitis corinthiaca* picked from the stalks.

THESE fruits are used as agreeable lubricating acescent sweets, in pectoral decoctions, and for obtunding the acrimony of other medicines and rendering them acceptable to the palate and stomach: the first sort inclines most to acidity, the sweetness of the latter being more of the mucilaginous kind. They both give out their sweetness and their pleasant flavour to water and spirit: the stones or seeds are supposed to communicate a disagreeable relish, and hence are generally directed to be taken out; but it did not appear on trial that they give any taste at all to water, proof spirit, or rectified spirit.

UVA URSI.

UVA URSI: Pharm. Lond. & Edinb. *Vitis idæa foliis carnofis & veluti punctatis, five idæa radix diofcoridis* C. B. *Arbutus (Uva Urfi) cauliculis procumbentibus, foliis integerrimis* Linn. BEARS WHORTLEBERRY: an evergreen trailing fhrubby plànt; with numerous fmall oblong oval leaves; monopetalous white flowers with a flefh-coloured edge cut into five fections; and red berries. It greatly refembles the common red whort-bufh; from which it may be diftinguifhed, by the leaves being more oblong, and by the flower having ten ftamina, and the berry five feeds; whereas the flower of the common whort has only eight ftamina, and the berry often twenty feeds. It is found on the fnowy hills of Auftria and Styria. but more plentifully on the Swedifh hills. It is alfo a native of the highlands of Scotland, and is now cultivated in fome of our gardens.

The leaves of this plant have a bitterifh aftringent tafte; without any remarkable fmell, at leaft in the dry ftate in which they have been brought to us from Germany. Infufions of them in water ftrike a deep black colour with folution of chalybeate vitriol, but foon depofite the black matter, and become clear: I do not recollect any other aftringent infufion, from which the blacknefs, produced by vitriol, feparate fo very fpeedily.

The leaves of uva urfi have of late been greatly celebrated in calculous and nephritic complaints, and other diforders of the urinary organs: the dofe is half a dram of the powder of the leaves, every morning, or two or three times

times a day. De Haen relates, after large experience of this medicine in the hofpital of Vienna, that fuppurations, though obftinate and of long continuance, in the kidneys, ureters, bladder, urethra, fcrotum and perinæum, where there was no venereal taint or evident marks of a calculus, were in general completely cured by it: that of thofe who had a manifeft calculus, feveral found permanent relief, fo that long after the medicine had been left off, they continued free from pain or inconvenience in making water, though the catheter fhewed that the calculus ftill remained: that others, who feemed to be cured, relapfed on leaving off the medicine, were again relieved on repeating its ufe, and this for feveral times fucceffively; while others obtained from it only temporary and precarious relief, the complaints being often as fevere during the continuance of the medicine as when it was not ufed. It may be obferved, that in feveral cafes which he relates, paregorics were joined to the uva urfi; and that other mild aftringent plants have been recommended for the fame intentions; from fome of which De Haen himfelf expects the fame good effects. The trials of the uva urfi, made in this country, have by no means anfwered expectation: in all the cafes that have come to my knowledge, it produced great ficknefs and uneafinefs, without any apparent benefit, though continued for a month.

* WINTERANUS CORTEX.

WINTERANUS CORTEX, Cortex magellanicus. WINTER'S BARK. The tree producing the Winter's bark *(Winterana aromatica Soland.)* is one of the largeft foreft trees on *Terra del Fuego*.

Fuego. Its leaves are ever-green, smooth, oval, and entire. Its flowers consist of seven petals, with from fifteen to thirty stamina, and from three to six germina, terminating in as many stigmata. Each germen becomes a seed-vessel, containing several seeds. The bark of the trunk of the tree is externally grey, and very little wrinkled.

The Winter's bark, which takes its name from Capt. Winter, who discovered it on the coast of Magellan in 1577, is brought to us in pieces of different degrees of thickness, from a quarter to three quarters of an inch. It is of a dark brown cinnamon colour, with an aromatic smell when rubbed, and of a pungent, hot, spicy taste, which is lasting on the palate, though imparted slowly. A watery infusion of it struck a black colour with a solution of green vitriol. From an infusion of two ounces of the bark, coarsely powdered, was obtained on evaporation an extract weighing two drams and twenty-four grains. The same quantity, treated with rectified spirit, yielded two drams of extract. A pound of the bark was infused in a proper quantity of water, and the liquor submitted to distillation. The distilled water was clear, of a pleasant taste, and somewhat of the cinnamon flavour. There was no appearance of essential oil. The residuum afforded six ounces of a soft extract, of a grateful aromatic taste.

A mixture of this bark seemed very effectually to cover and correct the disagreeable taste and smell of certain drugs; a property common to it with the canella alba.

Almost the only use hitherto made of the Winter's bark has been by the crews of ships navigating the streights of Magellan, as a preservative fervative from the fcurvy. It has been confounded in the fhops with the canella alba, from, which it is totally different.

An exact defcription of the plant, with a figure, is contained in a paper publifhed in the *Medical Obf. and Inq.* vol. v. from whence this account is extracted.

ZEDOARIA.

ZEDOARIA *Pharm. Lond. & Edinb.* Zedoaria longa & Zedoaria rotunda, *C. B.* Zedoary: the root of an Indian plant, *(Amomum fcapo nudo, fpica laxa truncata, Berg. Mat. Med.)* brought over in oblong pieces, about the thicknefs of the little finger and two or three inches in length; or in roundifh ones† about an inch in diameter; of an afh-colour on the outfide, and white within. The long fort is faid by fome to be the ftrongeft, but the difference, if any, is very inconfiderable, and hence the college allows both to be ufed indifcriminately.

† Zerumbeth *Ph. Parif.*

This root has an agreeable fmell, and a bitterifh aromatic tafte. It impregnates water with its fmell, a flight bitternefs, a confiderable warmth and pungency, and a yellowifh brown colour: the reddifh-yellow fpirituous tincture is in tafte ftronger, and in fmell weaker, than the watery. In diftillation with water, it yields a thick ponderous effential oil, fmelling ftrongly of the zedoary, in tafte very hot and pungent: the decoction, thus deprived of the aromatic matter, and concentrated by infpiffation, proves weakly and difagreeably bitter and fubacrid. A part of its odorous matter rifes alfo in the infpiffation of the fpirituous tincture: the remaining extract is a very warm, not fiery, moderately

ZIBETHUM.

moderately bitter aromatic, in flavour more grateful than the zedoary in substance.

Zedoary root is a very useful warm stomachic. It was employed by some as a succedaneum to gentian root; at a time when a poisonous article, mixed with the gentian brought from abroad, rendered its use hazardous: but from the above analysis it appears to be not entirely similar to that simple bitter; its warm aromatic part being the prevailing principle, in virtue of which, its spirituous extract (the most elegant preparation of it) is made an ingredient in the cordial confection of the London pharmacopœia.

ZIBETHUM.

CIVETTA. Civet: a soft unctuous odoriferous substance, about the consistence of honey or butter; of a whitish, yellowish, or brownish colour, and sometimes blackish; brought from the Brazils, the coast of Guinea, and the East Indies; found in certain bags situated in the lower part of the belly of an animal of the cat kind*(a). The bag has an aperture externally, by which the civet is shed or extracted.

This substance has a very fragrant smell, so strong as, when undiluted, to be disagreeable; and an unctuous subacrid taste. It is used chiefly in perfumes, rarely or never for medicinal purposes, though the singular effects which musk has been found to produce may serve as an inducement to the trial. It unites with oils, both expressed and distilled, and with animal fats: in watery or spirituous liquors it does not dissolve, but both menstrua may be strongly

(a) Or rather of the weasel kind.

impregnated with its odoriferous matter, water by diftillation, and rectified fpirit by digeftion: by trituration with mucilages, it becomes foluble in water.

ZINCUM.

ZINCUM Pharm. Lond. & Edinb. Zinc, or TUTENAG: a bluifh white metal; crackling, in being bent, like tin, and quickly breaking; about feven times fpecifically heavier than water; beginning to melt in a moderate red heat, and very flowly calcining on a continuance of the fire; in a moderate white heat flowing thin, burning, fulgurating, with a bright deep green or bluifh green flame, and fubliming into light white flowers, which concrete about the upper part of the veffel, or on the bodies adjacent, into thin crufts, or foft loofe filaments like down or cobwebs. In its metallic form, and in that of a calx or flowers, it diffolves readily in all acids, and precipitates from them almoft all the other metallic bodies.

The calces or flowers of zinc are difficultly revived into their metallic form. Though perfectly fixed in the fire fo long as they continue in a ftate of calx; yet, as calces in general require for their revival a greater heat than that in which the metal itfelf melts, and as a full melting heat is the greateft that zinc can fupport; the inftant they are revived, they burn and calcine again in open veffels, and efcape through the pores of clofe ones. Hence fome ores and preparations of this metal have been long kept in the fhops, and even chemically examined, without being difcovered to be fuch. The revival may be effected, by ufing compact veffels of fuch a ftructure, that the zinc, in proportion

ZINCUM.

portion as it is reftored to its metallic form by the charcoal powder, or other inflammable additions commonly made ufe of for thofe purpofes, may be fuffered to fublime or run off from the heat without being expofed to the outward air; or by adding fome other metallic fubftance to detain it, as copper, which is thus changed into brafs,

This metal has but lately been received into the fhops in its own form; in which it deferves a place, as affording preparations fuperiour to the ores or productions of it now made ufe of. A white vitriol made from pure zinc, by diffolution in the diluted vitriolic acid and cryftallization†, is doubtlefs preferable for medicinal ufe to the common impure white vitriol; and the white flowers, into which it is changed by deflagration‡, to the very impure calamine and tutty. Moderately pure white flowers, fublimed from it in the brafs or other furnaces, wherein zinc or its ores are melted with other metals, were formerly kept in the fhops, and diftinguifhed by the names of *pompholix* and *nihil album*.

† Vitriolum album *Ph.Ed.*

‡ Calx zinci *vulgo* flores zinci *Ph. Ed.* Zincum calcinat. *Ph. Lond.*

* The flowers of zinc were firft ufed as an internal medicine by the celebrated chemift Glauber, but were little known in practice till Dr. Gaubius, of Leyden, gave an account of their virtues in his *Adverfaria*. They have fince been much employed in convulfive and fpafmodic difeafes, and fometimes with good effects. Even obftinate epilepfies have been rendered much lefs violent by their ufe. Like all other medicines, however, in difeafes of this clafs, their good effects are often only temporary, and they often fail altogether. When the flowers are genuine, a grain or two generally at firft excites naufea or ficknefs, but by degrees a

confiderable

considerable dose may be taken with little or no sensible effect. As they are liable to be adulterated, it may be proper to mention, as tests of their purity, that they make no effervescence with acids; and that, when exposed to a strong heat, they become yellow, but on cooling, turn white again. An application for external use, made by mixing one part of flowers of zinc with six of the simple liniment of wax and oil, is directed in the Edinburgh pharmacopœia.

Unguent. e calce zinci Ph. Ed.

ZINGIBER.

ZINGIBER Pharm. Lond. & Edinb. & C. B.
GINGER: the root of a reed-like plant *(Amomum Zingiber Linn.)*, growing spontaneously in the East Indies, and cultivated in some part of the West; brought over in knotty branched flattish pieces, freed from the outer bark, of a pale colour and fibrous texture: that which is least fibrous is accounted the best.

THIS warm aromatic root, of common use as a spice in flatulent colics, &c. appears to be much less liable to heat the constitution than might be expected from the penetrating heat and pungency of its taste, and from the fixedness of its active principles. It gives out the whole of its virtue to rectified spirit, and great part of it to water, tinging the former of a deep, the latter of a pale yellow colour: the spirituous tincture, inspissated, yields a fiery extract, smelling moderately of the ginger: the watery infusion, boiled down to a thick consistence, dissolved a fresh in a large quantity of water and strongly boiled down again, retains still the heat and pungency of the root, though little or nothing

ZINGIBER.

thing of its smell: there does not seem to be any of the common spices whose pungency is of so fixed a kind. In the shops are kept a syrup made from an infusion of three† or four‡ ounces of the root in four† or three‡ pints of boiling water, which is agreeably impregnated with its warmth and flavour; and the candied ginger‖, brought from abroad, which is likewise moderately aromatic.

Syr. zingib.
† *Ph. Ed.*
‡ *Ph. Lond.*
‖ Zingiber conditum *Ph. Ed.*

ADDENDA.

ANGUSTURÆ CORTEX.

IN the year 1788 a confiderable quantity of a bark, not before known in this country, was imported from the Weft Indies, but as of African growth. The only account fent with it was, " that it had been found very fuperior to the Peruvian bark in the cure of fevers." In the fucceeding year, two letters were publifhed in the *London Medical Journal* for 1789, part ii. from Dr. J. Ewer, and Dr. Alexander Williams, phyficians at Trinidad in South America, containing a defcription of this bark under the name of Cortex Angufturæ, and giving an account of its medicinal effects. It is there faid to come from the Spaniards in Anguftura; and this is confirmed by the fubfequent importation of parcels of it from Cadiz and the Havanna. No accurate account, however, has yet been obtained of the place of its growth, nor does the name of Anguftura feem to belong to a particular diftrict, but rather to be the Spanifh term
for

for a narrow pafs between mountains. The fuppofition is, that the tree producing it grows on the banks of the river Oronoko. Not the leaft infight has been gained into the fpecies of vegetable whence this bark is derived; for although Mr. Bruce, who had been cured of a dyfentery in Abyffinia by the bark of a fhrub called the *Wooginoos*, now cultivated in Kew and other gardens under the name of *Brucea antidyfenterica* or *ferruginea*, declared that it appeared to him from recollection to be the fame; yet Dr. Duncan, in his *Medical Commentaries* for 1790, afferts, that upon comparifon, the two barks feem effentially different. At prefent, therefore, it muft be confidered as a drug of unknown origin, though its fenfible qualities and medical powers have been well afcertained by the experiments of various perfons.

Mr. Brande, apothecary to the queen, who publifhed an account of this bark firft in the *London Medical Journal*, and then in a feparate pamphlet, thus defcribes it. " There is a confider-
" able variety in the external appearance of the
" Anguftura bark, owing, however, probably,
" to its haying been taken from trees of differ-
" ent fizes and ages, or from various parts of
" the fame tree, as the tafte and other proper-
" ties perfectly agree. Some parcels which I
" have examined, confift chiefly of flips, torn
" from branches, which could not have ex-
" ceeded the thicknefs of a finger: thefe are
" often fmooth, three feet or more in length,
" and rolled up into fmall bundles. In others,
" the pieces have evidently been, for the greater
" part, taken from the trunk of a large tree,
" and

"and are wrinkled, and nearly flat, with quills
"of all sizes intermixed.

"The outer surface of the Anguftura bark,
"when good, is in general more or lefs wrinkled,
"and covered with a coat of a greyifh white,
"below which it is brown, with a yellow caft:
"the inner furface is of a dull brownifh-yellow
"colour. It breaks fhort and refinous. The
"fmell is fingular and unpleafant, but not very
"powerful: the tafte intenfely bitter, and
"flightly aromatic; in fome degree refembling
"bitter almonds, but very lafting, and leaving
"a fenfe of heat and pungency in the throat.
"This bark, when powdered, is not unlike the
"powder of Indian rhubarb. It burns pretty
"freely, but without any particular fmell."

With refpect to its habitude to menftrua, it yields its tafte and flavour to water, cold and hot, to rectified and proof fpirit, and to wine. The watery extract is large in quantity, bitter, but not acrid. After the action of water, the refiduum imparts colour, and great acrimony, with naufeoufnefs, to fpirit. The fpirituous extract is much lefs in quantity, and confifts of lefs than a fourth of refin, the reft being partly gum, and partly a greafy matter, in which the acrid tafte and unpleafant fmell of the fubject appear to refide. Water diftilled from the Anguftura bark bruifed had a fingular flavour, fomewhat refembling that of ftrong parfley water. A fmall portion of white effential oil fwam on the furface, which poffeffed the full fmell of the bark, was acrid, and left a glow in the mouth like camphor. The preparations of this bark are not affected in colour by the addition of vitriol of iron.

Mr.

ADDENDA.

Mr. Brande made various experiments to afcertain the comparative antifeptic power of the Anguftura bark; from the refult of which, it appears to rank very high among the vegetables poffeffing that quality, not one of the fubftances with which he compared it feeming to have the advantage of it.

With refpect to its medicinal qualities, from the teftimony of the gentlemen at Trinidad as well as thofe who have tried it in thefe climates, it appears to act as a very powerful tonic, and to be particularly efficacious in fevers of the intermittent kind, dyfenteries and diarrhœas. In large dofes it is apt to occafion naufea, or to purge; but in fmaller ones, it fits eafy on the ftomach, and is free from that common inconvenience of the Peruvian bark, of caufing a fenfe of weight and fulnefs. Indeed, the efficacy of moderate dofes is a peculiar advantage of the Anguftura bark; from ten to twenty grains of the powder, and from one ounce to one and a half of the infufion or decoction with a portion of the tincture, having been found fufficient, a few times repeated, to prevent the paroxyfms of an intermittent. In diarrhœas and dyfenteries, after the due exhibition of laxatives, its effects are ufually very fpeedy. Mr. Wilkinfon of Sunderland *(Lond. Med. Journ.* for 1790, part iv.) who has employed it extenfively, found it peculiarly efficacious in low or nervous fevers, and the irregular intermittents of children, ufually termed wormfevers. As a general tonic, Mr. Brande thinks it fuperior to every other medicine of that clafs; and this is the light in which Dr. Pearfon regards it, who rather compares it to the warm bitters, fuch as camomile, than to the Peruvian bark.

bark. Dr. Ewer mentions a cafe in which its external application in a mortification proved very effectual. On the whole, it appears not to be doubted that this bark is a valuable addition to the clafs of tonics of the higher order, and it is to be hoped that we fhall not long be left ignorant of its natural hiftory and botanical character.

BARYTES.

A SUBSTANCE of a fparry appearance found in mines, called *cauk* or *calk*, has by the later chemifts been difcovered to contain an earth, the properties of which entitle it to form a new genus among earthy bodies. From its remarkable fpecific gravity, it has obtained the name of *terra ponderofa*, or *barytes*; and is now found to exift in various combinations, particularly, united with the vitriolic, and with the aerial acids. It was from the firft fufpected to be of a metallic nature by that eminent chemift Profeffor Bergman; and Dr. Withering, in fome excellent obfervations and experiments on it publifhed in the *Philof. Tranfactions* (vol. lxxiv. part ii.) places it between the earths and the metallic calces. Its native combinations exert deleterious effects upon animals; and its artificial ones, though milder, are capable of acting with violence in moderate quantities; a farther prefumption of its metallic nature, fince no combinations of the fimple earths fhew any activity of that kind. From the aerated barytes an artificial combination has been made with the muriatic

muriatic acid, which has been introduced into medicine.

Dr. Crawford, in the year 1789, made several trials in St. Thomas's hospital of the muriated barytes, the result of which was published in the *Medical Communications*, vol. ii. The preparation he used was a saturated solution of the salt in water; but in part of the cases this salt was not the pure muriated barytes, but had a mixture of an eighth of muriated iron; the medical effects, however, of the pure and the compound salt were not found to be sensibly different. The cases in which it was used with the most striking success were scrofula in its different forms and combinations, with swelled glands, foul ulcers, enlarged joints, and general cachexy. Some of these which had resisted the usual remedies, were singularly relieved by the muriated barytes, either given alone, or in conjunction with mercurial and antimonial medicines and bark. The dose varied from two drops to twenty, twice a day; few patients, however, could bear more than from six to ten without nausea; and it did not appear that by habit the stomach was enabled to bear a considerable increase of dose, but rather the contrary. In a few instances this medicine appeared to increase the cuticular secretion; in most it occasioned an unusual flow of urine: and almost universally improved the appetite and general health. Sometimes it produced vertigo, an effect apparently connected with its nauseating quality. It is not to be doubted, Dr. Crawford observes, that if administered injudiciously, it is capable of producing deleterious effects, both by disordering the nervous system, and bringing on violent vomiting and purging. From trials made upon dogs, it appears

appears that a very large dose would prove fatal.

It may be proper to mention, that the aerated barytes is found in the lead mine of Anglezark, near Chorley in Lancashire, and as far as appears, there only in England. [See a paper by Mr. James Watt, junior, in the third volume of the *Manchester Society's Memoirs.*]

PIPER INDICUM.

A SINGULAR use of this substance is mentioned in two letters from John Collins, Esq. of the island of St. Vincent, inserted in the second volume of the *Medical Communications*. In a peculiar kind of angina maligna prevailing among children in that island, which began with blackness, sloughiness and ulceration of the fauces and tonsils *without fever*, and proved extremely fatal, he was induced, from a letter published by a Mr. Stewart of Grenada, to exhibit the following remedy. " Take two table-spoonfuls " of small red pepper, or three of the common " Cayenne pepper, and two tea-spoonfuls of fine " salt; beat them into a paste, and then add to " them half a pint of boiling water. Strain off " the liquor when cold, and add to it half a pint " of very sharp vinegar. Let a table-spoonful " of this liquor be taken every half hour as a " dose for an adult; diminishing it in proportion " for children." The extreme acrimony of this preparation rendered it difficult to be exhibited, and its effects were to inflame and excoriate the throat; but by this the sloughs were entirely cleansed away, the ulcers brought to a healing state,

ADDENDA.

ftate, and the difeafe removed. It is to be obferved, that fuccefs was to be expected chiefly when the medicine was adminiftered in its early ftage, before the fever had come on, while the power of fwallowing was little impaired, and the affection feemed nearly a local one. Its ufe is farther confirmed by a letter from Mr. James Stephens of St. Chriftopher's to Dr. Duncan, printed in the *Med. Comment.* for 1787.

It appears likewife that the capficum has been given with great fuccefs in the intermittents prevalent in Guiana, and for the fuppreffion of vomitings in putrid fevers.

INDEX

NOMINUM ET SYNONYMORUM.

A.

ABELMOSCH, i. 200.
Abies, i. 1.
    balsamea, i. 4.
    Canadensis, i. 4.
Abiga, i. 326.
Abrotanum femina, i. 6.
    mas, i. 5.
Absinthium, i. 8.
    Alpinum, i. 12.
    maritimum, i. 11.
    minus, i. 12.
    Ponticum, i. 12.
    Romanum, i. 8.
    Seriphium, i. 13.
    Valesiacum, i. 13.
Acacia Germanica, ii. 247.
    vera, i. 13.
Acajou, i. 79.
Acanthina mastiche, i. 287.
Acanthus, i. 15.
Acer, i. 16.
Acetabulum, i. 82.
Acetosa, i. 17.
Acetosella, ii. 75.
Acetum, i. 19.
Achillea, ii. 108.
    Ageratum, i. 192.
    Ptarmica, ii. 249.

Aconitum Anthora, i. 93.
    Napellus, i. 28.
    pardalianches, i. 407.
Acopa, V. Trifol. palud. ii. 440.
Acorus adulterinus, ii. 16.
    Calamus, i. 251.
    verus, i. 251.
    vulgaris, ii. 16.
Acte, ii. 325.
Acus moschata, i. 465.
Adeps, ii. 220.
Adianthum Canadense, i. 30.
    verum, i. 29.
    vulgare, V. Trichomanes, ii. 440.
Adipson, i. 469.
Aer fixus, i. 31.
Ærugo, i. 34.
Æther, ii. 145, 322, 471.
Æthiops antimonialis, i. 161.
    mineralis, i. 148.
    vegetabilis, ii. 259.
Æthusa Meum, ii. 107.
Agallochum, ii. 59.
Agaricus, i. 37.
    mineralis, i. 372.
    quercinus, i. 39.
Ageratum, i. 192.

VOL. II.   K k   Agnus

INDEX

Agnus castus, i. 41.
Agrimonia, i. 42.
    odorata, i. 43.
Agrioriganum, ii. 171.
Agripalma, i. 282.
Aigeiros, ii. 244.
Ajuga, i. 326.
    reptans, i. 242.
Aizoon, ii. 359.
Alcanna, i. 81.
Alchimilla, i. 43.
Alexandrina herba, i. 497.
Alisma, i. 405.
Alkekengi, i. 44.
Alkermes, ii. 26.
Alkohol, ii. 382.
Alleluja, ii. 75.
Alliaria, i. 46.
Allium, i. 47.
Alnus, i. 51.
    nigra, i. 440.
Aloe, i. 51.
Aloes lignum, ii. 59.
Alsine, i. 57.
Altercum, i. 499.
Althæa, i. 58.
Alumen, i. 60,
    plumosum, ii. 407.
    antiquorum, i. 60.
Alypum, ii. 438
Amaracus, *V*. Majorana, ii. 82.
Amaradulcis, ii. 375.
Amaranthus luteus, *V*. Stœchas, i. 411.
Amarella, ii. 243.
Ambarvalis, ii. 242.
Ambarum, i. 65.
    citrinum, ii. 394.
Ambra grisea, i. 65.
    liquida, ii. 72.
Ambrosia, i. 235.
Ambrosioides, i. 236.
Ambutua, ii. 186.
Amianthus, ii. 407.
Ammi, i. 68.
Ammoniacum gummi, i. 69.
Ammosteus, ii. 172.
Amomum Germanorum, ii. 226.

Amomum verum, i. 72.
    vulgare, i. 73.
Amygdalæ amaræ, i. 78.
    dulces, i. 75.
Amylum, i. 445.
Anacampseros, ii. 414.
Anacardium, i. 79.
Anagallis, i. 80.
    aquatica, i. 203.
Anatron, ii. 128.
Anchusa, i. 81.
    officinalis, i. 242.
Andropogon Schœnanthus, ii. 20.
Androsace, i. 82.
Androsæmum, i. 501.
Anemone Hepatica, i. 493.
    pratensis, ii. 252.
Anethum, i. 83.
    Fœniculum, i. 436.
Angelica, i. 85.
    montana, ii. 54.
    pratensis, ii. 348.
Anguria, i. 379.
Angusturæ cortex, ii. 488.
Anime, i. 88.
Anisum, i. 89.
    stellatum, i. 91.
Anodynum minerale, i. 101.
Anodynus liquor mineralis, ii. 468.
Anonis, ii. 160.
Anserina, *herba*, i. 133.
    semen, *V*. Ornithogloss.
    i. 441.
Anthelmia, ii. 377.
Anthemis, i. 324.
    Cotula, i. 325.
    nobilis, i. 321.
    Pyrethrum, ii. 253.
Anthora, i. 93.
Anthos, *V*. Rosmarinus, ii. 280.
Antimonium, i. 94.
Antirrhinum Elatine, i. 409.
    Linaria, ii. 69.
Antithora, i. 93.
Anthophyllus, i. 292.
Aparine, i. 112.
Aphronitrum, ii. 128.
    Apiastrum,

NOMINUM ET SYNONYMORUM.

Apiaftrum, ii. 99.
Apios, ii. 438.
Apis, i. 113.
Apium, i. 113.
    dulce, i. 115.
    hortenfe, ii. 214.
    Macedonicum, ii. 215.
    petræum, ii. 215.
    Petrofelinum, ii. 214.
Apollinaris, i. 499.
Aqua fortis. ii. 141.
    regia, ii. 142.
Aquæ communes, i. 116.
Aquæ medicinales, i. 123.
    alkalinæ, i. 123.
    catharticæ, i. 124.
    chalybeatæ, i. 127.
    cupreæ, i. 131.
    marinæ, i. 125.
    fulphureæ, i. 131.
Aquila alba, i. 160.
Aquilæ lignum, ii. 60.
Aquilegia, i. 132.
Aquileia, i. 132.
Aquilina, i. 132.
Arabicum gummi, i. 479.
Aracus aromaticus, ii. 446.
Aranearum telæ, i. 133.
Arantia, i. 183.
Arbutus Uva Urfi, II. 479.
Arcanum corallinum, i. 153.
    duplicatum, ii. 472.
    tartari, i. 25.
Archangelica, ii. 40.
Arcium, i. 200.
Arctium Lappa, i. 200.
Ardefia, i. 495.
Arefta bovis, ii. 160.
Argentina, i. 133.
Argentum, i. 134.
    vivum, i. 136.
Argilla alba, i. 342.
Ariftolochia, i. 163.
Armoracia, ii. 263.
Arnica, i. 405.
Arfenicum, i. 166.
Artemifia, i. 170.
    Abrotanum, i. 5.
    Abfinthium, i. 8.

Artemifia glacialis, i. 12.
    maritima, i. 11.
    pontica, i. 12.
    Santonicum, ii. 332.
Artemifiæ lanugo, ii. 116.
Arthanita, i. 172.
Arthritica, *V.* Chamæpitys, i. 326.
Arum, i. 173.
    Dracunculus, i. 408.
Afa dulcis, i. 206.
    foetida, i. 176.
Afarum, i. 178.
Afbeftos, ii. 407.
Afclepias Vincetoxicum, ii. 449.
Afelli, *V.* Millepedes, ii. 110.
Afpalathus, ii. 60.
Afparagus, i. 180.
Afphaltus, i. 224.
Afplenium, i. 317.
    Scolopendrium, ii. 69.
    Trichomanes, ii. 440.
Affis, i. 271.
Aftacus, i. 267.
After Atticus, *V.* Eryngium, i. 418.
    omnium maximus, i. 412.
Aftrantia, ii. 7.
Athamanta, ii. 107.
    cretenfis, i. 394.
Athanafia, ii. 409.
Atractylis gummifera, i. 287.
Atriplex olida, i. 181.
Atropa Belladonna, ii. 372.
Avellana purgatrix, ii. 272.
Avena, i. 445.
Aurantia Curaflavenfia, i. 186.
    Hifpalenfia, i. 183.
    Sinenfia, i. 186.
Aureliana Canadenfis, i. 467.
Auricula muris, ii. 217.
Auricularia, ii. 104.
Auripigmentum, i. 170.
Aurum, i. 187.
    fulminans, i. 189.
    mufivum, ii. 387.
Axungia, ii. 220.

Baccæ

INDEX

B.

Baccæ Bermudenses, ii. 339.
Badian, i. 91.
Balanus myrepsica, i. 206.
Balaustia, i. 190.
Balsamea, i. 4.
Balsamita femina, i. 192.
— mas, i. 191.
Balsamum, ii. 169.
— Americanum, i. 195.
— Braziliense, i. 193.
— Canadense, i. 4.
— Copaiba, i. 193.
— Gileadense, ii. 169.
— Indicum, i. 195.
— Mexicanum, i. 195.
— Peruvianum, i. 195.
— album, i. 197.
— nigrum, i. 195.
— Tolutanum, i 198.
Bamia moschata, i. 200.
Bancia, ii. 189.
Bangue, i. 271.
Banilia, ii. 446.
Baptisicula, *V.* Cyanus, i. 387.
Barba capræ, ii. 474.
— Jovis, ii. 359.
Bardana, i. 200.
Barilla, ii. 129.
Barytes, ii. 492.
Basilicum, ii. 156.
Batrachium, i. 465.
Baurach, ii. 128.
Bdellium, i. 202.
Becabunga, i. 203.
Bechium, ii. 442.
Belladona, ii. 372.
Bellis major, i. 204.
— minor, i. 205.
Ben, i. 206.
Benzoinum, i. 206.
Berberis, i. 210.
Beta, i. 212.
Betonica, i. 213.
— aquatica, ii. 358.
— Pauli, ii. 448.
Betula, i. 215.

Betula Alnus, i. 51.
Bezoar microcosmicum, i. 220.
— minerale, i. 107.
— occidentale, i. 219.
— orientale, i. 217.
— porcinum, i. 220.
— simiæ, i. 220.
Bilis, i. 422.
Bingalle, i. 309.
Bismuthum, i. 221.
Bistorta, i. 222.
Bitumen Barbadense, ii. 213.
— Judaicum, i. 224.
Blitum fœtidum, i. 181.
Bois de Coissi, ii. 256.
Boletus igniarius, i. 39.
Bolus Armena & aliæ, i. 225.
Bonus Henricus, i. 228.
Borago, i. 229.
Borax, i. 230.
Botrys, i. 235.
— Mexicana, i. 236.
Bovista, ii. 77.
Branca leonina, ii. 189.
— ursina, i. 15.
Brassica, i. 236.
— Eruca, i. 416.
— Erucastrum, i. 416.
— marina, i. 238.
— Napus, ii. 124.
— Rapa, ii. 264.
Britannica, ii. 43.
Brunella, ii. 245.
Bruscus, ii. 283.
Brutua, ii. 186.
Bryonia, i. 239.
Bubon macedonicum, ii. 215.
Buglossum, i. 242.
Bugula, i. 242.
Bumelia, *V.* Fraxinus, i. 441.
Bunias, ii. 124.
Buphthalmum, i. 204.
Bursa pastoris, i. 243.
Butomon, ii. 16.
Butua, ii. 186.
Butyrum antimonii, i. 106.
Buxus, i. 244.

Cabureiba

NOMINUM ET SYNONYMORUM.

C.

Cabureiba, i. 195.
Cacao, i. 245.
Cacara, i. 403.
Cadmia foſſilis, i. 247.
    lapidoſa, i. 247.
    metallica, i. 166.
Cadjuƈt, i. 403.
Cæruleum Berolinenſe, i. 424.
    nativum, ii. 50.
Caffe, i. 354.
Cajous, i. 79.
Calambac, ii. 59.
Calambour, ii. 60.
Calaminaris, i. 247.
Calamintha, i. 248.
    magno fiore, i. 250.
    montana, i. 250.
Calamus aromaticus, i. 251.
Calcanthum, ii. 457.
Calculus humanus, i. 220.
Calendula, i. 253.
    Alpina, V. arnica, i. 405.
Callitrichum, ii. 440.
Calomelas, i. 160.
Caltha, i. 253.
Calumba, i. 359.
Calx viva, i. 254.
Cambogia Gutta, i. 456.
Camphora, i. 259.
Camphoroſma, ii. 112
Cancrorum chelæ, i. 266.
    oculi, i. 267.
Candelaria, ii. 446.
Canella alba, i. 269.
    malavarica, i. 305.
Cannabina aquatica, V. Eupatorium, i. 419.
Cannabis, i. 270.
Cantharides, i. 272.
Caphura, i. 259.
Caphura baros, i. 348.
Capillus Veneris, i. 29.
Capſicum, ii. 227.
Carabaccium, i. 302.
Carabe, V. Succinum, ii. 394.
Caranna, i. 277.
Carcas, ii. 271.

Cardamantica, ii. 53.
Cardamine, i. 279.
Cardamomum majus, i. 281.
    majus Gallorum, i. 471.
    medium, i. 281.
    minus, i. 279.
    Siberienſe, i. 91.
Cardiaca, i. 282.
Cardopatium, i. 286.
Carduus benediƈtus, i. 282.
    pineus, i. 287.
Caricæ, i. 285.
Carlina, i. 285.
    gummifera, i. 287.
Carpathicum oleum, ii. 417.
Carpentaria, ii. 108.
Carpobalſamum, i. 287.
Carthamus, i. 288.
Carum, i. 289.
Caruon, i. 289.
Caryophylla aromatica, i. 291.
Caryophyllata, i. 294.
Caryophylloides cortex, i. 301.
Caryophyllus ruber, i. 295.
Caſcarilla, i. 297.
Caſia caryophyllata, i. 301.
    fiſtularis, i. 302.
    lignea, i. 305.
Caſtoreum, i. 306.
Caſumunar, i. 309.
Cataputia, ii. 438.
    major, ii. 271.
Cataria, ii. 130.
Catechu, ii. 420.
Cauda equina, i. 415.
    porcina, ii. 216.
Cauſticum antimoniale, i. 106.
    commune fortius, ii. 301.
    mitius, ii. 301, 336.
    lunare, i. 135.
Cedrinum lignum, i. 310.
    oleum, i. 349.
Cedronella, ii. 99.
Cedrus, i. 310.
Celeri, i. 115.
Centaurea benediƈta, i. 282.
Centaureum, i. 312.
    Cyanus, i. 387.

Centaurea

INDEX

Centaurea moschata, i. 388.
Centaurium majus, i. 312.
    minus, i. 311.
Centimorbia, ii. 146.
Centipedes, ii. 110.
Cepa, i. 313.
Cera alba, i. 315.
    di cardo, *V.* Carlina, i. 287.
    cinnamomi, i. 348.
    flava, i. 315.
Cerasa, i. 443.
Cerealia, i. 445.
Cerefolium, i. 318.
Cervispina, ii. 378.
Cerussa, ii. 234.
    antimonii, i. 98.
Ceterach, i. 317.
Cevadilla, i. 318.
Chacarilla, i. 297.
Chærefolium, i. 318.
Chalcitis, ii. 462.
Chalybs, i. 425.
Chamæacte, ii. 327.
Chamæcissus, i. 484.
Chamæclema, i. 484.
Chamæcyparissus, i. 6.
Chamædrys, i. 320,
    maritima, ii. 91,
    palustris, ii. 355.
    spuria, ii. 448.
Chamælæa, ii. 431.
Chamælæagnus, ii. 123.
Chamæleon, i. 285, 287.
Chamælinum, ii. 72.
Chamæmelum, i. 321.
    flore pleno, i. 324.
    fœtidum, i. 325.
Chamæpitys, i. 326.
Chamomilla, i. 321.
Cheiranthus Chieri, i. 327.
Cheiri, i. 327.
Chelæ cancrorum, i. 266.
Chelidonium majus, i. 328.
    minus, i. 329.
Chenopodium ambrosioides, i. 236.
    Bonus Henricus, i. 228.
    Botrys, i. 235.

Chenopodium fœtidum, i. 181.
Chermes, ii. 25.
Cherva, ii. 271.
Chibou gummi, i. 410.
Chiliophyllon, ii. 108.
China, i. 330.
China china, ii. 194.
Chrysanthemum Leucanthemum, i. 204.
Chrysocome, *V.* Elichrysum, i. 411,
Cichoreum, i. 332.
    Endivia, i. 412.
Cicla, i. 212.
Cicuta, i. 333.
    aquatica, i. 114.
Cimolia alba, i. 342
    purpurascens, i. 342.
Cinæ semen, ii. 331.
Cinara, i. 342.
Cinchona Carribæa, ii. 209.
    Jamaicensis, ii. 209,
    officinalis, ii. 194.
Cineres clavellati, ii. 297.
    Russici, ii. 297.
Cinnabaris antimonii, i. 108.
    factitia, i. 149.
    Græcorum, ii. 328.
    nativa, i. 343.
Cinnamomum, i. 345
Citrago, ii. 99.
Citraria, ii. 99.
Citrea, i. 349
Citrullus, i. 379,
Civetta, ii. 483.
Clematis recta, i. 435.
Clinopodium, ii. 92.
Cnicus, i. 282, 288.
Cobaltum, i. 166.
Coccagnidia, ii. 432.
Coccinella, i. 350.
Coccus baphica, ii. 25,
Coccus Cacti, i. 350.
Cochlea, ii. 66.
Cochlearia, i. 351.
    armoracia, ii. 263.
    marina, i. 351.
Codago-pala, i. 361.
Coffea, i. 354.
        Colchicum,

NOMINUM ET SYNONYMORUM.

Colchicum, i. 355.
Colcothar, i. 429. ii. 462.
Colocynthis, i. 357.
Colophonia, ii. 418.
Coluber Berus, ii. 454.
Colubrina, ii. 364.
Colubrinum lignum, ii. 152.
Columbinus pes, i. 465.
Columbo, i. 359.
Coma aurea, *V.* Elichryfum i. 411.
Coneffi, i. 361.
Conium maculatum, i. 333.
Confolida major, i. 362.
 media, i. 204, 242.
 minima, i. 205.
 minor, ii. 245.
 faracenica, ii. 455.
Contrayerva, i. 363.
 Germanica, ii. 450.
Convallaria, ii. 65.
 multiflora, ii. 367.
Copal, i. 260, 364.
Corallina, i. 366.
Corallium, i. 366.
Coriandrum, i. 368.
Cornu cervi, i. 368.
Corona terræ, i. 484.
Cortex cardinalis de Lugo, ii. 194.
 caryophylloides, i. 301.
 coneffi, i. 361.
 culilawan, i. 302.
 eleutheriæ, i. 297.
 granatorum, i. 473.
 Jefuiticus, ii. 194.
 Magellanicus, ii. 480.
 Peruvianus, ii. 194.
 • ruber, ii. 205.
 fimarouba, ii. 367.
 thuris, i. 297. ii. 433.
 thymiamatis, ii. 433.
 Winteranus, ii. 480.
Cortufa, ii. 91.
Coltus arabicus, i. 370.
 corticofus, i. 269.
 hortorum, i. 191.
  minor, i. 192.
 orientalis, i. 370.

Cotonea, i. 388.
Cotula fœtida, i. 325.
Cotyledon marina, i. 83.
Courbaril, i. 88.
Craffula, ii. 414.
Crepitus lupi, ii. 77.
Crefpinus, i. 210.
Creffio, ii. 126.
Creta, i. 372.
Crocus, i. 374.
 . antimonii, i. 100.
 Indicus, i. 386.
 martis, i. 105, 429.
 metallorum, i. 100.
Cryftallus, i. 377.
 mineralis, ii. 140.
Cubebæ, i. 378.
Cucullata, ii. 222.
Cucumis, i. 379.
 agreftis, i. 381.
 Colocynthis, i. 358.
Cucurbita, i. 380.
 Citrullus, i. 379.
 lagenaria, i. 379.
 Pepo, i. 380.
Culilawan, i. 302.
Cuminum, i. 390.
 pratenfe, i. 289.
Cunila bubula, ii. 171.
 fativa, ii. 344.
Cuprum, i. 382.
Curcuma, i. 386.
Curfuta, i. 387.
Cufcuta, i. 414.
Cyanus, i. 387.
 lapis, ii. 50.
 mofchatus, i. 388.
Cyclamen, i. 172.
Cydonia, i. 388.
Cyminum, i. 390.
Cynocrambe, ii. 107.
Cynogloffum, i. 390.
Cynorrhodon, ii. 279.
Cynofbatos, ii. 279.
Cynoforchis, ii. 345.
Cyperus longus, i. 392.
 rotundus, i. 393.
Cypira, i. 386.

Dactyli,

INDEX

D.

Dactyli, ii. 177.
Daphne Gnidium, ii. 431.
    Laureola, ii. 431.
    Mezereum, ii. 431.
Datura Stramonium, ii. 390.
Daucus Creticus, i. 394.
    silvestris, i. 395.
Delphinium Staphysagria, ii. 388.
Dens leonis, i. 396.
Dentaria, ii. 253.
Diagrydium, ii. 349.
Dianthus Caryophyllus, i. 295.
Diapensia, ii. 329.
Dictamnus albus, i. 398.
    Creticus, i. 399.
Digitalis, i. 401.
Dodecatheon, ii. 222.
Dolichos, i. 403.
Doria herba, ii. 455.
Doronicum Germanicum, i. 405.
    Romanum, i. 407.
Dorstenia, i. 363.
Draco silvestris, ii. 249.
Dracocephalum canariense, ii. 112.
Dracontium, i. 408.
Dracunculus pratensis, ii. 249.
Dragacantha, i. 481.
Drakena, i. 363.
Dulcamara, ii. 375.

E.

Ebulus, ii. 327.
Elaphoboscum, ii. 189.
Elaterium, i. 381.
Elatine, i. 409.
Elæagnus, ii. 123.
Electrum, ii. 394.
Elemi gummi, i. 410.
Elettari, i. 72, 280.
Eleoselinum, i. 113.
Eleutheriæ cortex, i. 297.
Elichrysum, i. 411.
Elleborum, i. 487.

Endivia, i. 412.
Enula campana, i. 412.
Epithymum, i. 414.
Equisetum, i. 415.
Erigerum, i. 415.
Eruca, i. 416.
Eryngium, i. 418.
Erysimum, i. 418.
    Alliaria, i. 46.
Erythrodanum, ii. 282.
Esula, ii. 438.
Eupatorium Arabum, i. 419.
    cannabinum, i. 419.
    Mesues, i. 192.
    odoratum, i. 43.
    verum, i. 42.
Euphorbia Cyparissias, ii. 438.
    Paralias, ii. 437.
Euphorbium, i. 420.
Euphrasia, i. 421.

F.

Faba crassa, ii. 414.
    febrifuga, ii. 153.
    Indica, ii. 153.
    purgatrix, ii. 271.
    sancti Ignatii, ii. 153.
    suilla, i. 499.
Fabaria, ii. 414.
Farfara, ii. 442.
Farinacea, i. 445.
Febrifuga, ii. 94.
Febrifugum Craanii, i. 102.
Fel, i. 422.
Ferrum, i. 424.
Ficaria, V. Chelidonium minus, i. 329.
Ficus, i. 285.
Filix, i. 433.
Flammula, ii. 262.
    Jovis, i. 435.
Fœniculum, dulce, i. 436.
    erraticum, ii. 348.
    porcinum, ii. 216.
    silvestre, ii. 216.
    Sinense, i. 91.
    tortuosum, ii. 366.
    vulgare, i. 437.

Fœnum

NOMINUM ET SYNONYMORUM.

Fœnum camelorum, ii. 20.
    Græcum, i. 438.
Folium Indum, ii. 84.
    orientale, ii. 361.
Formica, i. 439.
Fraga, i. 443.
Frangula, i. 440.
Fraxinella, i. 398.
Fraxinus, i. 441.
Fructus horæi, i. 443.
Frumenta, i. 445.
Fucus veficulofus, ii. 259.
Fuga dæmonum, i. 501.
Fuligo, i. 450.
Fumaria, i. 450.
Fungus arboreus, i. 39.
    igniarius, i. 39.
    laricis, i. 37.
    petræus marinus, i. 82.
    rotundus orbicularis, ii. 77.
    vinofus, i. 41.
Furfur, i. 445.

G.

Gabianum oleum, ii. 212.
Galanga, i. 452.
Galbanum, i. 453.
Gale, ii. 122.
Galeopfis, ii. 40.
Galerita, ii. 210.
Gallæ, i. 454.
Gallitrichum, i. 498.
Gallium, i. 455.
    Aparine, i. 112.
Gamandra, i. 456.
Gambogia, i. 456.
Gamma, i. 456.
Garofmum, i. 182.
Garyophyllus, i. 291.
Genifta, i. 458.
Gentiana, i. 460.
    Centaurium, i. 311.
    Indica, i. 462.
    lutea fylveftris, i. 387.
Geoffræa, i. 462.
Geranium, i. 464.
Geum urbanum, i. 294.

Gingiberi, ii. 486.
Gingidium, i. 318.
Ginfeng, i. 467.
Gladiolus luteus, ii. 16.
Glans unguentaria, i. 206.
Glechoma hederacea, i. 484
Glycypicros, ii 375.
Glycyrrhiza, i. 469.
Gnaphalium Stœchas, i. 411.
Gramen caninum, i. 471.
Grana cnidia, ii 432.
    paradifi, i. 471.
    regia, ii. 271.
    tiglia, ii 272.
    tinctoria, ii. 25.
Granata, i 472.
Granatus filveftris, i. 190.
Granum mofchi, i. 200.
Graphoy, i. 407.
Gratia Dei, i. 465, 473.
Gratiola, i. 473.
Guaiacum, i. 475.
Gummi acanthinum, i. 479.
    ammoniacum, i. 69.
    anime, i. 88.
    arabicum, i. 479.
    chibou, i. 410.
    courbaril, i. 88.
    elemi, i. 410.
    Gambienfe, ii. 27.
    guaiaci, i. 475.
    gutta, i. 456.
    hederæ, i. 484.
    juniperinum, ii. 23.
    lacca, ii. 35.
    rubrum aftringens, ii. 27.
    Senegalenfe, i. 480.
    Thebaicum, i. 479.
    tragacanthæ, i. 481.
Gutta gamba, i. 456.
Gypfum, ii. 360.

H.

Hæmatites, i. 482.
Hæmatodes, i. 464.
Halicacabum, i. 44.
Hardefia, i. 495.

Hedera

INDEX

Hedera arborea, i. 483.
    terrestris, i. 484.
Helenium, i. 412.
Heliochryfum, *V*. Stœchas, i. 411.
Helleborafter, i. 485.
Helleborus albus, i. 487.
    fœtidus, i. 485.
    niger, i. 489.
Helvetii pulvis, i. 63.
Helxine, *V*. Parietaria, ii. 188.
Hepar antimonii, i. 100.
    fulphuris, ii. 402.
Hepatica nobilis, i. 493.
    terrestris, i. 492.
Hepatorium, *V*. Eupatorium, i. 419.
Heptaphyllum, ii. 439.
Herba alba, i. 13.
    Alexandrina, i. 497.
    Apollinaris, i. 499.
    bafilica, ii. 156.
    Britannica, ii. 43.
    cephalalgica, ii. 447.
    doria, ii. 455.
    felis, ii. 130.
    fancti Jacobi, ii. 1.
    julia, i. 192.
    fanctæ Kunigundis, *V*. Eupatorium, i. 419.
    militaris, ii. 108.
    papillaris, ii. 41.
    pedicularis, ii. 388.
    pulicaris, ii. 248.
    regia, ii. 156.
    facra, ii. 447.
    falivaris, ii. 253.
    Roberti, i. 465.
    trinitatis, i. 493.
Hermodactylus, i. 494.
Hibernicus lapis, i. 495.
Hiera picra, i. 56.
Hieracium Pilofella, ii. 217.
Hierobotane, ii. 447.
Hippocaftanum, i. 496.
Hippolapathum, ii. 44.
Hippomarathrum, ii. 348.
Hippofelinum, i. 497.
Hippuris vulgaris, i. 415.

Hirundinaria, ii. 146, 449.
Hoitziloxitl, i. 195.
Holofteus, ii. 172.
Hordeolum, i. 318.
Hordeum, i. 445.
    caufticum, i. 318.
    perlatum, i. 447.
Horminum, i. 498.
Humulus Lupulus, ii. 76.
Hydrargyrus, i. 136.
Hydrolapathum, ii. 43.
Hydropiper, ii. 193.
Hyofcyamus, i. 499.
    luteus, ii. 135.
Hypericum, i. 501.
Hypociftis, i. 502.
Hyffopus, i. 503.
Hyftricis lapis, i. 220.

I.

Iberis, ii. 53.
Ibifcus, i. 58.
Ichthyocolla, ii. 6.
Igafur, ii. 153.
Illecebra, ii. 7.
Illicium anifatum, i. 92.
Imperatoria, ii. 7.
Inguinalis, *V*. Eryngium, i. 418.
Intybus, i. 412.
Inula Helenium, i. 412.
Ipecacoanha, ii. 8.
Iquetaia, ii. 358.
Iringus, *V*. Eryngium, i. 418.
Irio, i. 418.
Iris Florentina, ii. 15.
    noftras purpurea, ii. 14.
    paluftris lutea, ii. 16.
Iva arthritica, *V*. Chamæpitys, i. 326.
Ixine, i. 287.
Ixion, i. 287.

J.

Jacobæa, ii. 1.
Jalapium, ii. 2.
Jalappa alba, ii. 96.
    Japonica

NOMINUM ET SYNONYMORUM.

Japonica terra, ii. 420.
Jecoraria, i. 492.
Jemu, i. 456.
Jefuiticus cortex, ii. 194.
Juglans, ii. 18.
Jujuba, ii. 19.
Julia herba, i. 192.
Juncus odoratus, ii. 20.
Juniperus, ii. 21.
 Sabina, ii. 286.
Jupiter, *V*. Stannum, ii. 385.

K.

Kali, ii. 24, 299.
Keiri, i. 327.
Kermes, ii. 25.
 mineralis, i. 103.
Ketmia, i. 200.
Kiki, ii. 271.
Kina kina, ii. 194.
Kino, ii. 27.
Kunigundis herba, *V*. Eupator,
 i. 419.

L.

Labdanum, ii. 29.
Lac, ii. 31.
 lunæ, i. 372.
 fulphuris, ii. 403.
 virginis, i. 209.
Lacca, ii. 35.
Lachryma abiegna, ii. 417.
Lactucæ, ii. 37.
Ladanum, ii. 29.
Lamium album, ii. 40.
Lampfana, ii. 41.
Lanaria, ii. 446.
Lapathum acetofum, i. 17.
 acutum, ii. 41.
 aquaticum, ii. 43.
 hortenfe, ii. 44.
 unctuofum, i. 228.
Lapides cancrorum, i. 267.
Lapis bezoar, i. 216.
 cæruleus, ii. 50.
 calaminaris, i. 247.
 cyanus, ii. 50.

Lapis hæmatites, i. 482.
 Hibernicus, i. 495.
 hyftricis, i. 220.
 lazuli, ii. 50.
 Malacenfis, i. 220.
 porcinus, i. 220.
 fepticus, ii. 301.
 fimiæ, i. 220.
 fpecularis, ii. 360.
Lappa, i. 200.
Lapfana, ii. 41.
Larix, ii. 416.
Lafer, i. 176.
Laferpitium, i. 176.
 Chironium, ii. 190.
 Siler, ii. 365.
Lathyrus, ii. 438.
Laudanum, ii. 167.
Lavendula anguftifolia, ii. 45.
 latifolia, ii. 47.
 Stœchas, ii. 389.
Laver Germanicum, *V*. Becabunga, i. 203.
Laureola, ii. 431.
Laurocerafus, ii. 47.
Laurus, ii. 49.
Lazuli lapis, ii. 50.
Leontodon Taraxacum, i. 396.
Leonurus Cardiaca, i. 282.
Lentifcus, ii. 51.
Lepidium, ii. 53.
 Iberis, ii. 53.
 fativum, ii. 128.
Leucanthemum Diofcoridis, i. 324.
 odoratius, i. 321.
 vulgare, i. 204.
Leucoium luteum, i. 327.
Leuconymphæa, ii. 154.
Leucopiper, ii. 224.
Levifticum, ii. 54.
Libanotis, ii. 280.
Lichen cinereus terreftris, ii. 55.
 iflandicus, ii. 57.
 petræus, i. 492.
Lignum aloes, ii. 59.
 aquilæ, ii. 60.
 calambac, ii. 59.

Lignum

INDEX

Lignum Campechense, ii. 60.
   cedrinum, i. 310.
   spurium, ii. 23.
   colubrinum, ii. 152.
   guaiacum, i. 475.
   Indicum, ii. 60.
   juniperinum, ii. 23.
   lentiscinum, ii. 52.
   Moluccense, ii. 272.
   nephriticum, ii. 131.
   pavanum, ii. 272.
   Quassiæ, ii. 256.
   rhodium, ii. 62.
   sanctum, i. 475.
   santalum, ii. 329.
   sappan, ii. 61.
   sassafras, ii. 343.
   vitæ, i. 475.
Ligusticum, ii. 54.
Lilium album, ii. 63.
   convallium, ii. 65.
Limaces, ii. 66.
Limones, ii. 67.
Linaria, ii. 69.
   segetum, i. 409.
Lingua avis, i. 441.
   canina, i. 390.
   cervina, ii. 69.
Lini semen, ii. 70.
Linum catharticum, ii. 72.
Liparis, ii. 222.
Liquidambra, ii. 72, 393.
Liquiritia, i. 469.
Lithargyrus, ii. 233.
Lithospermum, ii. 73.
Lixivium causticum, ii. 300.
   saponarium, ii. 300.
   tartari, ii. 297.
Lobelia, ii. 73.
Lopeziana radix, ii. 260.
Lotus silvestris, ii. 98.
Lujula, ii. 75.
Lumbrici terrestres, ii. 76.
Lumbus Veneris, ii. 108.
Luna, i. 134.
Lupulus, ii. 76.
Lycoperdon, ii. 77.
Lysimachia Nummularia, ii. 146.

M.

Macerone, i. 497.
Macis, ii. 78.
Macropiper, ii. 225.
Magellanicus cortex, ii. 480.
Magisterium solubile, ii. 425.
Magistrantia, V. Imperatoria, ii. 7.
Magnesia alba, ii. 78.
   vitrariorum, ii. 81.
Maianthemum, ii. 65.
Majorana, ii. 82.
   Syriaca, ii. 91.
Malabathri oleum, i. 348.
Malabathrum, ii. 84.
Malicorium, i. 473.
Malva, ii. 85.
Malvaviscus, V. Althæa, i. 58.
Maniella kua, i. 386.
Maniguetta, i. 471.
Manna, ii. 86.
   Brigantiaca, ii. 87.
Marathrum, i. 435.
   silvestre, ii. 216.
Marathrophyllum, ii. 216.
Marcasita, i. 221.
Marcassita, ii. 254.
Marchantia polymorpha, i. 492.
Margaritæ, ii. 88.
Marrubium, ii. 89.
   Cardiaca, i. 282.
Mars, i. 425.
Marum Syriacum, ii. 91.
   vulgare, ii. 92.
Mastiche, ii. 93.
   acanthina, i. 287.
Mastichina, ii. 92.
Materia perlata, i. 98.
Matricaria, ii. 94.
   Chamomilla, i. 324.
Matrisylva, V. Ulmaria, ii. 474.
Mechoacanna, ii. 95.
   nigra, ii. 2.
Meconium, ii. 182.
Medica malus, i. 349.
Medulla saxi, i. 372.
Mel, ii. 96.

                Melampodium

NOMINUM ET SYNONYMORUM.

Melampodium, i. 489.
Melanopiper, ii. 224.
Meleguetta, i. 471.
Melilotus, ii. 98.
Meliſſa, ii. 99.
    Calamintha, i. 250.
    grandiflora, i. 250.
    Nepeta, i. 249.
    Turcica, ii. 112.
Meliſſophyllon, ii. 99.
Mellifolium, ii. 99.
Mellitis, ii. 99.
Meloe veſicatorius, i. 272.
Mentaſtrum, ii. 104, 250.
Mentha aquatica, ii. 104.
    cataria, ii. 130.
    cervina, ii. 250.
    corymbifera, i. 191.
    paluſtris, ii. 250.
    piperitis, ii. 104.
    Pulegium, ii. 250.
    Saracenica, i. 191.
    vulgaris, ii. 101.
Menyanthes, ii. 440.
Mercurialis, ii. 106.
    montana, ii. 107.
Mercurius, i. 136.
    vitæ, i. 107.
Metella nux, ii. 150.
Meum, ii. 107.
Mezereon, ii. 431.
Militaris herba, ii. 108.
Milium ſolis, ii. 73.
Millefolium, ii. 108.
Millemorbia, ii. 357.
Millepedæ, ii. 110.
Minium, ii. 233.
    Græcorum, i. 343.
Mirabilis Peruviana, ii. 6.
Moldavica, ii. 112.
Momordica Elaterium, i. 381.
Mora, i. 443.
Morſus gallinæ, V. Alſine, i. 57.
Moſchata nux, ii. 146.
Moſchus, ii. 113.
Moxa, ii. 116.
Muſcus maritimus, i. 366.
Myacantha, ii. 283.
Myoſotis, ii. 217.

Myrica Gale, ii. 133.
Myriophyllon, ii. 108.
Myriſtica nux, ii. 146.
Myrobalani, ii. 117.
Myrrha, ii. 119.
Myrtacantha, ii. 283.
Myrtilli, ii. 122.
Myrtus, ii. 122.
    Brabantica, ii. 122.
Myxa, ii. 20.
Myxaria, ii. 20.

N.

Napellus, i. 28.
Napha, i. 183.
Naphtha, ii. 211.
Napium, ii. 41.
Napus dulcis, ii. 124.
    ſilveſtris, ii. 124.
Nardus Celtica, ii. 125.
    Indica, ii. 126.
    ruſtica, i. 178.
Naſturtium aquaticum, ii. 126.
    hortenſe, ii. 128.
    pratenſe, i. 279.
Natron, ii. 128.
Nenuphar, ii. 154.
Nepeta, i. 249, ii. 130.
Nephriticum lignum, ii. 131.
Neroli eſſentia, i. 183.
Nicotiana, ii. 133.
    minor, ii. 135.
Nihil, ii. 485.
Ninzin, i. 468.
Nitrum, ii. 136.
    ammoniacale, ii. 143.
    antiquorum, ii. 128.
    calcareum, ii. 144.
    hydrologorum, ii. 313.
    cauſticum, i. 105.
    cubicum, ii. 143.
    flammans, ii. 143.
    vitriolatum, ii. 140.
    volatile, ii. 143.
Nuciſta, ii. 146.
Nuculæ ſaponariæ, ii. 339.
Nummularia, ii. 146.
Nux Barbadenſis, ii. 271.

Nux

INDEX

Nux ben, i. 206.
   Juglans, ii. 18.
   metella, ii. 150.
   moschata, ii. 146.
   pistacia, ii. 149.
   purgans, ii. 272.
   vomica, ii. 150.
Nymphæa, ii. 154.

O.

Ochra, ii. 155.
Ocularia, *V*. Euphrasia, i. 421.
Oculi cancrorum, i. 267.
Oculus bovis, i. 204.
Ocymum, ii. 156.
Œnanthe, ii. 157.
Offa alba, ii. 308.
Olea, ii. 158.
Oleum abietinum, ii. 417.
   animale, ii. 303.
   Carpathicum, ii. 417.
   Cedrinum, i. 349.
   Gabianum, ii. 212.
   laurinum, ii. 49.
   macis, ii. 149.
   malabathri, i. 348.
   Neroli, i. 183.
   olivarum, ii. 159.
   palmæ, ii. 177.
   petræ, ii. 211.
   susinum, ii. 64.
   Syriæ, ii. 113.
   templinum, i. 3.
   terræ, ii. 211.
   vini, ii. 469.
   vitrioli, ii. 463.
Olibanum, ii. 159.
Olivæ, ii. 159.
Olusatrum, i. 497.
Onisci, ii. 110.
Onitis, ii. 171.
Ononis, ii. 160.
Opium, ii. 161.
Opobalsamum, ii. 169.
Opopanax, ii. 170.
Orchis, ii. 345.
Origanum, ii. 171.
   Creticum, i. 399.

Origanum Dictamnus, i. 399.
   Majorana, ii. 82.
   Syriacum, ii. 91.
Ornithogalum, ii. 351.
Ornithoglossum, i. 441.
Oryza, i. 445.
Ossifragus, ii. 172.
Osteites, ii. 172.
Osteocolla, ii. 172.
Osteolithos, ii. 172.
Ostreum, ii. 173.
Ostrutium, ii. 7.
Osyris, ii. 69.
Ovum, ii. 174.
Oxalis, i. 17.
   Acetosella, ii. 75.
Oxyacantha, i. 210.
Oxylapathum, ii. 41.
Oxmyrsine, ii. 283.
Oxyphœnicon, ii. 408.
Oxys, ii. 75.
Oxytriphyllon, ii. 75.

P.

Pæonia, ii. 175.
Palea de Mecha, ii. 20.
Palimpissa, ii. 231.
Palma, ii. 177.
Palma Christi, ii. 271.
Palum sanctum, *V*. Guaiacum,
   i. 475.
Panava, ii. 273.
Panax, ii. 190.
   quinquefolium, i. 467.
Panis cuculi, ii. 75.
   porcinus, *V*. Arthanita,
   i. 172.
Papaver, ii. 161, 179.
   erraticum, ii. 183.
Papillaris, ii. 41.
Paradisi grana, i. 471.
Paralysis, ii. 185.
Pareira brava, ii. 186.
Parietaria, ii. 188.
Parthenium, ii. 94.
   mas, ii. 409.
Passerina, *V*. Ornithoglossum,
   i. 441.

Passulæ,

NOMINUM ET SYNONYMORUM.

Paſſulæ, ii. 478.
Paſtinaca, i. 395, ii. 189.
Patientia, ii. 44.
Patrum pulvis, ii. 194.
Pavana, ii. 273.
Pedicularia, ii. 388.
Pedro del porco, i. 220.
Pentaphylloides, i. 133.
Pentaphyllum, ii. 191.
Peplus, ii. 438.
Pepo, i. 380.
Perforata, i. 501.
Periſterium. ii. 447.
Perlæ, ii. 88.
Perſica, ii. 192.
Perſicaria, ii. 192.
Perſonata, i. 200.
Peruvianus cortex. ii. 194.
    cortex ruber, ii. 205.
Pes alexandrinus, ii. 253.
    columbinus, i. 465.
    leonis, i. 43.
Petaſites, ii. 210.
Petrapium, ii. 215.
Petroleum, ii. 211.
    Barbadenſe, ii. 213.
Petroſelinum, ii. 214.
    Macedonicum, ii. 215.
Petum, ii. 133.
Peucedanum, ii. 216.
    Silaus, ii. 348.
Philanthropus, i. 112.
Phu, *V.* Valeriana, ii. 443.
Phyllitis, ii. 69.
Phyſalis Alkekengi, i. 44.
Picea, i. 1.
Piloſella, ii. 217.
Pimenta, ii. 226.
Pimpinella, ii. 217.
Pinaſtellum, ii. 216.
Pineus purgans, ii. 271.
Pinguedo, ii. 220.
Pinguicula, ii. 222.
Pinhones Indici, ii. 271.
Pinus Abies, i. 1.
    balſamea, i. 4.
    Cedrus, i. 310.
    Larix, ii. 416.

Pinus Picea, i. 1.
    ſylveſtris, ii. 417.
Piper album, ii. 224.
    caudatum, i. 378.
    Hiſpanicum, ii. 227.
    Jamaicenſe, ii. 226.
    Indicum, ii. 227, 494.
    longum, ii. 225.
    Luſitanicum, ii. 227.
    murale, ii. 7.
    nigrum, ii. 224.
Piperitis, ii. 53.
Piſſaſphaltum, ii. 211.
Piſſelæum, ii. 211.
    Indicum, ii. 213.
Piſtacia, ii. 149.
Piſtolochia, i. 163.
Pityuſa, ii. 438.
Pix Burgundica, ii. 418.
    liquida. ii. 229.
    ſicca, ii. 231.
Plantago, ii. 231.
    Pſyllium, ii. 248.
Plumbago, ii. 193.
Plumbum, ii. 232.
    candidum, *V.* Stannum, ii. 385.
Polium montanum, ii. 239.
Polygala, ii. 240,
Polygonatum, ii. 367.
Polygonum Biſtorta, i. 222.
    Hydropiper, ii. 193.
    Perſicaria, ii. 193.
Polypodium, ii. 243.
    Filix mas, i. 433.
Polytrichum, ii. 440.
Pompholyx, ii. 485.
Populus, ii. 244.
    balſamifera, ii. 245, 406.
Potentilla, i. 133.
    reptans, ii. 191.
Priapeia, ii. 135.
Primula veris, ii. 185.
Pruna, ii. 246.
Prunella, ii. 245.
    Germanorum, i. 242.
Prunus Laurocerafus, ii. 47.
                   Prunus

INDEX

Prunus silveftris, ii. 247.
Pfeudoacorus, ii. 16.
Pfeudocoftus, ii. 190.
Pfeudoiris, ii. 16.
Pfeudonardus, ii. 45, 47.
Pfeudopyrethrum, ii. 249.
Pfychotria emetica, ii. 8.
Pfidium, i. 473.
Pfyllium, ii. 248.
Ptarmica, ii. 249.
Pulegium, ii. 250.
Pulicaris, ii. 248.
Pulmonaria maculofa, ii. 251.
Pulfatilla, ii. 252.
Punica, i. 190, 472.
Pyrethrum, ii. 253.
 filveftre, ii. 249.
Pyrites. ii. 254.

Q.

Quaffia, ii. 256.
Quercula, *V.* Chamædrys, i. 320.
Quercus, ii. 258.
 marina, ii. 259.
Quinquefolium, ii. 191.
Quinquina, ii. 194.

R.

Radix Brafilienfis, ii. 8.
 dulcis, *V.* Glycyrrhiza, i. 469.
 indica Lopeziana, ii. 260.
 rofea, ii. 415.
 rubra, ii. 282.
 urfina, ii. 107.
Raiis di Juan Lopez, ii. 260.
 di Mofambique, ii. 359.
Ranunculus, ii. 262.
 Ficaria, i. 329.
Raphanus rufticanus, ii. 263.
 filveftris, ii. 53.
Rapum, ii. 264.
Rapunculus virginianus, ii. 73.
Realgar, i. 170.
Regina prati, ii. 474.

Regulus antimonii, i. 97.
 martialis, i. 105.
 ftellatus, i. 105.
Remora aratri, ii. 160.
Refina fiava, alba, & nigra, ii. 418.
Refta bovis, ii. 160.
Rhabarbarum, ii. 265.
 album, ii. 96.
 antiquorum, ii. 269.
 monachorum, ii. 44.
Rhamnus catharticus, ii. 378.
 Frangula, i. 440.
 Zizyphus, ii. 19.
Rhaponticum, ii. 269.
 vulgare, i. 312.
Rheum, ii. 265.
Rhodia radix, ii. 415.
Rhodiola, ii. 415.
Rhodium lignum, ii. 62.
Rhododendron, ii. 270.
Rhus, ii. 405.
 belgica, ii. 122.
Ribefia, i. 443.
Ricinus, ii. 271.
Rifagon, i. 309.
Rifigal, i. 170.
Ros Calabrinus, ii. 86.
Rofæ, ii. 275.
Rofa filveftris, ii. 279.
Rofea radix, ii. 415.
Rofmarinus, ii. 280.
Rubia, ii. 282.
Rubrica, ii. 155.
Rubi Idæi fructus, i. 443.
Rumex, i. 17.
 acutus, ii. 41.
 aquaticus, ii. 43.
 Patientia, ii. 44.
Rufcus, ii. 283.
Ruta, ii. 284.

S.

Sabadilla, i. 318.
 Sabina,

NOMINUM ET SYNONYMORUM.

Sabina, ii. 286.
Saccharum, ii. 287.
  Canadenfe, i. 16.
  lactis, ii. 34.
  Saturni, ii. 235.
Sagapenùm, ii. 290.
Sago, ii. 178.
Sal alkalinus fixus, ii. 292.
  foffilis, ii. 24, 128.
  volatilis, ii. 301.
Sal ammoniacus, ii. 310.
  fixus, ii. 320.
  catharticus amarus, ii. 313.
  Glauberi, ii. 314.
  communis, ii. 315.
  diureticus, i. 25.
  enixus, ii. 472.
  gemmæ, ii. 315.
  marinus, ii. 316.
  microcofmicus, ii. 477.
  mirabilis, ii. 314.
  muriaticus calcareus, ii. 320.
  petræ, ii. 136.
  prunellæ, ii. 140.
  polychreftus, ii. 140.
  Rupellenfis, ii. 25, 413.
  fedativus, i. 233.
  vegetabilis, ii. 412.
Salep, ii. 346.
Salivaris herba, ii. 253.
Saliunca, ii. 125.
Salix, ii. 322.
  amerina, i. 41.
Salfola Kali, ii. 24.
Salvia, ii. 323.
  Sclarea, i. 498.
Sambucus, ii. 325.
Sampfuchus, ii. 92.
Sandaracha Arabum, ii. 23.
  Græcorum, i. 170.
Sanguis draconis, ii. 328.
Sanicula, ii. 329.
  Eboracenfis, ii. 222.
Santalum lignum, ii. 329.
Santolina, i. 6.
Santonicum, ii. 331.
Sapo, ii. 332.

Sapo volatilis, ii. 336.
Saponaria, ii. 338.
  nucula, ii. 339.
  terra, *V*. Cimolia, i. 342.
Sappan lignum, ii. 61.
Sarcocolla, ii. 340.
Sarfaparilla, ii. 341.
Saffafras, ii. 343.
Satureia, ii. 344.
Saturnus, ii. 232.
Satyrion, ii. 345.
Savina, ii. 286.
Saxifraga alba, ii. 347.
  vulgaris, ii. 348.
Scammonium, ii. 348.
Scandix Chærefolium, i. 318.
Scariola, *V*. Endivia, i. 412.
  Gallorum, ii. 38.
Schœnanthus, ii. 20.
Scilla, ii. 351.
Scincus, ii. 355.
Sclarea, i. 498.
Scolopendria, i. 317.
Scolopendrium, ii. 69.
Scolymus, i. 343.
Scopa regia, ii. 283.
Scordium, ii. 355.
Scoria reguli antimonii fuccinea, i. 105.
Scorzonera, ii. 356.
Scrophularia, ii. 357.
  minor, *V*. Chelid. min. i. 329.
Sebadilla, i. 318.
Sebeften, ii. 20.
Sedum, ii. 359.
  acre, ii. 7.
  Telephium, ii. 415.
Selenites, ii. 360.
Semen contra, ii. 331.
  fanctum, ii. 331.
Sementina, ii. 331.
Sempervivum, ii. 359.
  acre, ii. 7.
Sena, ii. 361.
Senecio, i. 415.
  Jacobæa, ii. 1.
Seneka, ii. 240.

        Serapias,

INDEX

Serapias, ii. 346.
Serapinum, ii. 290.
Seriola, *V.* Endivia, i. 412.
Seriphium, i. 13.
Seris, *V.* Endivia, i. 412.
Serpentaria Gallorum, i. 408.
  Hispanica, ii. 356.
  Virginiana, ii. 364.
Serpyllum, ii. 435.
  citratum, ii. 435.
Seseli, ii. 365.
  Massiliense, ii. 366.
  pratense, ii. 348.
  tortuosum, ii. 366.
Sevum, ii. 220.
Sicula, i. 212.
Sidium, i. 473.
Sigillum Salomonis, ii. 367.
Sil, ii. 155.
Silaus, ii. 348.
Siler montanum, *V.* Seseli. ii. 365.
Siliquastrum, ii. 227.
Simarouba, ii. 367.
Simiæ lapis, i. 220.
Sinapi, ii. 369.
Sisarum montanum, i. 468.
Sison, i. 73.
  Ammi, i. 68.
Sisymbrium, ii. 104.
  aquaticum, ii. 126.
Sium, ii. 371.
  aromaticum, *V.* Amomum, i. 73.
  foliis serratis, i. 468.
Smectis, *V.* Cimolia purpurascens, i. 342.
Smyrnion, ii. 7.
Smyrnium Olusatrum, i. 497.
Soda, ii. 24, 129.
Sol, i. 187.
Solanum, ii. 372.
  fœtidum, ii. 390.
  lignosum, ii. 375.
  nigrum, ii. 372.
  vesicarium, i. 44.
Soldanella, i. 238.
Solidago Virga aurea, ii. 455.
Spartium scoparium, i. 458.

Specularis lapis, 360.
Sperma ceti, ii. 376.
Spica, ii. 45, 47.
  Celtica, ii. 125.
  nardi, ii. 126.
Spigelia, ii. 377
Spina acida, i. 210.
  cervina, ii. 378.
  infectoria, ii. 378.
Spiræa Ulmaria, ii. 474.
Spiritus mindereri, i. 26.
  vinosus rectificatus, ii. 379.
  tenuior, ii. 381.
Spongia, ii. 384.
Squilla, ii. 351.
Squinanthus, ii. 20.
Stalactitæ, i. 372.
Stannum, ii. 385.
Staphisagria, ii. 388.
Staphylinus, i. 395.
Stellaria, *V.* A$_l$chimi$_{ll}$a, i. 43.
Stelochites, ii. 172.
Sternutamentoria, ii. 249.
Stibium, i. 94.
Stizolobium, i. 403.
Stœchas, ii. 389.
  citrina, i. 411.
Stramonium, ii. 390.
Stratiotes, ii. 108.
Struthium, ii. 7.
Styrax alba, i. 197.
  calamita, ii. 391.
  liquida, ii. 393.
Succinum, ii. 394.
  cinereum, i. 65.
  griseum, i. 65.
Sulphur, ii. 397.
Sumach, ii. 405.
Supercilium Veneris, ii. 108.
Susinum oleum, ii. 64.
Sylphium, i. 176.
Symphytum, i. 362.
  maculosum, ii. 251.
  minus, ii. 245.
Syriæ oleum, ii. 113.

T.

Tabacum, ii. 133.
    Tacamahaca,

NOMINUM ET SYNONYMORUM.

Tacamahaca, ii. 406.
Talcum, ii. 407.
Tamalapatra, ii. 84.
Tamarindus, ii. 408.
Tanacetum, ii. 409.
    Balfamita, i. 191.
    hortenfe, i. 191.
Tanafia, ii. 409.
Tapfus barbatus, ii. 446.
Taraxacum, i. 396.
Tarchon filveftris, ii. 249.
Tartarum, ii. 410.
    emeticum, i. 108.
    regeneratum, i. 25.
    folubile, ii. 412.
Tegula Hibernica, i. 495.
Telephium, ii. 414.
Templinum oleum, i. 3.
Terebinthinæ, ii. 415.
Terra foliata tartari, i. 25.
    Japonica, ii. 420.
    merita, i. 386.
    ponderofa, ii. 492.
    faponaria, *V.* Cimolia
        purp. i. 343.
    figillata, i. 226.
Terræ oleum, ii. 211.
Terrea abforbentia, ii. 423.
Tefticulus caninus, ii. 345.
Teucrium, ii. 440.
    capitatum, ii. 239.
    Chamædrys, i. 320.
    Chamæpitys, i. 326.
    creticum, ii. 239.
    Scordium, ii. 355.
Thea, ii. 427.
    Germanica, ii. 448.
Theriaca rufticorum, i. 49.
Thermæ, i. 131.
Thlafpi, ii. 429.
    Burfa paftoris, i. 243.
Thuris cortex, i. 297, ii. 433.
Thus, ii. 430.
    Judæorum, i. 297,
    ii. 433.
    mafculum, *V.* Oliba-
        num, ii. 159.
Thymbra, ii. 344.
    Hifpanica, ii. 92.

Thymelæa, ii. 430.
Thymiama, i. 297, ii. 433.
Thymus, ii. 434.
    citratus, ii. 435.
    maftichina, ii. 92.
    Serpyllum, ii. 435.
Tiglia grana, ii. 272.
Tilia, ii. 436.
Tincal, i. 231.
Tithymali, ii. 437.
Tormentilla, ii. 439.
Tobacco anglicum, ii. 135.
Tota bona, i. 228.
Tragacantha, i. 481.
Tragofelinum, ii. 218.
Trichomanes, ii. 440.
Trifolium acetofum, ii. 75.
    aquaticum, ii. 440.
    hepaticum, i. 493.
    Melilotus, ii. 98.
    odoratum, ii. 98.
    paludofum, ii. 440.
Trigonella Fœnum-græcum,
    i. 438.
Trinitatis herba, i. 493.
Triffago, *V.* Chamædrys, i.
    320.
    paluftris, ii. 355.
Triticum, i. 445.
    repens, i. 471.
Turpethum, ii. 441.
    minerale, i. 154.
Tuffilago, ii. 442.
    Petafites, ii. 210.
Tutia, ii. 442.

U.

Ulmaria, ii. 474.
Ulmus, ii. 475.
Umbilicus marinus, i. 83.
Ungula caballina, ii. 442.
Uniones, ii. 88.
Urina, ii. 476.
Urinaria, ii. 69.
Urfina radix, ii. 107.
Urtica, ii. 478.
    mortua, ii. 40.
Uvæ paffæ, ii. 478.
Uva urfi, ii. 479.

Veleriana,

INDEX

V.

Valeriana, ii. 443.
    Celtica, ii. 125.
Vanilla, ii. 446.
Venus, i. 382.
Veratrum album, i. 487.
    nigrum, i. 489.
Verbasculum, ii. 185.
Verbascum, ii. 446.
Verbena, ii. 447.
    femina, *V.* Erysimum, i. 418.
Vermicularis, ii. 7.
Vermis terrestris, ii. 76.
Vernix, ii. 23.
Veronica, ii. 447.
    aquatica, i. 203.
    Becabunga, i. 203.
    femina, *V.* Elatine, i. 409.
    Teucrium, ii. 448.
Vetonica, i. 213.
Vincetoxicum, ii. 449.
Vinum, ii. 450.
Viola, ii. 452.
    lutea, i. 327.
    palustris, ii. 222.
    tricolor, ii. 453.
Violaria, ii. 452.
Vipera, ii. 454.
Viperaria, ii. 356.
Viperina, ii. 364.
Virga aurea, ii. 455.
Viride æris, i. 34.
Viscus, ii. 456.
Vitex Agnus castus, i. 41.
Vitis alba, i. 239.
Vitriolum album, ii. 457, 485.
    cæruleum, ii. 459.
    Romanum, ii. 459.
    viride, ii. 461.
Vitrum antimonii, i. 99.
    ceratum, i. 111.
Vomica nux, ii. 150.
Vulgago, i. 178.
Vulvaria, i. 181.

W.

Wanhom, i. 452.
Winteranus cortex, ii. 480.

X.

Xyloaloes, ii. 59.

Z.

Zagarilla, *V.* Eleutheria, i. 297.
Zarza, ii. 341.
Zedoaria, ii. 482.
Zerumbeth, ii. 482.
Zibethum, ii. 483.
Zincum, ii. 484.
Zingi, i. 91.
Zingiber, ii. 486.
Zizyphus, ii. 19.

NOMINUM ET SYNONYMORUM.

INDEX

Of ENGLISH NAMES.

A.

ABSORBENT earths, ii. 425.
Acacia, i. 13.
　German, ii. 247.
Aconite, i. 28, 93.
Acorus, *baftard,* ii. 17.
Adder, ii. 455.
Agallochum, ii. 59.
Agaric, i. 37.
　of the oak, i. 39.
Agnus-caftus, i. 41.
Agrimony, i. 42.
　Hemp, i. 419.
　fweet-fcented, i. 43.
　Water, i. 419.
Air, *fixed,* i. 31.
Alder, i. 51.
　black, i. 449.
Alecoft, i. 191.
Alehoof, i. 484.
Alexanders, i. 497.
Alkanet, i. 81.
Allheal, ii. 190.
Allfpice, ii. 226.
Almonds, i. 74.
Aloes, i. 51.
Aloes-wood, ii. 59.
Alum, i. 60.
　plumous, ii. 407.
Amber, ii. 394.
Ambergris, i. 65.
Amethyft, i. 377.
Amianthus, ii. 407.
Ammoniac falt, ii. 310.
Ammoniacum gum, i. 69.
Amomum, i. 72.
　common, i. 73.

Anacardium, i. 79.
Anemone, ii. 252.
Angelica, i. 85.
Anguftura bark, ii. 488.
Anime, i. 88.
Anife, i. 89.
　ftarry-headed, i. 91.
Anodyne liquor of Hoffman, ii. 468.
Antimony, i. 94.
Ants, i. 439.
Aqua-fortis, ii. 141.
Aqua-regis, ii. 142.
Arabic gum, i. 479.
Archangel, ii. 40.
Armenian bole, i. 225.
Arrach, *ftinking,* i. 182.
Arfenic, i. 166.
Arfmart, ii. 192.
　ftinging, ii. 193.
Artichoke, i. 343.
Afafetida, i. 176.
Afarabacca, i. 178.
Afhtree, i. 441.
Afhes, *Ruffia* ii. 297.
Afhes of vegetables, ii. 79.
Afparagus, i. 180.
Avens, i. 294.

B.

Bacher's pills, i. 491.
Balauftines, i. 190.
Baldmoney, *V.* Gentian, i. 460.
Balm, ii. 99.
　of Gilead, herb, ii. 112.
　　fir, i. 4.
　Turkey ii. 112.

Balfam

INDEX

Balsam of Canada, i. 4.
   of Copaiba, i. 193.
   of Gilead, ii. 169.
   of Peru, i. 195.
   white of Peru, i. 197.
   of Tolu, i. 198.
Barbadoes nut, ii. 271.
   tar, ii. 213.
Barberry, i. 210.
Bark, *Carribæan*, ii. 209.
   Jamaica, ii. 209.
   Peruvian, ii. 194.
   red ii. 205.
Barley, i. 445.
   cauſtic, i. 318.
Basil, ii. 156.
Bauldmony, ii. 107.
Bay, ii. 49.
   Cherry, ii. 47.
Bdellium, i. 202.
Bean, *Malaca*, i. 79.
   St. Ignatius's, ii. 153.
Bears-breech, i. 15.
Bears-foot, i. 485.
Bears-whortleberry, ii. 479.
Bedstraw, *Ladies*, i. 456.
Beech, *Sea-ſide* ii. 209.
Bees, i. 113.
Beets, i. 212.
Ben nuts, i. 206.
Benit herb, i. 294.
Benzoine, i. 206.
Bermudas berries, ii. 340.
Betony, i. 213.
   Paul's, ii. 447.
   Water, ii. 358.
Bezoar microcosmic, i. 220.
   mineral, i. 107.
   occidental, i. 219.
   oriental, i. 217.
   of the ape, i. 220.
   of the porcupine, i. 221.
Bile, i. 422.
Birch, i. 215.
Birdsnest, i. 395.
Birdstongue, i. 441.
Birthworts, i. 163.
Bishopsweed, i. 68.
Bismuth, i. 221.

Bistort, i. 222.
Bittergourd, *V.* Colocynth, i. 357.
Bitterſweet, ii. 375.
Blisters, i. 273.
Bloodstone, i. 482.
Blue-stone, ii. 459.
Bluebottle, i. 387.
Blue, *Pruſſian*, i. 424.
Bois de Coiffi, ii. 256.
Bolar earths, i. 225.
Bones, i. 370.
Bonebinder, ii. 172.
Borage, i. 229.
Borax, i. 230.
Box tree, i. 244.
Bran, i. 448.
Brankurſine, i. 15.
Braſs, i. 385.
Bread, i. 445.
Briar, *wild* ii. 279.
Brignole plum, i. 246.
Brimstone, ii. 397.
Brooklime, i. 203.
Broom, i. 458.
Bruiſewort, ii. 338.
Bryony, i. 239.
Buckbean, ii. 440.
Buckthorn, ii. 378.
Bugle, i. 242.
Buglofs, i. 242.
Bulgewater tree, 462.
Bullfiſt, *V.* Boviſta, ii. 77.
Burdock, i. 200.
Burgundy pitch, ii. 418.
Burnet-ſaxifrage, ii. 217.
Butcherſbroom, ii. 283.
Butter of antimony, i. 106.
Butterbur, ii. 210.
Butterflower, ii. 262.
Butterwort, ii. 222.

C.

Cabbages, i. 236.
Cabbage bark, i. 462.
Cacao nut, i. 245.
Calamine, i. 247.
Calamints, i. 248.

Calambac

NOMINUM ET SYNONYMORUM.

Calambac, ii. 59.
Calamus, i. 251.
Calamy, i. 247.
Calomel, i. 160.
Camels-hay, ii. 20.
Cammock, *V.* Ononis, ii. 161.
Camomile, i. 320.
   ſtinking, i. 325.
Campeachy wood, ii. 61.
Camphor, i. 259.
   of cinnamon, i. 348.
   of peppermint, ii. 105.
Canella, i. 269.
Cantharides, i. 272.
Capſicum, ii. 227, 494.
Caranna, i. 277.
Caraway, i. 289.
Cardamom ſeeds, i. 279.
   greater, i. 281, 472.
Cardinal-flower, *blue*, ii. 73.
Carduus, i. 282.
Carline thiſtle, i. 286.
Carpobalſam, i. 287.
Carrot of Candy, i. 394.
   wild, i. 395.
Caſcarilla, i. 297.
Caſhew nut, i. 79.
Caſia fiſtula, i. 302.
   lignea, i. 305.
Caſſidony, *V.* Stœchas, i. 411.
Caſtor, i. 306.
Caſtor oil, ii. 273.
Caſumunar, i. 309.
Catmint, ii. 130.
Cauliflower, i. 236.
Cauſtic alkalies, ii. 300.
   barley, i. 318.
   Antimonial, i. 106.
   Lunar, i. 135.
Cayenne pepper, ii. 228.
Cedar, i. 310.
Celandine, i. 328.
Celeri, i. 115.
Centaury, i. 311.
   greater, i. 312.
Ceruſſe, ii. 234.
   of antimony, i. 98.
Ceterach, i. 317.

Chalk, i. 372.
Chamemel, i. 320.
Chaſte-tree, i. 41.
Cheeferennet, i. 456.
Cherries, i. 443.
   Winter, i. 44.
Cherry-bay, ii. 47.
Chervil, i. 318.
Chibou *Gum*, i. 410.
Chickweed, i. 57.
China root, i. 330.
Chocolate, i. 245.
Chriſtmas-flower, *V.* Black.
   Helleb. i. 489.
Cichory, i. 332.
Cinnabar of antimony, i. 108.
   factitious, i. 149.
   native, i. 343.
Cinnamon, i. 345.
Cinquefoil, ii. 191.
Citrons, i. 349.
Citrul, i. 379.
Civet, ii. 483.
Clary, i. 498.
Clay, i. 342.
Cleavers, i. 112.
Clotbur, i. 200.
Clove bark, i. 301.
Clove gilliflower, i. 296.
Clove ſpice, i. 291.
Cobalt, i. 166.
Cobweb, i. 133.
Cochineal, i. 350.
Coffee, i. 354.
Colcothar, i. 429, ii. 462.
Cold ſeeds, i. 380.
Coleworts, i. 236.
   Sea, V. Soldanella, i. 238.
Colophony, ii. 418.
Coloquintida, i. 357.
Coltsfoot, ii. 442.
Columbine, i. 132.
Columbo, i. 359.
Comfry, i. 362.
Coneſſi bark, i. 361.
Confound, greater, *V.* Comfry, i. 362.
   middle, i. 242.
             Confound,

INDEX

Confound, small, ii. 245.
—— Saracens, ii. 455.
Contrayerva root, i. 363.
Copaiba balsam, i. 193.
Copal gum, i. 364.
Copper, i. 382.
Copper waters, i. 131.
Copperas, ii. 461.
Coral, i. 366.
Coralline, i. 366.
Coriander, i. 368.
Cornelian, i. 377.
Cornflower, *V.* Cyanus, i. 387.
Cornrose, ii. 183.
Costmary, i. 191.
Costus root, i. 370.
Couch-grass, i. 471.
Couhage, i. 403.
Counterpoison, *V.* Contrayerva, i. 363.
—— Monkshood, *V.* Anthora, i. 93.
Cowitch, i. 403.
Cowslips, ii. 185.
—— of Jerusalem, ii. 251.
Crabs-claws, i. 266.
Crabs-eyes, i. 267.
Cranesbill, i. 464.
Creme of tartar, ii. 410.
Cresses, ii. 128.
—— Bank, *V.* Erysimum, i. 418.
—— Dock, ii. 41.
—— Sciatica, ii. 53.
—— Water, ii. 126.
Crowfoot, ii. 262.
Crystal, i. 377.
Crystal mineral, ii. 140.
Cubebs, i. 378.
Cuckow-flower, i. 279.
Cuckowpint, i. 173.
Cucumber, i. 379.
—— *wild* i. 381.
Cummin, i. 390.
Curassoa apples, i. 186.
Currants, i. 443; ii. 478.
Cutch, ii. 421.
Cyperus root, i. 392.

D.

Daisy, i. 205.
—— *Ox-eye,* i. 204.
Damsons, ii. 246.
Dandelion, i. 396.
Danewort, ii. 327.
Dates, ii. 177.
Devils-dung, *V.* Asafetida, i. 176.
Dill, i. 83.
Dittander, ii. 53.
Dittany of Crete, i. 399.
—— *white* i. 398.
Dock, ii. 41.
—— *sour,* i. 17.
—— *Water,* ii. 43.
Dock-cresses, ii. 41.
Dodder, i. 414.
Dog-rose, ii. 279.
Dogs-grass, i. 471.
Dogs-mercury, ii. 107.
Dogstones, *V.* Orchis, ii. 345.
Dovesfoot, i. 465.
Dragant gum, i. 481.
Dragons, i. 408.
Dragons blood, ii. 328.
Dropwort *Hemlock,* ii. 157.
Dwale, ii. 372.

E.

Earths, i. 117.
—— absorbent, ii. 425.
—— aluminous, i. 60.
—— animal calcareous, i. 266.
—— not calcareous, i. 370.
—— argillaceous, i. 225, 342.
—— bolar, i. 225.
—— mineral calcareous, i. 372.
—— vegetable, i. 79.
Earth *Fullers,* i. 342.
—— *Heavy,* ii. 492.
—— *Japan,* ii. 420.
Earths *sealed,* i. 226.
Earthworms, ii. 76.

Earwort,

OF ENGLISH NAMES.

Earwort, ii. 104.
Eatons ftyptic, ii. 462.
Eau de luce, ii. 309.
Eau de Rabel, ii. 466.
Eggs, ii. 174.
Elaterium, i. 381.
Elder, ii. 325.
   Dwarf, ii. 327.
Elecampane, i. 412.
Elemi *Gum*, i. 410.
Eleutheria bark, i. 297.
Elm, ii. 475.
Emerald, i. 377.
Emulfions, i. 76.
Endive, i. 412.
Epfom falt, i. 313.
Eryngo, i. 418.
Eternal-flower, *V.* Stœchas, i. 411.
Ether, ii. 471.
   nitrous, ii. 145.
   marine, ii. 322.
Ethiops antimonial, i. 161.
   mineral, i. 148.
   vegetable, ii. 259.
Euphorbium, i. 420.
Eyebright, i. 421.

F.

Fats, ii. 220.
Felwort, *V.* Gentian, i. 460.
Fennel, i. 435.
   Hogs, ii. 216.
Fenugreek, i. 438.
Fern, *male*, i. 433.
Feverfew, ii. 94.
Figs, i. 285.
Figwort, ii. 357.
Finckle, i. 437.
Fir tree, i. 1.
   Balm-of-Gilead, i. 4.
   Canada, i. 4.
Fifhglue, ii. 6.
Fiftick-nut, *V.* Piftacia, ii. 149.
Five-leaved grafs, ii. 191.
Flag, *sweet*, i. 251.
Flax, *purging*, ii. 72.
Fleawort, ii. 248.

Flies, *Spanifh*, i. 272.
Flower-de-luce, ii. 14.
Fluellin, i. 409.
Foxglove, i. 401.
Frankincenfe, ii. 430.
Fruits, *Summer*, i. 443.
Fumitory, i. 450.
Fullers earth, i. 342.
Furze, *Ground*, *V.* Ononis, ii. 161.

G.

Galangal, i. 452.
   Englifh, i. 392.
Galbanum, i. 453.
Gall of animals, i. 422.
Gall-ftones, i. 221.
Galls, i. 454.
Gamboge, i. 456.
Garlic, i. 47.
Garnet-ftone, i. 378.
Gas, *fixed*, i. 31.
Gaule, ii. 123.
Gelly of fruits, i. 389, 444.
   horns, i. 369.
Gentian, i. 460.
   Indian, i. 462.
Germander, i. 320.
   Water, ii. 355.
Gill-go-by-ground, *V.* Groundivy, i. 484.
Gilliflower, i. 296.
Ginger, ii. 486.
Ginfeng root, i. 467.
Glafs of antimony, i. 99.
   cerated, i. 111.
Glafswort, ii. 24.
Glaubers falt, ii. 314.
Gold, i. 187.
   Mofaic, ii. 387.
Goldcup, ii. 262.
Golden rod, ii. 455.
Goldilocks, i. 411.
Goofegrafs, i. 112.
Gourd, i. 379.
   bitter, *V.* Colocynth, i. 357.

Grains

INDEX

Grains of paradife, i. 472.
Grana-tilia, ii. 272.
Grafs, *Dogs*, i. 471.
Gromwell, ii. 73.
Groundivy, i. 484.
Groundpine, i. 326.
Groundfel, i. 415.
Guaiacum, i. 475.
Gum ammoniacum, i. 69.
    anime, i. 88.
    arabic, i. 479.
    elemi, i. 410.
    gambia, ii. 27.
    guaiacum, i. 475.
    hederæ, i. 484.
    juniper, ii. 23.
    lac, ii. 35.
    red aftringent, ii. 27.
    fenica, i. 480.
    tragacanth, i. 481.
Gypfum, ii. 360.

H.

Hartfhorn, i. 368.
Hartftongue, ii. 69.
Hartwort, ii. 365.
    of Marfeilles, ii. 366.
Hay *Camels*, ii. 20.
Heartfeafe, ii. 453.
Hedgehyffop, i. 473.
Hedgemuftard, i. 418.
Hellebore *black*, i. 489.
    white, i. 487.
Hellweed, i. 414.
Helmet-flower, i. 93.
Helvetius's powder, i. 63.
    ftyptic, ii. 462.
Hemlock, i. 333.
    Dropwort, ii. 157.
    Water, i. 114.
Hemp, i. 270.
    Water, i. 419.
Hemp-Agrimony, i. 419.
Henbane, i. 499.
Hepatica, i. 493.
Herb-benit, i. 294.
Herb-of-Grace, *V.* Gratiola, i. 473.

Herb-maftich, ii. 92.
    Syrian, ii. 91.
Herb-Robert, i. 465.
Herb-trinity, i. 493.
Hercules's allheal, ii. 190.
Hermodactyl, i. 494.
Hipps, ii. 279.
Hogs-fennel, ii. 216.
Holly *Sea*, i. 418.
Holy Thiftle, i. 282.
Honey, ii. 97.
Hops, ii. 76.
Horehound, ii. 89.
Horeftrong, ii. 216.
Horns, i. 368.
Horfecheftnut, i. 496.
Horferadifh, ii. 263.
Horfetail, i. 415.
Houndftongue, i. 391.
Houfeleek, ii. 359.
Hurtfickle, *V.* Cyanus, i. 387.
Hyacinth ftone, i. 378.
Hypociftis, i. 502.
Hyffop, i. 503.

I.

Jack-by-the-hedge, i. 46.
Jalap, ii. 2.
Jamaica bark, ii. 209.
    pepper, ii. 226.
James's wort, *St. V.* Ragwort, ii. 1.
Japan-earth, ii. 420.
Jerufalem cowflips, ii. 251.
    oak, i. 235.
    fage, ii. 251.
Jefuits bark, ii. 194.
Jews-pitch, i. 224.
Ignatius's bean, ii. 153.
Ilathera bark, i. 297.
Indian-leaf, ii. 84.
    pink, ii. 377.
John's wort, *St.* i. 501.
Ipecacoanha, ii. 9.
Irifh flate, i. 495.
Iron, i. 424.
Ifinglafs, ii. 6.
    mineral, ii. 360.

Jujubes,

OF ENGLISH NAMES.

Jujubes, ii. 19.
July flowers, i. 296.
Juniper, ii. 21.
Ivy, i. 483.

K.

Kelp, ii. 24.
Kermes, ii. 25.
   mineral, i. 103.
Kernels of fruits, i. 444.
Kernelwort, ii. 358.
Keyfer's pills, i. 161.
Kino, ii. 27.
Kneeholly, ii. 283.

L.

Labdanum, ii. 29.
Lac *Gum*, ii. 35.
Ladies bedftraw, i. 456.
   mantle, i. 43.
   fmock, i. 279.
Lakeweed, ii. 193.
Larch tree, ii. 416.
Lard, ii. 221.
Laudanum, ii. 167.
Lavender, ii. 44.
   French, ii. 389.
Lavender Cotton, i. 6.
Laurel, ii. 47.
   Spurge, ii. 431.
Lazule-ftone, ii. 50.
Lead, ii. 232.
Leaf, *Indian*, ii. 84.
Lemnian Earth, i. 226.
Lemons, ii. 67.
Lemon-thyme, ii. 435.
Lentifk wood, ii. 52.
Leopards bane, i. 405.
Lettuces, ii. 37.
Lignaloes, ii. 59.
Lily of the valley, ii. 65.
   May, ii. 65.
   Water, ii. 154.
   white, ii. 63.
Lime, i. 254.
Lime ftone, i. 372.
Lime tree, ii. 436.

Linden tree, ii. 436.
Linfeed, ii. 70.
Liquidambar, ii. 72.
Liquid-fhell, ii. 321.
Liquorice, i. 469.
Litharge, ii. 233.
Liverwort, i. 492.
   Ground, ii. 55.
   Iceland, ii. 57.
   noble, i. 493.
Logwood, ii. 61.
Lopez root, ii. 260.
Lovage, ii. 54.
Lunar cauftic, i. 135.
   pills, i. 136.
Lungwort, ii. 251.
Lye of tartar, ii. 297.
Lyes, *Soap*, ii. 300.

M.

Mace, ii. 78.
   Oil of, ii. 149.
Madder, ii. 282.
Magnefia, ii. 78.
Maidenhair, i. 29.
   Canada, i. 30.
   Englifh, ii. 440.
Malaca-bean, i. 79.
Mallow, ii. 85.
Manganefe, ii. 81.
Manna, ii. 86.
   of Briançon, ii. 86.
Maple, i. 16.
Marble, i. 372.
Marcafite, ii. 254.
Marigold, i. 253.
Marjoram, ii. 82.
   wild, ii. 171.
Marle, i. 372.
Marfhmallow, i. 58.
Marfhtrefoil *or* buckbean, ii. 440.
Marvel of Peru, ii. 6.
Marum, ii. 91.
Mafterwort, ii. 8.
Maftich, ii. 93.
Maftich-wood, ii. 52.
   Herb, ii. 92.

INDEX

Maftich *Syrian Herb*, ii. 91.
Maudlin, i. 192.
May lily, ii, 65.
Mayweed, i. 325.
Meadow-faffron, i. 355.
Meadowfweet, ii. 474.
Mechoacan, ii. 96.
Melilot, ii. 98.
Mercury, i. 136.
   Dogs, ii. 107.
   Englifh Herb, i. 228.
   French Herb, ii. 106.
Mexico-feed, ii. 271.
Mexico tea, i. 236.
Mezereon, ii. 431.
Microcofmic bezoar, i. 220.
   falt, ii. 477.
Milfoil, ii. 108.
Milk, ii. 31. 4.
Milkwort, ii. 240.
Millepedes, ii. 110.
Millmountain, ii. 72.
Miltwafte, i. 317.
Mindererus's fpirit, i. 26.
Mineral anodyne liquor, ii. 468.
Mineral waters, i. 123.
Mint, ii. 101.
   Pepper, ii. 105.
   Water, ii. 104.
Miffeltoe, ii. 456.
Mithridate-muftard, ii. 429.
Mollipuff, *V*. Lycoperdon, ii. 77.
Monkfhood, *V*. Anthora, i. 93.
Monk's rhubarb, ii. 44.
Moneywort, ii. 146.
Mofaic gold, ii. 387.
Mother-of-thyme, ii. 435.
Motherwort, i. 282.
Moufear, ii. 217.
Moxa, ii. 116.
Mugwort, i. 171.
Mulberries, i. 443.
Mullein, ii. 446.
Mufcovy-glafs, ii. 360.
Mufhroom, *dufty*, *V*. Lycoperdon, ii. 77.
Mufk, ii. 113.
Mufk-cranefbill, i. 465.

Mufk-cyanus, i. 388.
Mufkfeed, i. 200.
Muftard, ii. 369.
   Hedge, i. 418.
   Mithridate, ii. 429.
   Treacle, ii. 429.
Myrobalans, ii. 117.
Myrrh, ii. 119.
Myrtle, ii. 122.
   Dutch, ii. 123.

N.

Nard, *Celtic*, ii. 125.
   Indian, ii. 126.
Natron, ii. 128.
Navew, ii. 123.
   wild, ii. 124.
Nep, ii. 130.
Nephritic wood, ii. 131.
Nettle, ii. 478.
   dead, ii. 40.
Nightfhade, ii. 372.
   deadly, ii. 372.
   Garden, ii. 372.
   Woody, ii. 375.
Ninzin, i. 468.
Nipplewort, ii. 41.
Nitre, ii. 136.
   ammoniacal, ii. 143.
   calcareous, ii. 144.
   falfely fo called, ii. 313.
   cubical, ii. 143.
   fixt or alkalized, ii. 136.
   volatile, ii. 143.
Nitrous acid, ii. 141.
   dulcified, ii. 145.
Nitrous ether, ii. 145.
Nutmeg, ii. 146.
Nut, *Barbadoes*, ii. 271.
   Piftachio, ii. 149.
   purging, ii. 272.
   Vomic, ii. 150.

O.

Oak, ii. 258.
   of Jerufalem, i. 235.
                Oak,

OF ENGLISH NAMES.

Oak, *Sea*, ii. 259.
Oats, i. 445.
Ochre, ii. 155.
Oil animal, ii. 303.
  of bays, ii. 49.
  of mace, ii. 149.
  olive, ii. 159.
  British, ii. 213.
  Castor, ii. 273.
  Palm, ii. 177.
  Rock, ii. 211.
Olibanum, ii. 159.
Olives, ii. 158.
Olive, *Spurge*, ii. 431.
Onion, i. 313.
  Sea, ii. 351.
Opium, ii. 161.
Opobalsamum, ii. 169.
Opopanax, ii. 171.
Oppodeldoch, ii. 336.
Orache, *stinking*, i. 182.
Orange, *Seville*, i. 183.
  sweet, i. 186.
Orchis, ii. 345.
Origany, ii. 171.
Orpiment, i. 170.
Orpine, ii. 415.
Orris, ii. 14.
Osteocolla, ii. 172.
Ox eye daisy, i. 304.
Oyster-shells, ii. 173.

P.

Paigil, ii. 185.
Palmachristi, ii. 271.
Palm oil, ii. 177.
Pansies, ii. 453.
Pareira brava, ii. 186.
* Parsley, ii. 214.
  Macedonian, ii. 215.
Parsnep, ii. 189.
  water, ii. 371.
Pasque flower, ii. 252.
Patience, ii. 44.
Paul's betony, ii. 447.
Peach, ii. 192.
Peagle, ii. 185.
Pearl, ii. 88.

Pearl-ashes, ii. 298.
Pearl-barley, i. 447.
Pedro del porco, i. 221.
Pellitory of Spain, ii. 253.
  of the wall, ii. 180.
  bastard, ii. 249.
Pennyroyal, ii. 250.
Peony, ii. 175.
Pepper, ii. 223.
  Guinea, ii. 227, 494.
  Jamaica, ii. 226.
  long, ii. 225.
  Poor man's, ii. 53.
  Wall, ii. 7.
  Water, ii. 193.
  white, ii. 224.
Peppermint, ii. 105.
Pepperwort, ii. 53.
Peruvian bark, ii. 194.
  red, ii. 205.
Pestilentwort, ii. 210.
Petrefactions, i. 372.
Petroleum, ii. 211,
Petty-whin, *V*. Ononis, ii. 161.
Pilewort, i. 329.
Pimento, ii. 226.
Pimpernel, i. 80.
  Water, i. 203.
Pine tree, ii. 417.
Pine thistle, i. 287.
Pink, *Indian*, ii. 377.
Piony, ii. 175.
Pismire, i. 439.
Pissabed, *V*. Dandelion, i. 396.
Pistachio nut, ii. 149.
Pitch, ii. 231.
  Burgundy, ii. 418.
  Jews, i. 224.
  tree, i. 1.
Plantain, ii. 281.
Plaster-of-Paris, ii. 360.
Plums, ii. 245.
Poley-mountain, ii. 239.
Polypody, ii. 243.
Pomegranate, i. 472.
Poor-man's pepper, ii. 53.
Poplar, ii. 244.
Poppy, ii. 161, 179.
  wild, ii. 183.

Potash

Potash, ii. 297.
Primrose, ii. 185.
Prunelloes, ii. 246.
Prunes, ii. 246.
Prussian blue, i. 424.
Puffball, ii. 77.
Pumpion, i. 380.
Purging-flax, ii. 72.
   nuts, ii. 272.
   salt, ii. 312.
   waters, i. 124.
Pyrites, ii. 254.

Q.

Quassy, ii. 256.
Queen of the meadows, ii. 474.
Quickgrass, i. 471.
Quicklime, i. 254.
Quicksilver, i. 136.
Quince, i. 389.

R.

Ragwort, ii. 1.
Raisins, ii. 478.
Rape, ii. 124.
Raspberries, i. 443.
Rattlesnake root, ii. 240.
Realgar, i. 170.
Reddle, ii. 155.
Regulus of antimony, i. 97.
     martial, i. 105.
     stellated, i. 105.
Resin, ii. 418.
Restharrow, ii. 161.
Rhapontic, ii. 269.
Rhodium wood, ii. 62.
Rhubarb, ii. 265.
  Monks, ii. 44.
Ribwort, ii. 232.
Rice, i. 445.
Risigal, i. 170.
Rochelle salt, ii. 413.
Rocket, i. 416.
Rock-oil, ii. 211.
Rose, *Damask*, ii. 275.

OF ENGLISH NAMES.

Salts, Seignette's, ii. 413.
Saltpetre, ii. 136.
Saltwort, ii. 24.
Sandarach of the Arabians, ii. 23.
   of the Greeks, i. 170.
Sanicle, ii. 329.
   Yorkshire, ii. 222.
Sapphire, i. 377.
Saracens confound, ii. 455.
Sarcocol, ii. 340.
Sarsaparilla, ii. 341.
Saffafras, ii. 343.
Satyrion, ii. 345.
Sauce-alone, i. 46.
Savin, ii. 286.
Savoury, ii. 344.
Savoy, i. 236.
Saunders woods, ii. 329.
Saxifrage, *Burnet*, ii. 217.
   English, ii. 348.
   Meadow, ii. 348.
   white, ii. 347.
Scammony, ii. 348.
Sciatica-cresses, ii. 53.
Scordium, ii. 355.
Scorzonera, ii. 356.
Scurvygrafs, i. 351.
  Sirich, V. Solda nella i. 238.
Sea-holly, i. 418.
Sea-moss, i. 366.
Sea-oak, ii. 259.
Sea-onion, ii. 351.
Sea-water, i. 125.
Sea-wrack, ii. 259.
Sealed earths, i. 226.
Sebestens, ii. 20.
Sedative salt, i. 233.
Sedge, ii. 17.
Seggrum, *V.* Jacobæa ii. 1.
Sel de Seignette, ii. 413.
Selenite, ii. 360.
Self heal, ii. 245.
Sena, ii. 361.
Seneca gum, i. 480.
Senegaw rattlesnake root, ii. 240.

Sengreen, ii. 359.
Septfoil, ii. 439.
Sermountain, ii. 365.
Setterwort, i. 485.
Shepherd's purse, i. 243.
Silesian earth, i. 226.
Silver, i. 134.
Silverweed, i. 133.
Simaruba bark, ii. 367.
Skink, ii. 355.
Slate, *Irish*, i. 495.
Slaters, *V.* Millepedes, ii. 110.
Sloes, ii. 247.
Smallage, i. 113.
Snails, ii. 66.
Snakeroot, ii. 364.
Snakeweed, i. 222.
Snakewood, ii. 152.
Sneezewort, ii. 249.
Soap, ii. 333.
Soaps, *Volatile*, ii. 336.
Soap-lyes, ii. 300.
Soap-berries, ii. 339.
Soap-wort, ii. 338.
Soldanella, i. 238.
Solomon's-seal, ii. 367.
Soot, i. 450.
Sorrel, i. 17.
  Wood, ii. 75.
Soude, ii. 84.
Soude blanche, ii. 128.
Sour-dock, i. 17.
Southernwood, i. 5.
Sowbread, i. 172.
Spanish flies, i. 272.
Spar, i. 372.
Spearmint, ii. 101.
Spearwort, ii. 262.
Speedwell, *Female*, i. 409.
   Male, ii. 448.
   Mountain, ii. 448.
Spermaceti, ii. 376.
Spignel, ii. 107.
Spike, ii. 47.
Spikenard, ii. 126.
Spirits, *Vinous*, ii. 379.
   Volatile, ii. 308.
Spleenwort, i. 317.
Sponge, ii. 384.

Spunk,

INDEX

Spunk, i. 39.
Spurges, ii. 437.
Spurge-flax, ii. 431.
Spurge-laurel, ii. 431.
Spurge-olive, ii. 431.
Squill, ii. 351.
Squinanth, ii. 20.
Starch, i. 445.
Stavesacre, ii. 388.
Stechas, ii. 389.
    yellow i. 411.
Steel, i. 425.
Stonecrop, ii. 7.
Stone-parsley, *bastard*, i. 73.
Stones, *precious*, i. 377.
Storax, ii. 391.
    Liquid, ii. 393.
    white, i. 197.
Strawberries, i. 443.
Sublimate corrosive, i. 156.
Succory, i. 332.
Suet, ii. 221.
Sugar, ii. 287.
    of milk, ii. 34.
    of lead, ii. 235.
    Maple, i. 16.
Sulphur, ii. 397.
Sulphurwort, ii. 216.
Sultan-flower, i. 388.
Sumach, ii. 405.
Swallowwort, ii. 449.
Sweet-rush, ii. 20.
Sweet-sultan, i. 388.
Sweet willow, ii. 123.
Sycamore, i. 16.

T.

Tacamahaca, ii. 406.
Talc, ii. 407.
    English, ii. 360.
Tamarind, ii. 408.
Tamepoison, ii. 449.
Tansy, ii. 409.
    wild, i. 133.
Tar, ii. 229.
    Barbadoes, ii. 213.
Tartar, ii. 410.
    emetic, i. 108.

Tartar regenerated, i. 25.
    soluble, ii. 412.
    vitriolated, ii. 472.
Tea, ii. 427.
    German, ii. 448.
    Mexico, i. 236.
Thistle, *Carline*, i. 286.
    holy, i. 282.
    pine, i. 287.
Thorn-apple, ii. 390.
    black, ii. 247.
Thyme, ii. 434.
    Lemon, ii. 435.
    Mother of, ii. 435.
Tin, ii. 385.
Tincal, i. 231.
Tinglass, i. 221.
Toadflax, ii. 69.
Tobacco, ii. 133.
Tobacco-pipe clay, i. 342.
Tormentil, ii. 439.
Touchwood, i. 39.
Tragacanth, i. 481.
Treacle-mustard, ii. 429.
Trefoil, *Marsh*, ii. 440.
Trinity-herb, i. 493.
Tunhoof, *V.* Ground-ivy, i. 484.
Turbith mineral, i. 154.
Turbith root, ii. 441.
Turmerick, i. 386.
Turnep, ii. 264.
Turpentines, ii. 415.
Tutenag, ii. 484.
Tutty, ii. 442.

V.

Valerian, ii. 444.
Vanelloes, ii. 446.
Verdegris, i. 35.
Vermilion, i. 149.
Vervain, ii. 447.
Vine, *wild*, i. 239.
Vinegar, i. 19.
Violets, ii. 452.
Viper, ii. 455.
Viper-grafs, ii. 356.
Virgins-bower, *upright*, i. 435.
    Virgins-

OF ENGLISH NAMES.

Virgins-milk, i. 209.
Vitriol, *blue*, ii. 459.
   green, ii. 461.
   Roman, ii. 459.
   white, ii. 457, 485.
Vitriolic acid, ii. 463.
   volatile, ii. 464.
   dulcified, ii. 467.
Vitriolated falts, ii. 472.
Vomic nut, ii. 150.

Urine, ii. 476.

W.

Wakerobin, i. 173.
Wallflower, i. 327.
Wallpepper, ii. 7.
Walnut, ii. 18.
Waters, *common*, i. 116.
   Mineral, i. 123.
   alkaline, i. 123.
   chalybeate, i. 127.
   cupreous, i. 131.
   purging, i. 124.
   fea, i. 125.
   fulphureous, i. 131.
Water-creffes, ii. 126.
Water-germander, ii. 355.
Water-flag, ii. 16.
Water-hemp, i. 419.
Water-lily, ii. 154.
Water-parfnep, ii. 371.
Water-pepper, ii. 193.
Wax, i. 314.
Wheat, i. 445.
Whey, ii. 33.

Whin, *Petty*, *V.* Ononis, ii. 161.
Whortleberry, *Bears*, ii. 479.
Widow-wail, ii. 431.
Willow, ii. 322.
   fweet, ii. 123.
Wine, ii. 450.
Winter-cherries, i. 44.
Winter's bark, ii. 480.
Wolffbane, i. 28.
   blue, i. 28.
Woodlice, ii. 110.
Wood-forrel, ii. 75.
Wormbark, i. 462.
Wormfeed, ii. 331.
   white, *V.* Coralline, i. 366.
Wormwood, i. 8.
   Mountain, i. 12.
   of Valais, i. 13.
   Roman, i. 12.
   Sea, i. 11.
Wort, i. 448.
Woundwort, ii. 190.
Wrack, *Sea*, ii. 259.

Y.

Yarrow, ii. 108.

Z.

Zarnich, i. 170.
Zedoary, ii. 482.
Zerumbeth, ii. 482.
Zinc, ii. 484.

THE END.

Library

DO NOT
REMOVE
THE
CARD
FROM
THIS
POCKET

Acme Library Card Pocket
Under Pat "Ref. Index File"
Made by LIBRARY BUREAU